METHODS IN MOLECULAR BIOLOGY

Series Editor
John M. Walker
School of Life and Medical Sciences
University of Hertfordshire
Hatfield, Hertfordshire, UK

For further volumes:
http://www.springer.com/series/7651

For over 35 years, biological scientists have come to rely on the research protocols and methodologies in the critically acclaimed *Methods in Molecular Biology* series. The series was the first to introduce the step-by-step protocols approach that has become the standard in all biomedical protocol publishing. Each protocol is provided in readily-reproducible step-by step fashion, opening with an introductory overview, a list of the materials and reagents needed to complete the experiment, and followed by a detailed procedure that is supported with a helpful notes section offering tips and tricks of the trade as well as troubleshooting advice. These hallmark features were introduced by series editor Dr. John Walker and constitute the key ingredient in each and every volume of the *Methods in Molecular Biology* series. Tested and trusted, comprehensive and reliable, all protocols from the series are indexed in PubMed.

Somatic Cell Nuclear Transfer Technology

Edited by

Marcelo Tigre Moura

Chemical Biology Graduate Program, Federal University of São Paulo — UNIFESP, Campus Diadema, Diadema - SP, Brazil

Editor
Marcelo Tigre Moura
Chemical Biology Graduate Program,
Federal University of São Paulo –
UNIFESP, Campus Diadema
Diadema - SP, Brazil

ISSN 1064-3745   ISSN 1940-6029   (electronic)
Methods in Molecular Biology
ISBN 978-1-0716-3063-1   ISBN 978-1-0716-3064-8   (eBook)
https://doi.org/10.1007/978-1-0716-3064-8

© The Editor(s) (if applicable) and The Author(s), under exclusive license to Springer Science+Business Media, LLC, part of Springer Nature 2023, Corrected Publication 2023
This work is subject to copyright. All rights are solely and exclusively licensed by the Publisher, whether the whole or part of the material is concerned, specifically the rights of translation, reprinting, reuse of illustrations, recitation, broadcasting, reproduction on microfilms or in any other physical way, and transmission or information storage and retrieval, electronic adaptation, computer software, or by similar or dissimilar methodology now known or hereafter developed.
The use of general descriptive names, registered names, trademarks, service marks, etc. in this publication does not imply, even in the absence of a specific statement, that such names are exempt from the relevant protective laws and regulations and therefore free for general use.
The publisher, the authors, and the editors are safe to assume that the advice and information in this book are believed to be true and accurate at the date of publication. Neither the publisher nor the authors or the editors give a warranty, expressed or implied, with respect to the material contained herein or for any errors or omissions that may have been made. The publisher remains neutral with regard to jurisdictional claims in published maps and institutional affiliations.

This Humana imprint is published by the registered company Springer Science+Business Media, LLC, part of Springer Nature.
The registered company address is: 1 New York Plaza, New York, NY 10004, U.S.A.

Dedication

This book is dedicated to my beloved father Romero Moura, who is no longer with us. His love, friendship, and guidance were fundamental to my personal and professional lives.

Preface

Animal ontogeny has facinated researchers for centuries. The early embryo begins as a single cell (the zygote) that initiates a highly orchestrated sequence of events that ultimately leads to an organism with multiple organs and hundreds of cell types. Since early embryology research relied upon observation alone, quite simple questions remained unanswered. Do cells lose genes during development or retain the exact same genome? This question motivated early work of experimental biology and led to the conceptualization of nuclear transfer almost a century ago. The cloning of amphibians in the 1950s demonstrated that somatic cells undergo nuclear reprogramming (reversal to an embryo-like state) after nuclear transfer into enucleated oocytes. This reprogramming allows the acquisition of a cellular state equivalent to totipotency, proven by the birth of clones, albeit from a small percentage of cloning attempts.

The birth of Dolly was the most remarkable achievement of cloning technology and was widely publicized a quarter of a century ago. This scientific report was special because Dolly was a clone originated from somatic cells of an adult ewe, thus demonstrating the amenability of mammalian adult cells to nuclear reprogramming. This discovery led to widespread attention in the scientific community and the general public about the potential uses of cloning technology in animals and humans. Cloning by nuclear transfer proved versatile for dissecting the epigenetic basis of cellular identity and assisted in numerous commercial applications. The mechanisms by which oocytes reprogram somatic cells remain poorly understood but an attractive research goal. Such understanding of reprogramming-associated processes will likely improve cloning efficiency and the applicability of the technology.

Our intention with the book *Somatic Cell Nuclear Transfer Technology* was to celebrate the science and associated technologies inspired by the birth of Dolly and to reinforce the potential of cloning technology to continuously unravel the molecular basis of cellular states and their transitions.

São Paulo, Brazil   *Marcelo Tigre Moura*

Contents

Dedication . . . . *v*
Preface . . . . *vii*
Contributors . . . . *xi*

1 Cloning by SCNT: Integrating Technical
   and Biology-Driven Advances 1
   Marcelo Tigre Moura

2 Epigenetic Reprogramming and Somatic Cell Nuclear Transfer 37
   Luna N. Vargas, Márcia M. Silveira, and Maurício M. Franco

3 Early Cell Specification in Mammalian Fertilized
   and Somatic Cell Nuclear Transfer Embryos 59
   Marcelo D. Goissis and Jose B. Cibelli

4 Mitochondrial Inheritance Following Nuclear Transfer:
   From Cloned Animals to Patients with Mitochondrial Disease 83
   Jörg P. Burgstaller and Marcos R. Chiaratti

5 Stem Cells as Nuclear Donors for Mammalian Cloning 105
   Carolina Gonzales da Silva and Carlos Frederico Martins

6 Animal Transgenesis and Cloning: Combined Development
   and Future Perspectives 121
   Melissa S. Yamashita and Eduardo O. Melo

7 Mouse Cloning Using Outbred Oocyte Donors
   and Nontoxic Reagents 151
   *Sayaka Wakayama, Yukari Terashita, Yoshiaki Tanabe,
   Naoki Hirose, and Teruhiko Wakayama*

8 Somatic Cell Nuclear Transfer in Rabbits 169
   Pengxiang Qu, Wenbin Cao, and Enqi Liu

9 Production of Cloned Pigs by Handmade Cloning 183
   Gábor Vajta, Wen Bin Chen, and Zoltan Machaty

10 Somatic Cell Nuclear Transfer in Pigs 197
   Werner G. Glanzner, Vitor B. Rissi, and Vilceu Bordignon

11 Somatic Cell Nuclear Transfer Using Freeze-Dried
   Protaminized Donor Nuclei 211
   *Luca Palazzese, Marta Czernik, Kazutsugu Matsukawa,
   and Pasqualino Loi*

12 Cattle Cloning by Somatic Cell Nuclear Transfer 225
   *Juliano Rodrigues Sangalli, Rafael Vilar Sampaio, Tiago Henrique
   Camara De Bem, Lawrence Charles Smith, and Flávio Vieira Meirelles*

13 Production of Water Buffalo SCNT Embryos by Handmade Cloning 245
   Prabhat Palta, Naresh L. Selokar, and Manmohan S. Chauhan

14 Bovid Interspecies Somatic Cell Nuclear Transfer with Ooplasm Transfer 259
   L. Antonio González-Grajales and Gabriela F. Mastromonaco

ix

x   Contents

15   Horse Somatic Cell Nuclear Transfer Using Zona Pellucida-Enclosed
     and Zona-Free Oocytes... 269
     Daniel Salamone and Marc Maserati

16   A Modified Handmade Cloning Method for Dromedary Camels 283
     Fariba Moulavi and Sayyed Morteza Hosseini

17   Derivation of Bovine Primed Embryonic Stem Cells from Somatic
     Cell Nuclear Transfer Embryos 305
     Delia A. Soto, Micaela Navarro, and Pablo J. Ross

Correction to: Cloning by SCNT: Integrating Technical and Biology-Driven
Advances C1
Marcelo Tigre Moura

Index *317*

Contributors

VILCEU BORDIGNON • *Department of Animal Science, McGill University, Sainte-Anne-de-Bellevue, QC, Canada*

JÖRG P. BURGSTALLER • *Institute of Animal Breeding and Genetics, University of Veterinary Medicine, Vienna, Austria*

WENBIN CAO • *Laboratory Animal Center, Xi'an Jiaotong University Health Science Center, Xi'an, Shaanxi, China; Key Laboratory of Environment and Genes Related to Diseases, Ministry of Education of China, Xi'an, Shaanxi, China*

WEN BIN CHEN • *VitaVitro Biotech Co., Ltd., Shenzhen, Guangdong, China*

MARCOS R. CHIARATTI • *Departamento de Genética e Evolução, Universidade Federal de São Carlos, São Carlos, Brazil*

JOSE B. CIBELLI • *Department of Animal Science, Michigan State University, East Lansing, MI, USA*

MARTA CZERNIK • *Institute of Genetics and Animal Biotechnology of the Polish Academy of Sciences, Jastrzebiec, Poland; Faculty of Veterinary Medicine, University of Teramo, Teramo, Italy*

CAROLINA GONZALES DA SILVA • *Federal Institute of Education, Science and Technology of Bahia, Campus Xique-Xique, Xique-Xique, Bahia, Brazil*

TIAGO HENRIQUE CAMARA DE BEM • *Departamento de Medicina Veterinária, Faculdade de Zootecnia e Engenharia de Alimentos, Universidade de São Paulo, Pirassununga, SP, Brazil*

MAURÍCIO M. FRANCO • *Laboratory of Animal Reproduction, Embrapa Genetic Resources and Biotechnology, Brasília, Distrito Federal, Brazil; Institute of Biotechnology, Federal University of Uberlândia, Uberlândia, Minas Gerais, Brazil; School of Veterinary Medicine, Federal University of Uberlândia, Uberlândia, Minas Gerais, Brazil*

WERNER G. GLANZNER • *Department of Animal Science, McGill University, Sainte-Anne-de-Bellevue, QC, Canada*

MARCELO D. GOISSIS • *Department of Animal Reproduction, School of Veterinary Medicine and Animal Science, University of Sao Paulo, Sao Paulo, SP, Brazil*

L. ANTONIO GONZÁLEZ-GRAJALES • *Institut für Fortpflanzung landwirtschaftlicher Nutztiere Schönow e.V, Bernau bei Berlin, Germany*

NAOKI HIROSE • *Faculty of Life and Environmental Sciences, University of Yamanashi, Yamanashi, Japan*

SAYYED MORTEZA HOSSEINI • *Department of Embryology, Camel Advanced Reproductive Technologies Centre, Government of Dubai, Dubai, United Arab Emirates*

ENQI LIU • *Laboratory Animal Center, Xi'an Jiaotong University Health Science Center, Xi'an, Shaanxi, China; Key Laboratory of Environment and Genes Related to Diseases, Ministry of Education of China, Xi'an, Shaanxi, China*

PASQUALINO LOI • *Faculty of Veterinary Medicine, University of Teramo, Teramo, Italy*

ZOLTAN MACHATY • *Department of Animal Sciences, Purdue University, West Lafayette, IN, USA*

CARLOS FREDERICO MARTINS • *Brazilian Agricultural Research Corporation (Embrapa Cerrados), Brasília, Federal District, Brazil*

xii Contributors

MARC MASERATI • *Embryo Production In Vitro Clonagem/In Vitro Equinos, Mogi Mirim, São Paulo, Brazil*

GABRIELA F. MASTROMONACO • *Reproductive Sciences Unit, Toronto Zoo, Toronto, ON, Canada*

KAZUTSUGU MATSUKAWA • *Faculty of Agriculture and Marine Science, Kochi University, Kochi, Japan*

FLÁVIO VIEIRA MEIRELLES • *Departamento de Medicina Veterinária, Faculdade de Zootecnia e Engenharia de Alimentos, Universidade de São Paulo, Pirassununga, SP, Brazil*

EDUARDO O. MELO • *Embrapa Genetic Resources and Biotechnology, Brasília, Distrito Federal, Brazil; Graduation Program in Biotechnology, University of Tocantins, Gurupi, Tocantins, Brazil*

FARIBA MOULAVI • *Department of Embryology, Camel Advanced Reproductive Technologies Centre, Government of Dubai, Dubai, United Arab Emirates*

MARCELO TIGRE MOURA • *Chemical Biology Graduate Program, Federal University of São Paulo – UNIFESP, Campus Diadema, Diadema - SP, Brazil*

MICAELA NAVARRO • *Department of Animal Science, University of California, Davis, CA, USA; Instituto de Investigaciones Biotecnológicas, CONICET, Universidad Nacional de San Martín, Buenos Aires, Argentina*

LUCA PALAZZESE • *Institute of Genetics and Animal Biotechnology of the Polish Academy of Sciences, Jastrzebiec, Poland*

PRABHAT PALTA • *Animal Biotechnology Centre, National Dairy Research Institute, Karnal, India*

PENGXIANG QU • *Laboratory Animal Center, Xi'an Jiaotong University Health Science Center, Xi'an, Shaanxi, China; Key Laboratory of Environment and Genes Related to Diseases, Ministry of Education of China, Xi'an, Shaanxi, China*

VITOR B. RISSI • *Faculty of Veterinary Medicine, Federal University of Santa Catarina, UFSC, Curitibanos, SC, Brazil*

PABLO J. ROSS • *Department of Animal Science, University of California, Davis, CA, USA*

DANIEL SALAMONE • *Universidad de Buenos Aires, Facultad de Agronomía, Departamento de Producción Animal, Buenos Aires, Laboratorio Biotecnología Animal (LabBA), Buenos Aires, Argentina; Instituto de Investigaciones en Producción Animal (INPA), CONICET–Universidad de Buenos Aires, , Buenos Aires, Argentina*

RAFAEL VILAR SAMPAIO • *Departamento de Medicina Veterinária, Faculdade de Zootecnia e Engenharia de Alimentos, Universidade de São Paulo, Pirassununga, SP, Brazil; Centre de Recherche en Reproduction et Fértilité, Faculty of Veterinary Medicine, University of Montreal, Saint-Hyacinthe, QC, Canada*

JULIANO RODRIGUES SANGALLI • *Departamento de Medicina Veterinária, Faculdade de Zootecnia e Engenharia de Alimentos, Universidade de São Paulo, Pirassununga, SP, Brazil*

NARESH L. SELOKAR • *Animal Biotechnology Centre, National Dairy Research Institute, Karnal, India*

MÁRCIA M. SILVEIRA • *Laboratory of Animal Reproduction, Embrapa Genetic Resources and Biotechnology, Brasília, Distrito Federal, Brazil; Institute of Biotechnology, Federal University of Uberlândia, Uberlândia, Minas Gerais, Brazil*

LAWRENCE CHARLES SMITH • *Centre de Recherche en Reproduction et Fértilité, Faculty of Veterinary Medicine, University of Montreal, Saint-Hyacinthe, QC, Canada*

DELIA A. SOTO • *Department of Animal Science, University of California, Davis, CA, USA*

YOSHIAKI TANABE • *Biotechnical Center, Japan SLC, Inc., Shizuoka, Japan*

YUKARI TERASHITA • *Integrated Clinical Education Center, Kyoto University Hospital, Kyoto, Japan; Department of Cardiovascular Surgery, Takamatsu Red Cross Hospital, Takamatsu, Japan*

GÁBOR VAJTA • *VitaVitro Biotech Co., Ltd., Shenzhen, Guangdong, China; RVT Australia, Cairns, QLD, Australia*

LUNA N. VARGAS • *Laboratory of Animal Reproduction, Embrapa Genetic Resources and Biotechnology, Brasília, Distrito Federal, Brazil; Institute of Biotechnology, Federal University of Uberlândia, Uberlândia, Minas Gerais, Brazil*

SAYAKA WAKAYAMA • *Advanced Biotechnology Center, University of Yamanashi, Yamanashi, Japan*

TERUHIKO WAKAYAMA • *Advanced Biotechnology Center, University of Yamanashi, Yamanashi, Japan; Faculty of Life and Environmental Sciences, University of Yamanashi, Yamanashi, Japan*

MELISSA S. YAMASHITA • *Embrapa Genetic Resources and Biotechnology, Brasília, Distrito Federal, Brazil; Graduation Program in Animal Biology, University of Brasília, Brasília, Distrito Federal, Brazil*

Chapter 1

Cloning by SCNT: Integrating Technical and Biology-Driven Advances

Marcelo Tigre Moura ⓘ

Abstract

Somatic cell nuclear transfer (SCNT) into enucleated oocytes initiates nuclear reprogramming of lineage-committed cells to totipotency. Pioneer SCNT work culminated with cloned amphibians from tadpoles, while technical and biology-driven advances led to cloned mammals from adult animals. Cloning technology has been addressing fundamental questions in biology, propagating desired genomes, and contributing to the generation of transgenic animals or patient-specific stem cells. Nonetheless, SCNT remains technically complex and cloning efficiency relatively low. Genome-wide technologies revealed barriers to nuclear reprogramming, such as persistent epigenetic marks of somatic origin and reprogramming resistant regions of the genome. To decipher the rare reprogramming events that are compatible with full-term cloned development, it will likely require technical advances for large-scale production of SCNT embryos alongside extensive profiling by single-cell multi-omics. Altogether, cloning by SCNT remains a versatile technology, while further advances should continuously refresh the excitement of its applications.

Key words Cellular reprogramming, Enucleation, Nuclear transplantation, Nuclear remodeling, Pluripotency, Totipotent

---

1 Introduction

Mammalian ontogeny begins with fertilization, a process that culminates with the fusion of two fully differentiated haploid cells, namely the oocyte and the spermatozoon. The resulting embryo will inherit this novel genome and faithfully replicate it during mitotic cell divisions throughout development and the animal's lifespan [1]. There are very few exceptions to this rule, thus including chromosome crossing-overs in the germline [2], and both immunoglobulin class switching and somatic hypermutation during lymphocyte maturation [3]. Errors in DNA metabolism [4] or

---

The original version of this chapter was revised. The correction to this chapter is available at https://doi.org/10.1007/978-1-0716-3064-8_18

exposure to environmental (e.g., pollutants) and intracellular insults (e.g., oxidative stress) may lead to DNA damage and mutations [4, 5]. However, the genetic fidelity of mammalian genomes is ensured by the DNA damage response (DDR) signaling pathway [4]. Upon DNA damage, the DDR senses such lesions and applies one of its multiple enzymatic tools for repairing the affected DNA. Very few DNA lesions escape the surveillance of the DDR pathway, and this mutation burden may contribute to aging or the onset of diseases such as cancer [6, 7].

Despite this static nature of the genome sequence, cells undergo drastic phenotypic changes during mammalian development. Immediately after fertilization, the early embryo displays rapid mitotic cell divisions with gradual decrease of blastomere size, underscoring the period of cleavage or pre-compaction development [1, 8]. Embryo compaction increases cell-cell interactions between blastomeres preceding the morula stage and initiates the polarization of outer cells, which flatten their apical surfaces and become epithelial-like [1, 8, 9]. Inner cells in the compacting embryo remain unchanged phenotypically and a fluid-filled cavity coined blastocoel begins to form. The initial segregation of cell populations resolves at the blastocyst stage, when inner cells form the inner cell mass (ICM) and cells in the periphery give rise to the trophectoderm (TE) [1, 8, 10].

The phenotypic changes in embryonic cells during early development are the product of multiple cellular and molecular processes. During the pre-compaction period, the development is under maternal control since it relies on oocyte-derived constituents (organelles, mRNAs, proteins, and other cellular components) accumulated during the oocyte growth phase [1, 8, 10]. For instance, mouse oocytes hold ~20,000 mitochondrial DNA (mtDNA) copies, although descendent embryonic cells will carry ~4,000 mtDNAs around implantation [11]. The processes of polarization and blastocoel formation require cytoskeletal dynamics that include filamentous actin participating in forming intercellular junctions [12]. After a few cell divisions, embryos undergo embryonic genome activation (EGA) and newly synthesized gene products begin to replace those of maternal origin [1, 8, 10]. This oocyte to embryo transition is gradual, varies between species and gene transcripts, and ultimately ends around gastrulation [13, 14]. From the zygote stage until EGA, the embryo is totipotent, which is the capacity to progressively give rise to all cell types in the fetus, and extraembryonic tissues including the placenta [15, 16]. At the blastocyst stage, ICM cells are pluripotent and hold the capacity to give rise to all cell types of the fetus. In turn, the TE cells that are multipotent will differentiate into extraembryonic tissues and the placenta [1, 8, 10].

The early embryo is self-organizing, since it grows ex vivo under conditions of minimum nutrient availability [9, 10]. Part of this "developmental programming" is embedded in the genome. Transcription factors (TFs) are key players in early embryogenesis and its first cell fate decisions [9, 10]. TFs modulate numerous target genes by interacting with cis-regulatory elements (CREs; promoters and enhancers) across the genome via their DNA-binding domains. The interplay between TFs and CREs modulates gene activity by recruiting the transcriptional apparatus for gene activation or repressor complexes for gene silencing. This genome-wide regulatory role gave TFs the reputation of as master regulators of cell identity and coordinators of cell fate transitions. In the early embryo, the starting gene transcript abundance of key TFs is of maternal origin, such as *Oct4(Pou5f1)* and *Sox2* [17, 18]. Core pluripotency-associated TFs (*Oct4*, *Sox2*, *Nanog*) maintain the viability of pluripotent cells in early embryos and TE-inducing TFs (*Cdx2*, *Eomes*, *Tead4*) suppress pluripotency gene expression to inform the extraembryonic program for placenta formation [9, 10]. Therefore, TFs dictate a substantial part of both temporal and spatial gene expression programs in the developing embryo and are necessary for its developmental potential [9, 10, 17, 18].

Mammalian development is a unidirectional process. As mammalian cells respond to differentiation cues such as hormones or growth factors and commit to a specific cell lineage, their progeny cannot revert to the previous state, and thus, tissues are maintained by tissue-specific stem/progenitor cells and mitotic divisions of mature cells [19, 20]. There may be few context-specific exceptions in organs such as the liver and pancreas, where cell replenishment may benefit from the dedifferentiation of mature cell types that later commit to hepatocytes and ß-cell phenotypes [21, 22]. The inability to switch cell fates originates from the fact that cells acquire non-genetic memory from the exposure to developmental cues. This memory may contemplate extensive transcriptional rewiring [23–25] and genome-wide TF binding reallocation [23, 25]. A comprehensive example comes from the in vitro differentiation of preadipocytes into mature adipocytes in mice and humans [23]. To explore TF dynamics, authors mapped the TF binding sites (TFBS) of PPARɣ in both species during four steps in the differentiation protocol. PPARɣ is a TF mostly associated with gene activation and required for adipogenesis. The TFBS of PPARɣ overlapped with genes associated with adipogenesis and there was a strong correlation between PPARɣ binding and target gene expression [23]. Moreover, the profile of TFBS of PPARɣ changed substantially during cellular differentiation [23]. Likewise, profiling 38 TFs during differentiation of human pluripotent cells into the three germ layers provided more details on such TF rewiring [25]. For instance, the TFBS profile of *NANOG* in endoderm was much similar to pluripotent cells, while *GATA4* showed pronounced rewiring in the same developmental context [25].

Transitions in cellular states during development also rely on epigenetic remodeling [10, 23–25]. Epigenetics encompasses inheritable changes in gene activity that do not depend on the DNA primary sequence [26]. Epigenetics include DNA methylation, post-translational modifications of histone tails, and non-coding RNAs. DNA methylation represents the deposition of a methyl group in the carbon five of a cytosine by DNA methyltransferases 1, 3a, and 3b (DNMT1, DNMT3A, DNMT3B) in a symmetrical CpG context [27]. DNA methylation in CREs associates with gene repression, while its deposition in gene bodies correlates with gene activity. Despite much of DNA methylation being constant across cell types, DNA methylation levels increase during cell differentiation [26, 27]. Moreover, histones are basic proteins with high affinity for DNA [28]. There are four core types of histones (H2A, H2B, H3, and H4) that form globular structures known as nucleosomes, which is the basic unit of packing genomes. Each nucleosome consists of eight histones (two copies of each histone type) and tightly packs ~150 base pairs of DNA [28]. This interaction between the nucleosome and bound DNA is influenced by post-translational modifications of histone tails that impact the accessibility of the bound DNA. There are over 25 types of histone modifications [28]. For instance, histone acetylation associates with euchromatin formation, while histone methylation may be in both euchromatin and heterochromatin [26]. During cellular differentiation, DNA methylation and epigenetic modifications act in concert to make more stable chromatin states [24, 25]. For example, facultative heterochromatin in somatic cells shows enrichment for DNA methylation and histone 3 trimethylation at lysine 9 (H3K9me3) [29].

There are two developmental time windows when epigenetic patterns undergo a genome-wide reprogramming [26, 27]. Few hours after fertilization, sperm cells experience substantial nuclear remodeling, which includes global DNA demethylation [30] and exchange of sperm-specific protamines by histones stored in the oocyte [31]. Further, both paternal and maternal genomes experience extensive reprogramming of DNA methylation and histone modifications during preimplantation development [32, 33]. This reprogramming occurs as a passive loss of epigenetic marks up to the blastocyst stage and gradual reestablishment during fetal and placental developments. The second stage of intense epigenetic reprogramming occurs during gametogenesis, when primordial germ cells display very low levels of epigenetic marks [34] that will be established in a gender-specific fashion [35, 36].

The complexity of cellular states and the underlying mechanisms driving differentiation motivates the following question: Could one reverse these differentiation-associated mechanisms? And to what extent? The definitive answer to the aforementioned question came from the development of nuclear transfer (NT) technology. This chapter aims to review seminal developments in the field of animal cloning, thus revealing technical and biology-driven advances that systematically pushed the technology forward and refreshed the excitement of its applications in basic science and commercial endeavors.

2 Nuclear Equivalency and Animal Cloning

The determinants of cellular differentiation were elusive a century ago, despite the knowledge of developmental phenomena. Amphibians (mostly frogs and salamanders) were preeminent animal models in embryology research during the first half of the twentieth century. This preference for amphibians was due to multiple reasons, which included the large size of oocytes and eggs (immature and mature gametes, respectively), ex vivo development, and resistance to physical damage [37]. Based on rudimentary tools and technical skill, embryologists showed that blastomeres remained totipotent when separated or spatially rearranged in early embryos. The limitation of this experimental approach was that it becomes technically infeasible when embryos reach gastrulation and cells become too small and have stronger cell-cell adhesions. Nonetheless, these findings showed that early blastomeres were undifferentiated, but it remained unknown what happened during later stages of development [38].

Initial investigations on the genetic basis of cellular differentiation relied mostly on cytogenetics [39]. There was an understanding that somatic cells have few euchromatic regions and that such open chromatin state correlated positively with gene activity. However, cytogenetics at that time did not offer adequate resolution to distinguish between potential changes in gene activity from DNA rearrangements during cell differentiation. Therefore, the logic of such "adaptations" by which genes undergo during cell lineage commitment was unknown [39]. To unequivocally test the theory of nuclear equivalency (i.e., cells retain the exact same genome throughout their lifespan), the ultimate test would be to introduce a differentiated cell nucleus into an enucleated oocyte (egg) and determine its ability to choreograph development [38–40]. Despite the simplicity of this concept, it demanded effort to evolve into the NT technology known today.

Initial evidence suggesting the feasibility of NT in vertebrates came from early NT experiments in amoeba [41] and demonstrations that partial enucleation of zygotes (i.e., removal of the female pronucleus) impaired embryonic development in frogs and salamanders [42, 43]. Further, enucleated eggs allow testing the potency of embryonic cells of the leopard frog *Rana pipiens* [38, 44]. One initial concern on developing a NT protocol was to avoid damage to oocytes and donor cells during micromanipulation [38]. Eggs were subject to pricking that caused their activation (Fig. 1). Enucleation occurred after activation by aspirating the nucleus out of the animal pole with a glass micropipette under a stereomicroscope [38, 40, 44, 45]. Donor cells of late blastula had roughly the same size of somatic cells [38]. The donor cell sucked into a slightly larger glass micropipette led to the rupture of the

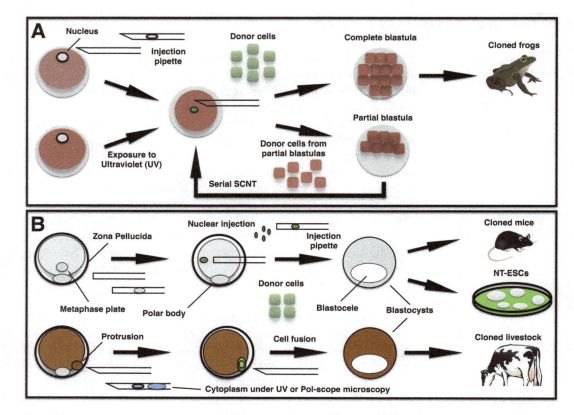

Fig. 1 Somatic cell nuclear transfer (SCNT) technology in amphibians and mammals. (**a**) SCNT in amphibians begins with egg enucleation using micromanipulation (Rana pipiens) or exposure to ultraviolet (UV) light (Xenopus laevis), which also causes oocyte parthenogenetic activation. Enucleated eggs undergo reconstruction by nuclear injection. Complete blastula may lead to cloned tadpoles and metamorphosed frogs. Partial blastulae cannot support full-term development but may acquire such ability by serial SCNT. (**b**) Mouse SCNT relies on oocyte enucleation by aspiration of the metaphase plate and reconstruction by nuclear injection using a piezo unit. Livestock SCNT uses oocyte enucleation by cytoplasm aspiration and exposure to UV (for assessing enucleation). Alternatively, oocytes are cultured with compounds that induce the formation of cytoplasmic protrusions holding the maternal chromosomes. These protrusions guide oocyte enucleation without UV exposure. Pol-scope microscopy also permits visualizing oocyte chromosomes without DNA dyes or UV. Reconstruction of livestock enucleated oocytes occurs by cell fusion (electrofusion or Sendai virus). Both mouse and livestock reconstructed oocytes undergo parthenogenetic activation and in vitro culture. Cloned blastocysts allow full-term development and the derivation of NT-derived embryonic stem cells (NT-ESCs)

cellular membrane and underwent injection into the enucleated egg [38, 40, 44]. This NT protocol in *R. pipiens* generated cloned blastulae at high frequencies (~40–50% reconstructed eggs) from blastula donor cells and many metamorphosed into healthy frogs [40]. Therefore, this represented the first NT protocol and initiated a series of developments in cloning technology (Fig. 2). In contrast, somatic cell NT (SCNT) using endoderm nuclei showed much lower cloning efficiency (17–33% blastulae) and very few clones

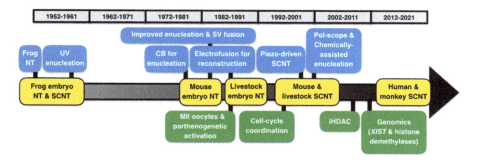

Fig. 2 Timeline of technical and biology-driven advances contributing to the development and applications of mammalian cloning by somatic cell nuclear transfer (SCNT) technology. Technical advances in blue, biology-driven advances in green, and key cloning achievements in yellow. CB: cytochalasin B, iHDAC: inhibitors of histone deacetylases, MII: metaphase II, NT: nuclear transfer, SV: Sendai virus, UV: ultraviolet light

reached the larval stage [46]. This was strong evidence that cloning efficiency was inversely proportional to the differentiation status of the donor cell [40, 46, 47].

Cloning technology was soon applied to the crawled frog *Xenopus laevis* [48, 49], albeit requiring technical adaptations [40, 50, 51]. The *X. laevis* egg has an impenetrable jelly which frustrated attempts to inject them [51]. The exposure of eggs to ultraviolet (UV) abolished their developmental capacity thus skipping enucleation by micromanipulation (Fig. 1), and triggered parthenogenetic activation [40, 51]. Surprisingly, the UV exposure dissolved the jelly and facilitated nuclear injections [51]. Although cloning of *X. laevis* was less efficient than *R. pipiens* using blastula cells (~15–30% reconstructed eggs formed blastulae), it fostered reprogramming of endoderm cells from pre-hatching tadpoles (0.35% reconstructed eggs) up to metamorphosed frogs [49]. The discovery of mutant *X. laevis* carrying one nucleolus (wild-type animals have two nucleoli) allowed tracking the fate of mutant donor cells and ensuring that clones originated from them [49, 51, 52]. Since few somatic cells were amenable to nuclear reprogramming, it remained unsolved if these cells harbored phenotypic plasticity (e.g., somatic stem cells) and did not reflect the phenotype of most somatic cells [49]. The use of intestine epithelium, keratin-positive skin cells, and spleen lymphocytes as nuclear donors gave more appealing evidence that fully differentiated cells are amenable to nuclear reprogramming [52–54], albeit these experimental settings did not convincingly demonstrate their origin from fully differentiated cells. These works did not apply lineage tracing approaches for tracking differentiated donor cells, some donor cells were negative for differentiation markers, and DNA rearrangements in lymphocytes were not yet known [52–54]. Since few reconstructed eggs using intestine epithelium developed into complete blastulae (~7%), abnormal embryos used as nuclear donors for a second round of SCNT (known as serial

SCNT) led to viable feeding tadpoles [53]. These results from serial SCNT strongly suggested that non-genetic factors were the underlying blocks to cloned development [55]. Despite correct interpretation of hindering constraints to nuclear reprogramming, the advantage of serial SCNT in *X. laevis* was due to its ability to enhance DNA replication in donor cells [56].

The pioneer work on NT technology [38] inspired its adaptation to other species, which included the *Drosophila melanogaster* [57] and the teleost fish *Misgurnus fossilis L.* [58]. Initial NT experiments in Drosophila injected multiple donor cells per enucleated egg [57, 59], thus limiting its ability to determine the developmental potency of donor cells. Upon stepwise adaptation of NT protocol, blastoderm cells injected into unfertilized eggs resulted in 1–2% clones developing up to larval stages [60]. Gasaryan et al. [58] took advantage that teleost eggs placed in tap water undergo parthenogenetic activation but fail to cleave. Eggs were exposed to X-ray irradiation of 38–49 rads to ensure enucleation, while donor cells came from uncultured blastulae [58]. From a total of 1,111 NTs to enucleated ($n = 791$) and non-enucleated ($n = 320$) eggs, reconstructed eggs cultured in Niu-twitty solution formed 200 gastrulae (18%), and 15 (1.3%) reached the feeding larvae stage [58]. In contrast, no cleavage was found after activation of 1,000 eggs, which included enucleated and intact gametes. The genetic contribution of teleost blastula cells to cloned development came from cytogenetics (e.g., 66% diploid embryos) and carboxylesterase isoenzyme genotyping [58].

The initial three decades of non-mammalian cloning showed that somatic nuclei are amenable to nuclear reprogramming, acquire the ability to orchestrate embryogenesis, and generate clones that reach adulthood and prove to be fertile. In turn, cloning efficiency was low, accompanied by substantial losses during cloned development, and even serial SCNT did not abrogate such inherent biology-driven limitations.

3 Mammalian Cloning from Embryonic Nuclei

The development of NT technology in mammals lagged behind until the 1970s (Fig. 2). Unlike their amphibian counterparts, mammalian oocytes (both immature and mature gametes) are smaller (1,000-fold) and less resistant to physical damage [61, 62]. Other caveats in mammals included limited knowledge on oocyte activation and embryo in vitro culture [61–64]. In turn, two early developments contributed to early NT works in mammals, namely initial demonstrations of partial zygote enucleation by removing one pronucleus [62] and oocyte reconstruction by cell fusion via inactivated Sendai virus (SV) [65].

The first successful description of NT in mammals was done in rabbits, which led to cloned embryonic development [61]. Oocytes underwent reconstruction by nuclear injection or SV-mediated fusion and the majority (70–100%) survived micromanipulation. The author avoided oocyte enucleation, most likely to avoid their lysis. There were two technical innovations that included donor cell synchronization in metaphase with nitrous oxide and their lineage tracing with tritiated thymidine [61]. Nonetheless, ~2% (15 of 694) of reconstructed oocytes cleaved and only four developed into embryos harboring 18–26 cells [61]. The low embryonic development was likely due to poor oocyte activation, which was insufficiently stimulated by oocyte cooling to 5 °C during NT. In mice, the injection of morulae nuclei carrying T6 chromosomes to intact (non-enucleated) zygotes led to low survival (9%) but more than half of reconstructed embryos formed morulae and blastocysts in vivo [66]. The fate of donor nuclei was suggested by T6 chromosomes in cloned blastocysts, albeit one drawback was the slower developmental kinetics [66].

A key discovery for mammalian NT came from an unexpected place. Hoppe and Illmensee [67] cultured mouse zygotes in medium containing a cell-permeable and reversible inhibitor of actin filaments called cytochalasin B (CB), thus intending the diploidization of single pronucleus in partially enucleated zygotes. Curiously, CB treatment increased zygote survival after micromanipulation [67]. Another report replicated these findings and showed that pre-treatment with CB before enucleation nearly doubled the survival of partially enucleated zygotes [68]. This tweak by CB allowed more sophisticated NT experiments (also known as microsurgery) for testing the developmental potency of ICM and TE nuclei [69, 70]. The injection of ICM nuclei into intact zygotes led to blastocysts after in vitro culture, while zygotes reconstructed with TE cells did not develop beyond the 8-cell stage [69]. The enucleation of CB-treated zygotes alongside reconstruction with ICM and TE gave reasonable survival rates (~40%) and supported development up to blastocysts [70]. These reports suggested that mouse ICM cells are more amenable to reprogramming than their TE counterparts [69, 70].

Despite some progress in animal cloning using embryonic donor cells, NT protocols remained partial (without enucleation) [69] or were difficult to replicate independently [64, 70, 71]. To circumvent such roadblocks, the use of CB and large injection pipettes to enucleate zygotes removed oocyte cytoplasm containing both pronuclei without plasma membrane lysis [63]. A pronuclear karyoplast (cytoplasm bridge carrying two pronuclei) from another zygote was inserted into the perivitelline space of the enucleated zygote for cell fusion using SV (Fig. 1b). These two steps allowed the survival of >90% of reconstructed zygotes that formed blastocysts in vitro and gave rise to newborn mice at similar rates than non-manipulated controls [63]. This novel NT protocol was accurate, reliable, and easy to master [64, 71]. In the following year, two independent groups applied this NT protocol to demonstrate the non-equivalence of male and female pronuclei during mouse development [72–74], thus acting as strong evidence for genomic imprinting. The reconstruction of androgenetic and gynogenetic embryos (carrying two male or female pronuclei, respectively) led to high preimplantation development [74]. The transfer of androgenetic or gynogenetic embryos to recipient female mice culminated in reasonable implantation rates (17–35%), albeit embryos were unviable and inevitably underwent reabsorption or displayed severe growth retardation [72–74].

Early NT experiments relied on zygotes as recipients since there was no efficient method for oocyte activation in mammals [66, 69, 70]. This useful experimental setting for NT with embryonic donor cells also led to a premature conclusion. The reconstruction of enucleated zygotes with blastomeres of cleavage-stage embryos gave intriguing results. Unlike pronuclear exchange that led to nearly all reconstructed zygotes reaching the blastocyst stage (~95%), the efficiency dropped to 19% when using 2-cell stage blastomeres as donors [75]. The use of donor cells from more advanced developmental stages (4-cell, 8-cell, and ICM) did not affect reconstruction rates, albeit very few embryos reached the morula stage [75]. Additional experiments showed that enucleation (but not nuclear injury or toxicity caused by the injection medium) was the cause of developmental arrest of cloned embryos. This progressive loss of developmental potential led to the conclusion that mammalian cloning was impossible [75]. Likewise, rabbit ICM cells showed limited potential to develop into cloned blastocysts, while cloned embryos from TE cells faced developmental arrest at the 8-cell stage [76].

Cloning work in amphibians motivated the field to overcome the hurdles described above [64, 71]. The first NT in large mammals provided more technical and biology-driven improvements [77]. Sheep metaphase II (MII) oocytes once enucleated by bisection using a pulled glass pipette underwent reconstruction by SV-mediated fusion or electrofusion. The later reconstruction approach proved much faster (1 h instead of 4 h) and supported high (42–48%) blastocyst development [77]. Reconstructed oocytes with 8-cell or 16-cell stage nuclei developed into full-term lambs and viable 60-day pregnancies (recipients killed to assess the development of clones), respectively [77]. This work was soon followed by several reports of mouse and livestock cloning using donor cells from early embryos [71, 78, 79].

Besides reinforcing the amenability of reprogramming embryonic nuclei to totipotency [70, 77, 80], these works collectively gave important insights into a key factor impacting cloning efficiency [71, 78, 79]. The use of MII oocytes (or 2-cell stage blastomeres) proved more efficient than zygotes as recipients for mammalian cloning [71, 79]. The drawback of MII oocytes is the high levels of maturation/meiosis/mitosis-promoting factor (MPF) and thereby requires cell cycle co-ordination with donor cells [78]. Most blastomeres in non-synchronized preimplantation embryos are under interphase of the cell cycle (S-phase) (80–90%) due to short (<1 h) G1/G2 phases [76, 78]. The high MPF level ruptured the nuclear membrane of blastomeres causing a metaphase-like premature chromosome condensation (PCC) followed by pronucleus formation and nuclear swelling [76, 78]. However, PCC damaged the chromosomes of S-phase nuclei and caused developmental arrest [78, 81, 82]. Embryonic nuclei at G1 phase remained intact after PCC and gave rise to high yields of cloned embryos [78, 81]. The activation of oocytes before reconstruction lowers MPF levels and creates a cytoplasmic environment that supports non-synchronized blastomeres (G1, S, or G2 phases) as nuclear donors [78]. This MPF-low "universal recipient" oocyte did not trigger PCC and nuclear swelling to the magnitude of MII oocytes but may lead to high blastocyst yields, at least under some experimental conditions [76, 78].

A prelude to SCNT came from cloning work using non-cultured ICM cells and ICM-derived cell lines. Sheep enucleated oocytes receiving ICM donor cells formed 38% blastocysts upon transfer to ligated oviducts of recipient ewes in diestrous [80]. The cloned lamb obtained did not survive most likely due to dystocia because there were no gross malformations [80]. In cattle, NT with ICM nuclei gave few blastocysts (3–7%) but reasonable (13–15%) calving rates [83, 84]. Suspension culture of ICM cells for NT translated into ~15% blastocysts and ~12% of transferred blastocysts gave rise to cloned calves [85]. Alternatively, sheep embryonic disk (ED) cells from day-9 blastocysts allowed the derivation of a cell line with flattened morphology [86]. The reconstruction of enucleated activated oocytes (universal recipients) with unsynchronized cultured ED cells (up to passage three) led to cloned lambs. Upon passage six, ED cells (renamed to TNT4) expressed the markers of differentiation cytokeratin and nuclear lamin A/C [86]. To facilitate the reprogramming of gene expression [87], authors cultured TNT4 cells under serum deprivation (0.5% serum) conditions instead of the usual supplementation with 10% fetal calf serum. The serum deprivation forces cells to exit the cell cycle and remain quiescent at the G0 phase [86]. The use of quiescent TNT4 cell donors offered similar blastocyst rates to ED cells, while recipient conditions (pre-activated, simultaneous fusion and activation, or post-activated MII oocytes) did not affect blastocysts yields. Five lambs were born from the transfer of cloned blastocysts obtained from TNT4 cells at passages 6–13 and the three oocyte recipient conditions [86]. ICM-derived cells were not equivalent to embryonic stem cells (ESCs) because there was no protocol for maintaining cattle pluripotent cells under in vitro culture [85]. Moreover, these reports were exciting because they gave definitive proof that partially differentiated cells are amenable

12   Marcelo Tigre Moura

to reprogramming [85, 86]. Furthermore, it paved the way for genetically modifying cells in culture before NT to generate cloned transgenic animals [86, 88, 89].

4 The Development of SCNT in Mammals

SCNT in mammals did not await the unfolding of cloning using embryonic nuclei. Rather, it began as cytological studies for investigating nuclear-cytoplasmic interactions [65, 90] and as an initial assay before testing SCNT [91]. These studies took advantage of fused couplets from somatic cells and oocytes, thereby allowing the inspection of two nuclei of different origins simultaneously. Although oocyte chromosomes remained unaltered upon cell fusion, mouse cumulus or thymus cells underwent nuclear membrane breakdown and a reversible PCC (lasted a few hours) with clustered or scattered chromosomes within the oocyte [90, 91]. Somatic nuclei behaved differently in activated oocytes, displaying a "pronucleus-like" morphology [90, 91]. Under such context, somatic cells held swelled nuclei (200-fold enlargement), decondensed chromosomes, and visible nucleoli [90, 91]. The decondensation of somatic nuclei reached similar levels and spatial dispersion than of maternal chromosomes with almost simultaneous (<1 h interval) fusion and activation [92]. Further, these oocytes extinguished gene transcription of somatic nuclei, thus mirroring the female pronucleus. SCNT into pre-activated oocytes (cell fusion >3 h after activation) did not abolish gene transcription of somatic nuclei [92]. When cell fusion occurred several hours after oocyte activation, somatic nuclei kept their naive chromatin conformation [91]. Due to technical constraints, these studies faced spontaneous oocyte activation [90] or fusion between multiple somatic cells to single oocytes [91]. Nonetheless, these studies showed that mammalian somatic nuclei undergo nuclear remodeling similar to embryonic cells. However, the biological relevance of these cytogenetic studies still required proof of embryogenesis after SCNT [90].

Accumulating evidence of mammalian cloning from embryonic nuclei and suggestive evidence of similar nuclear remodeling at the cellular level with thymocyte nuclei set the stage for cloning using somatic cells. Initial efforts revealed that mouse, rabbit, and cattle somatic nuclei (thymocyte, fetal fibroblasts, and granulosa, respectively) undergo reprogramming upon SCNT and orchestrate cloned preimplantation development [83, 93]. The transfer of 19 cloned blastocysts derived from granulosa cells did not evolve into implantation in cattle [83]. A common fact of these studies was that donor cells were not subject to cell cycle synchronization (i.e., no description available) [83, 93]. Since driving donor cells to a quiescent state favored sheep cloning [86], authors relied on this approach to use fetal fibroblasts as donor cells for SCNT [94]. Reconstructed oocytes were subject to in vivo embryo culture (oviduct of recipient ewes) or in vitro culture. From a total of 144 reconstructed oocytes, 47 became morulae or blastocysts [94]. Upon transfer of 40 embryos to synchronized recipients, three cloned lambs were born but one died soon after delivery. Based on the number of embryos transferred to recipients, the cloning efficiency using fetal fibroblasts (7.5%) was similar to using ED cells (5.6%) [94].

An additional experiment was carried out using donor cells from mammary epithelium of a 6-year-old ewe carrying a last trimester pregnancy [94]. Quiescent mammary cells used for SCNT formed fewer blastocysts than the other donor cell types (ED cells and fetal fibroblasts), albeit the transfer of 29 embryos to recipient ewes led to the birth of one cloned lamb [94], better known as Dolly.

5 Applications of Mammalian Cloning by SCNT

The cloning of an adult mammal caused much excitement and brought attention to its potential applications [88, 89, 95]. The full reprogramming of somatic nuclei of adult mammals gave more appealing evidence that the genome remains unchanged during the animal's lifespan. In a few years, mammalian cloning by SCNT under intense investigation led to cloning of 20 mammalian species [96]. Most proof-of-principle cloning studies relied on similar experimental conditions, albeit some species required protocol adaptations to match to their unique physiology, such as the blockade of spontaneous activation in rat oocytes [97] and shortening oocyte activation associated with non-synchronous embryo transfer in rabbits [98]. Despite low cloning efficiencies, the technology proved replicable, and thousands of clones (mostly cattle and other livestock) were born in the following years [99].

The most straightforward application of cloning by SCNT is the replication of desirable genomes. Appealing individuals eligible for cloning may come from rare livestock populations, endangered species, elite livestock, or companion animals [100–103]. Cloning also offers the potential to bring back genomes stored for long periods [104] and thus assisting conservation programs of genetic resources. The genomes of non-viable cells may be suitable for cloning [105], thereby expanding the repertoire of approaches to preserve rare genetic material. For instance, freeze-drying (also known as lyophilization) may become a more appealing approach for the conservation of genetic resources when combined with SCNT protocols [106].

The potential of animal cloning in the context of transgenesis was a driving force in the pursuit of SCNT in mammals

[88, 89]. Before cloning, rare transgenic animals were the end-product of injecting thousands of zygotes and the random integration of exogenous DNA [89, 107]. In contrast, somatic cells during in vitro culture are eligible to genetic modifications (addition or removal of DNA), selection of genetically modified cells (transgenic without off-target mutations), and expansion of transgenic cell clones before SCNT [89, 108]. This strategy ensures that all clones born harbor the genetic modification made on the donor cell line. The use of recloning and gene editing technologies made possible the generation of healthy clones carrying multiple site-specific editions in their genome [109]. The cloned transgenic animals serve as founder animals to generate herds of transgenic animals by natural breeding or assisted reproductive technologies [108]. Transgenesis in livestock led to novel disease models [110], large-scale production of recombinant proteins [111], organs with potential for xenotransplantation [112], among other applications.

Cloning by SCNT also contributes to unrevealing biological processes. For instance, the epigenetic reprogramming during cloning offered the possibility of distinguishing between the genetic and epigenetic control of a given phenomenon. Tumorigenesis is the culmination of stepwise accumulation of both genetic lesions and epigenetic alterations [113]. SCNT using cancer cells as donors would reset their epigenetic state and thereby generate clones with the sole burden of the genetic lesions. In other words, the phenotypic analyses of clones from cancer cells would uncouple the contribution of genetic mutations from acquired epigenetic alterations during cancer initiation and progression. The use of mouse cells from multiple cancer models led to few reconstructed oocytes developing into blastocysts (67 out of 2,201; 3%) [114]. From 57 cloned blastocysts explanted in ESC culture conditions, two gave rise to stable NT-ESC lines (3.5%). Both NT-ESC lines came from a melanoma model driven by doxycycline-inducible *Ras* oncogene combined with the deletion of the tumor suppressor Ink4a/Arf [114]. There were two reasons for not attempting to generate full-term clones directly from these SCNT embryos. Firstly, to circumvent the developmental abnormalities commonly found in cloned pregnancies. Secondly, to explore the potential of chimeras, which have wild-type cells that support development with another cell population carrying phenotypes not compatible with full-term development. The injection of NT-ESCs into wild-type host embryos ultimately resulted in mouse chimeras that developed cancers with shorter latency [114], greater penetrance, and expanded tumor spectrum than the original disease model. As a more stringent test of developmental potency, NT-ESCs were injected into tetraploid blastocysts (tetraploid complementation assay). Since polyploid cells progressively become largely restricted to extraembryonic tissues, injected pluripotent cells must orchestrate fetal development. Nonetheless, the fetal development from NT-ESCs inevitably arrested on embryonic day 9.5 [114]. Authors argued that secondary changes accumulated during cancer progression do not abrogate mouse development but predispose them to more aggressive tumors.

Another example of exploring the role of epigenetics during development came from the quest for elucidating the mystery of olfactory receptor choice. The mouse genome has >1,000 genes encoding olfactory receptors and each neuron expresses a specific receptor from this vast repertoire [115]. To exclude the possibility of DNA rearrangements locking receptor choice, two independent groups used olfactory sensory neurons (OSN) as donor cells for SCNT [116, 117]. The cellular identity of OSNs was based on a knock-in reporter line in which the green fluorescent protein (GFP) was expressed under the control of the olfactory marker protein (exclusively expressed in mature OSNs). Dissociated tissue allowed the identification of GFP-positive OSNs under epifluorescence before their use for SCNT. These two reports described 27 NT-ESC lines from nearly 2000 reconstructed oocytes altogether [116, 117]. NT-ESC lines generated chimeras with widespread tissue contribution, suggestive of pluripotency in vivo. Clonal mice were born from tetraploid complementation [116], and recloning using NT-ESC donor cells gave rise to cloned pups [117]. Moreover, OSNs expressing specific olfactory receptors (i.e., P2, M71) for SCNT ultimately generated chimeric and clonal mice with alternative receptors [116, 117]. Hence, epigenetic mechanisms govern olfactory receptor choice in OSNs and post-mitotic cells may undergo extensive nuclear reprogramming.

The utilization of T lymphocytes as donor cells put to rest the quest to demonstrate the ability to reprogram terminally differentiated cells [118]. Although lymphocytes could be reprogrammed by SCNT and making clonal mice from NT-ESCs [95, 119], it did answer if oocytes alone supported full epigenetic resetting. To test their potential for nuclear reprogramming, authors retrieved natural killer T (NKT) cells and helper T cells (hTCs) from mice and used them for SCNT [118]. While hTCs led to few morulae and blastocysts (12%), NKT supported their development at much higher rates (71%). While hTCs failed to support full-term development, NKT cells gave rise to clones at similar rates to other cell types (~1.5%) [118]. DNA rearrangements at the TCRVa14 and TCRVb loci in clones and their placentas gave definitive proof that terminally differentiated cells are amenable to nuclear reprogramming into totipotent cells [118].

The birth of Dolly was soon followed by the derivation of human ESCs [120]. This latter discovery prompted the idea of combining these two technologies for regenerative medicine, which became known as therapeutic cloning [121]. Instead of attempting to clone humans (i.e., reproductive cloning), the idea was to generate genetically matched NT-ESCs from patients and differentiate them into the desired cell type which is affected by disease [121, 122]. These disease-specific cells obtained by reprogramming could be used for in vitro models for understanding disease and screening therapeutics [123]. The derivation of mouse NT-ESCs proved feasible [114, 116, 117, 124, 125] and these cells were functionally similar to ESCs as demonstrated by in vitro differentiation potential, and extensive contribution to embryonic development alongside germline transmission [116, 117, 126]. Detailed molecular analyses of numerous cell lines demonstrated that NT-ESCs are indistinguishable to ESCs based on transcriptomes, DNA methylation maps, and differentiation potential [127, 128]. Proof-of-principle studies in the mouse suggested the potential of therapeutic cloning by ameliorating the effects of immunodeficiency (coupled with restoring *Rag2* expression) and Parkinson's disease models [129, 130]. The derivation of NT-ESCs in Rhesus monkeys (*Macaca mulatta*) [131] and humans [132] were additional steps toward the exploration of therapeutic applications.

It became evident that constraints to obtaining human oocytes and the complexity of SCNT would complicate the realization of therapeutic cloning [133, 134]. An alternative to oocytes could come from surplus zygotes from human-assisted reproduction. Due to the inability of interphase zygotes to reprogram embryonic and somatic cells [66, 69, 70, 135], one report described the synchronization of mouse zygotes in metaphase before enucleation [136]. The reasoning of metaphase arrest was that pronuclei enclose reprogramming factors and impede their interaction with the somatic nucleus [137]. Indeed, enucleation of zygotes arrested at metaphase induced nuclear reprogramming, thus allowing mouse cloning from donor ESCs and derivation of NT-ESCs from somatic cells [136, 138]. Nonetheless, the use of zygotes did not show promising results in humans because SCNT embryos failed to undergo EGA and faced developmental arrest at the morula stage [139].

The demonstration that expression of key TFs in somatic cells reprogrammed them into induced pluripotent cells (iPSCs), which are equivalent to ESCs and do not depend on oocytes or SCNT [140–142]. The ethical and technical advantages of iPSCs generated much enthusiasm for its development and applications. It circumvented the moral issues of making human embryos for deriving NT-ESCs, the potential risks of leading to human reproductive cloning, and the constraints of relying on human oocytes for SCNT [134]. Comparison of genetically matched human iPSCs and NT-ESCs showed similar transcriptomes, DNA methylation maps, and de novo mutations [143], thus suggesting that reprogramming methods generate pluripotent cell lines with similar molecular profiles and differentiation potential. Nowadays, the iPSC technology dominates the field of regenerative medicine and

Cloning by SCNT: Integrating Advances   17

numerous clinical trials of cell therapy rely on iPSC-derived cells [144].

6 Technical Adaptations of SCNT Protocols in Mammals

The protocol of SCNT is complex and time-consuming (Fig. 1). Beyond biology-driven adaptations to mammalian species mentioned above, the SCNT protocol remains essentially the same as the pioneer report in amphibians [38, 71]. The sole substantial simplification was the alternate oocyte enucleation approach developed for SCNT in *X. laevis* [50, 51]. The exposure to UV light did not affect oocyte developmental competence but impaired the maternal chromosomes [50]. This functional oocyte enucleation method by UV skipped the physical removal of oocyte chromosomes and was an actual shortcut to SCNT [48, 49]. Nonetheless, there were few technical adaptations that facilitated SCNT in mice and livestock (Fig. 2).

Attempts to adapt the UV-mediated enucleation to mammals failed due to developmental arrest most likely due to DDR pathway activation [145, 146]. Unlike mammals, amphibian pre-blastula development cannot activate the DDR pathway [147] and thereby supports embryonic development with damaged chromosomes [48, 49]. DNA intercalating drugs such as actinomycin D and mitomycin C have potential as alternative methods for oocyte functional enucleation [148, 149]. These approaches impeded parthenogenetic development after chemical activation of bovine oocytes but gave rise to blastocysts after SCNT, albeit at low efficiencies. Nonetheless, the removal of damaged oocytes fully recovers developmental potential [148, 149], further suggesting that the roadblock was restricted to DDR activation. Perhaps chemical inhibition of DDR [150] after oocyte reconstruction may support cloned development at high efficiency using functional enucleation with DNA intercalating drugs.

Mammalian oocytes incubated in DNA dyes and briefly exposed to UV reveal the location of the oocyte spindle and ensure enucleation by micromanipulation [151]. This approach became commonplace for oocyte enucleation in livestock because of their dark oocyte cytoplasm that impedes visual checking of successful enucleation [80, 152]. Two approaches became popular as alternatives to UV for ensuring oocyte enucleation. Pol-scope microscopy allows the visualization of oocyte spindles due to their birefringence [153]. Under the Pol-scope, oocytes from several species (mouse, golden hamster, bovine, and human) underwent enucleation at high efficiency and it did not affect preimplantation development of cloned mouse embryos [153]. Some compounds (demecolcine, nocodazole, and sucrose) cause the formation of cone-like protrusions in >70% of treated oocytes [154–

156]. These protrusions contain maternal chromosomes in ~100% oocytes and easily guide the enucleation without UV exposure [154–156]. These alternatives do not affect the oocyte developmental competence (similar cleavage rates, blastocyst rates, and blastocyst cell numbers) and support the production of cloned livestock [156, 157]. Therefore, chemically assisted enucleation methods improve oocyte enucleation but still require micromanipulation [154–157].

The first mouse cloning from somatic cells described a new strategy for oocyte reconstruction [158]. The piezo-impact pipette drive unit facilitated the rupture of donor cell membrane and release of the donor nucleus while discarding most of its cytoplasm. Upon oocyte reconstruction, the piezo unit allowed penetrating the oocyte oolemma and injecting the somatic nucleus with minimum cytoplasm [158]. Direct comparisons between cell fusion and piezo-driven nuclear injection in livestock showed conflicting results [159, 160]. One study in pigs found similar cleavage and blastocyst rates between oocyte reconstruction methods [160] but the other found comparable cleavage rates but much lower blastocyst rates using the piezo unit in cattle (35–39% vs. 11–18%) [159]. One similar study in mice revealed that SCNT embryos made using the piezo unit display more apoptotic cells and lower developmental potential than those from electrofusion [161]. Despite these facts, the vast majority of mouse and some livestock SCNT protocols use piezo units for micromanipulation.

An alternative SCNT protocol came from the development of handmade cloning (HMC), a SCNT protocol without micromanipulators [162, 163]. The traditional SCNT requires micromanipulators and micropipette-making equipment, which makes the experimental cloning setup quite expensive. The HMC protocol utilizes oocyte enucleation by bisection with a sharp blade controlled manually [162]. Halved oocytes are checked under UV to select enucleated ones and further subject them to two rounds of electrofusion. The first cell fusion combines the donor cell with a halved oocyte, while the second round fuses the reconstructed oocyte with another halved oocyte [162, 163]. One limitation of HMC is the need to perform embryo in vitro culture individually since it relies on starting zona pellucida-free oocytes [163]. HMC is particularly popular in livestock and numerous reports described clones using this protocol [164].

7 Improving SCNT-Mediated Reprogramming

Despite robust embryonic development after SCNT, molecular and cellular analyses showed incomplete reprogramming even at early stages of embryo development. Immunofluorescence analysis for epigenetic marks showed persistent DNA methylation levels and repressive histone marks concomitant with limited acquisition of euchromatic marks [165, 166]. This epigenetic memory (epigenetic states associated with the cell-of-origin) correlated with limited reactivation of embryonic genes [167], persistent expression of somatic genes [168], skewed X chromosome reactivation [169], and loss of gene imprinting [170]. At the cellular level, cloned embryos may display energy metabolism resembling somatic cells [171], with an increased number of apoptotic cells [172], among other detrimental phenotypes.

This plethora of epigenetic abnormalities in cloned embryos inevitably leads to substantial losses, mostly during peri-implantation development. Cloned pregnancies frequently fail to develop functional placentas [173], which begins with impaired TE differentiation [174], inadequate embryo-maternal signaling [175], and abnormally large placentas [176]. Likewise, full-term clones might display an array of abnormal phenotypes [177], thus including overgrowth [178], obesity [179], among others. Similar developmental constraints to those found in cloned development come from the phenotype of mice deficient [180] or overexpressing epigenetic regulators [181], which frequently display preimplantation development at mendelian rates but inevitably die in utero after implantation [180, 181]. These similarities suggest that gastrulation and organogenesis pose a greater challenge to embryos with epigenetic perturbations.

Mammalian cloning using various somatic cell types culminated in a view that its efficiency was inversely proportional to the differentiation status of the donor cell [47]. Since the epigenetic state of donor cells represents the main factor determining reprogramming potential, much work intended to understand the molecular roadblocks to safeguarding the cellular identity of somatic cells. Based on the notion that epigenetics governs cloning efficiency, treatment of donor cells with epigenetic modulators should facilitate reprogramming. Bovine somatic cells treated with 5-aza-2-′-deoxycytidine (5-aza; cytidine analog that prevents DNA methylation) and trichostatin A (TSA; deacetylase inhibitor) displayed lower DNA methylation and higher acetylation levels, respectively [182]. TSA-treated cells were more amenable to nuclear reprogramming as shown by higher blastocyst rates, albeit 5-aza was toxic and lowered embryo yields. Further, treatment of mouse reconstructed oocytes with TSA gave a threefold to fourfold increase in blastocyst rates [183, 184] and gave a more than fivefold (2.5–6.5% of transferred embryos) increase in live births [183]. Mechanistically, TSA recovers global levels of histone 3 dimethylation at lysine 4 (H3K4me2) and partially diminishes the H3K9me3 of SCNT embryos in comparison to fertilized embryos [185]. These modulations of histone marks in SCNT embryos were also accompanied by chromatin decondensation and increased DNA replication [185] alongside improved activation of EGA-associated genes [186]. Mouse SCNT with post-activation TSA treatment allowed serial cloning for 25 generations with reasonable efficiency (2–25%) [187], while clones had similar postnatal viability, telomere length, and transcriptomes (liver and brain). The effect of TSA was steady across generations and proved better than non-treated SCNT [187], which did not allow serial cloning for more than six generations [187, 188].

8 Genomics Revealed Roadblocks to Nuclear Reprogramming

The initial identification of key mechanisms associated with epigenetic memory enabled improvements in cloning efficiency, although it did not reveal the identity of genes or genomic regions needing remodeling during nuclear reprogramming. Genome-wide approaches offered this opportunity to explore nuclear reprogramming at much greater resolution that ultimately uncovered contextual chromatin states that act as roadblocks to reprogramming and that govern large portions of the donor cell identity (Fig. 2).

Transcriptomic analyses demonstrated that cloned embryos display aberrant expression of hundreds of genes during the EGA [189] and at the blastocyst stage [190, 191]. This flawed gene expression pattern derives from the combination of an inability to reactivate embryonic genes [167] and to properly silence genes active in donor cells [168]. For example, one genome-wide study using mouse cloned blastocysts found that the majority of dysregulated genes were found in the X chromosome [190]. Despite substantial variation among cloned blastocysts, a set of 129 genes were systematically aberrantly expressed in comparison with in vitro fertilization (IVF) embryo controls. Cloned embryos showed higher expression of the long non-coding RNA *Xist*, which is a major effector of X chromosome inactivation [190]. Curiously, both male and female clones displayed ectopic *Xist* expression in active X chromosomes. Upon SCNT using somatic cells with one *Xist*-deficient X chromosome ($X^{\Delta xist}$), cloned embryos had 73–85% fewer dysregulated genes, which included both X-linked and autosomal genes. X-linked genes with enrichment for the repressive mark H3K9me2 did not respond to *Xist* ablation [190]. More importantly, cloning efficiency using $X^{\Delta xist}$ cumulus and Sertoli cells improved eightfold to ninefold to 12.7% and 14.4%, respectively. Of note, the impact of correcting *Xist* expression was independent of TSA treatment. This conclusion lies in the inability of TSA to ameliorate the expression of X-linked genes of cloned embryos and the additive effect of TSA and Xist RNAi-mediated knockdown on boosting mouse cloning efficiency to ~20% [192].

An independent study found >1,200 differentially expressed genes between mouse IVF and SCNT embryos at the 2-cell stage

[189]. A total of 372 somatic genes remained active and 301 embryonic genes failed to become active in SCNT embryos. By mapping the genomic location of differentially expressed genes between zygotes and 2-cell IVF embryos, this work revealed 811 genomic regions associated with EGA [189]. From those, 222 genomic regions were refractory to reprogramming in SCNT embryos and referenced as "reprogramming resistant regions (RRRs)." Few genes overlap with RRRs because these are gene-poor genomic regions but are abundant for repeat sequences [192]. In silico analyses revealed that RRRs overlap with H3K9me3 regions in multiple somatic cell types. The injection of reconstructed oocytes with mRNA of the H3K9me3-specific histone demethylase *Kdm4d* greatly diminished H3K9me3 levels in SCNT embryos, while a catalytic-defective *Kdm4d* mRNA did not affect such embryos [192]. The *Kdm4d* injection corrected the expression of 737 dysregulated genes (~60%) in 2-cell SCNT embryos. This strategy also had a profound impact on blastocyst rates because SCNT using multiple cell types (cumulus cells, Sertoli, and embryonic fibroblasts) improved from 10–26% to 80–88% blastocysts [192]. After embryo transfer, cloning efficiency increased from 0% and 1.0% to 7.6% and 8.7% of live pups using cumulus and Sertoli as donor cells, respectively. Finally, embryonic fibroblasts deficient for two H3K9me3 methyltransferases *Suv39h1* and *Suv39h2* when used for SCNT recapitulated the improvements in cloning efficiency up to the blastocyst stage [192].

The greater resolution provided by genome-scale approaches made clearer the settlebacks made by incomplete reprogramming of DNA methylation profiles [193, 194]. For instance, repetitive DNA elements were refractory to demethylation [193], which may be due to the cumulative deposition of H3K9me3. Notwithstanding the reprogramming potential of oocytes, the DNA methylation maps of mouse somatic cells were much resistant and remained largely different from fertilized zygotes [194]. Mouse cloned blastocysts hold similar global DNA methylation levels to IVF controls, but genomics found thousands of differently methylated regions (DMRs) between them. Blastomere biopsying in 2-cell and 4-cell cloned embryos provided an assay to correlate DNA methylation maps and developmental potential [195, 196]. Clones displayed epigenetic memory as evidenced by persistent DMRs (resistant to reprogramming) and sequences that regained DNA methylation (re-methylation) during embryonic development [194]. Those re-methylated regions were mostly promoters and genomic regions harboring repeat elements. Upon injection of short interfering RNAs (siRNAs) for *Dnmt3a/3b* into enucleated oocytes, cloned preimplantation development was modestly ameliorated although full-term development increased from 0.88% to 5.33% [194]. Combined injection of *Dnmt3a/3b* siRNAs and *Kdm4d/Kdm5b* mRNAs further increased cloning efficiency to 17.21% [194], thereby suggesting non-redundant roles of DNA methylation and repressive histone marks in difficulting reprogramming. Vitamin C (VC) inhibits DNA methylation by blocking the conversion of Fe^{3+} into Fe^{2+} and therefore limiting Fe^{2+} and oxoglutarate mediated reactions necessary for this epigenetic mechanism [197]. Post-activation treatment of SCNT embryos with TSA and VC increased mouse cloning efficiency from 0.0% to 15.2%. Moreover, TSA + VC improved the transcriptional status of ~200 genes at the 2-cell stage, albeit it did not most affect genes located in RRRs [197].

The demonstration of major challenges during EGA, there was a search for factors driving EGA and that may become dysregulated during reprogramming. The chromatin conformation plays a major role during the onset of the EGA by allowing TFs and the transcription apparatus to engage with target genes. Histone 3 acetylation at lysine 9 (H3K9ac) correlates with gene activity during EGA and the failure to increase H3K9ac in SCNT embryos may pose another barrier to reprogramming [198]. TSA recovered most H3K9ac peaks in SCNT embryos (not in RRRs) although it did not translate into a more similar transcriptome to fertilized controls. Prospection of TF factors driving EGA found enrichment for *Dux* and *GATA3* binding motifs [198]. Gene expression analysis found that only *Dux* was downregulated in SCNT embryos. Remarkably, *Dux* mRNA injection rescued most SCNT embryos from the 2-cell block while cloned embryos from $Dux^{-/-}$ cells failed to develop [198]. Exogenous *Dux* increased cloning efficiency from 1.01% to 10.71% and was higher than the injection of *Kdm4d* mRNA (7.81%) [199]. The combination of *Dux* or *Kdm4d* mRNA with *Dnmt3a/3b* siRNAs further increased cloning efficiency to 18.6% and 12.24%, respectively. Cloning using *Dux* mRNA alongside *Dnmt3a/3b* siRNAs also rescued the large placenta phenotype [199]. Moreover, the effects of both TSA and *Kdm4d* mRNA strategies depend on *Dux* function.

Other histone modifications have important roles in gene transcriptional control and challenge reprogramming. The histone 3 trimethylation at lysine 4 (H3K4m3) and histone 3 trimethylation at lysine 27 (H3K27me3) modifications are associated with gene activity and repression, respectively. When found together, these histone marks form bivalent domains that pose genes for potential activation or long-term repression during cellular differentiation. Genes active in somatic cells may remain active or experience precocious activation in SCNT embryos [168]. This "gene-on" epigenetic memory was shown to largely depend on H3K4me3 and the histone variant H3.3 in both *X. laevis* and human SCNT embryos [200, 201]. A study that performed mouse SCNT with $X^{\Delta xist}$ donor cells combined with *Kdm4d* mRNA found that despite higher cloning efficiencies, the large placenta phenotype persisted in these clones. A transcriptomic analysis revealed that mouse SCNT embryos showed aberrant expression of multiple imprinted genes (e.g., *Sfmbt2*, *Jade1*, *Gab1*, *Somc1*), irrespectively of the $X^{\Delta xist}$ status and injection of *Kdm4d* mRNA [202]. Chromatin analysis found widespread loss of H3K27me3 at CpG islands and gene promoters of active genes in somatic cells but also caused the erasure of non-canonical imprinting [203]. This subset of imprinting marks contemplates 76 domains of H3K27me3 accumulated during oogenesis that remain in extraembryonic tissues but are lost during organogenesis [204]. In the mouse, the recovery of *Slc38a4* gene transcription rescued the large placenta phenotype in cloned pregnancies [205] and *Sfmbt2* deletion improved cloning efficiency [203].

Single-cell transcriptomic analysis showed that *Kdm4b* has higher expression in SCNT embryos that reach the blastocyst stage, and *Kdm4b* injection was more efficient to cloned development in vitro [196]. Combined injection of *Kdm4b* and *Kdm5b* mRNAs improved blastocyst yields, NT-ESC derivation efficiency, and full-term development. In cattle, the injection of a H3K27me3-specific demethylase *KDM6A* in reconstructed oocytes leads to morulae with transcriptomes more like fertilized counterparts [206], thus including EGA genes. Further, KDM6A injection increases cloned blastocyst yields and ICM cell numbers. Nonetheless, there are species-specific differences regarding epigenetic marks hindering reprogramming. Pig SCNT embryos showed profound differences in gene expression at the EGA (1,230 dysregulated genes) and RRRs contributed to this partial block of the EGA [207]. Curiously, RRRs also displayed enrichment for H3K9me3 and H3K27me3 but gene-on transcriptional memory did not overlap with H3K4me3 deposition. Injection of *KDM4A* mRNA increased blastocyst development but exogenous *KDM6A* mRNA was toxic and caused embryonic arrest [207]. The combination of injected *KDM4A* mRNA and GSK126 treatment (*Enhancer of zeste homolog 2—Ezh2* inhibitor) increased blastocysts and recovered ~50% of dysregulated genes. Thymine DNA glycosylase (TDG) is a pig-specific epigenetic regulator during preimplantation development and was not reactivated upon H3K9me3 and H3K27me3 removal [207]. Ectopic expression of TDG in SCNT embryos assisted in genome-wide DNA demethylation.

Epigenetic mechanisms act in concert to modulate chromatin structure and accessibility to transcriptional machinery. Therefore, addressing chromatin accessibility would favor a more reliable estimation of reprogramming outcome than assessing few histone marks or DNA methylation maps. DNase I hypersensitive sites (DHSs) are chromatin regions susceptible to DNA cleavage by DNAse I since protein-bound nucleic acids resist cleavage [208]. Somatic cells lose most DHSs after SCNT into oocytes, despite some accessible chromatin remaining open and H3K9me3-rich regions not acquiring DHSs [208]. Part of DHSs lost during nuclear remodeling suggest displacement of somatic cell-specific TFs due to closure of CREs enriched for TFs. Additional details came from an alternative reprogramming assay. The injection of mammalian somatic cells into *X. laevis* GV oocytes (100–300 cells per oocyte) represents an attractive assay to assess nuclear reprogramming in a DNA replication-independent fashion [209]. Mouse embryonic fibroblasts injected into *X. laevis* oocytes displayed closure of DHSs in enhancers. In turn, novel DHSs displayed motifs for pluripotency-associated TFs (e.g., *Utf1*, *Rara*) [209]. Modulation of Rara signaling affected reprogramming and TSA treatment increased the reactivation of *Oct4* and *Utf1* by 19- and 30-fold, respectively. Nearly half of reactivated embryonic genes had an accessible chromatin configuration [209], thus reinforcing the notion that the epigenetic state of donor cells contributes to the reprogramming outcome. Mapping the deposition of nucleosomes also revealed details of such remodeling of chromatin accessibility [210, 211]. Porcine somatic cells after SCNT undergo widespread nucleosome depletion [210], particularly in the X chromosome. Nucleosomes adjacent to the transcription start site were lost after SCNT, which suggest chromatin relaxation and propensity for gene activation. Housekeeping genes were not affected [210] and reinforce the notion of orchestrated, non-stochastic reprogramming mechanism. There were somatic genes that became silent but retained an open nucleosome-free chromatin configuration [210], thus representing evidence of epigenetic memory. Further work in mice elucidated that some nucleosome dynamics were due to the progression of cell cycle in SCNT embryos [211]. SCNT embryos display different patterns of nucleosome-depleted regions at the 2-cell stage, which may contribute as another constraint to adequate EGA. Furthermore, RRRs retain nucleosome-rich regions and reinforce the notion of epigenetic memory [211]. Enhancers were more faithfully remodeled than promoters and nucleosome-depleted regions overlapped with histone acetylation in SCNT embryos [211].

Nuclear reprogramming upon SCNT must also reset the three-dimensional configuration of high-order chromatin to resemble the state found in early embryos [212, 213]. Upon SCNT, reconstructed oocytes displayed metaphase-like chromosome configuration with stepwise dissolution of chromatin interactions [212, 213]. Short-distance interactions (<1 megabase pairs—Mbp) and intermediate-distance interactions (1-10Mbp) decreased and increased, respectively in SCNT embryos. Longer-distance interactions took much longer to remodel in cloned embryos [212]. Nonetheless, these findings support a fast and genome-wide restructuring of the somatic epigenome. Topologically associating domains (TADs) are high-order chromatin structures that isolate certain CRE interactions (enhancers and promoters) and represent building blocks of genome organization [214]. After SCNT, donor cell-associated TADs were lost and progressively re-established mirroring the configuration found in fertilized embryos [212, 213]. The more evident delayed high-order chromatin setting in cloned embryos during cleavage stages corroborates with inadequate EGA. H3K9me3 was an epigenetic barrier to reprogramming the donor cell at the high-order chromatin level and injection of *Kdm4d* mRNA could partially rescue the reorganization of H3K9me3-enriched TADs of somatic origin [212]. Depletion of cohesin, a chromatin architectural protein also favored TAD resetting and improved the yields of cloned blastocysts [213]. Further, cohesin-depleted cells (both donor cells and cloned embryos) showed the upregulation of nearly half of genes associated with the minor EGA [213] that anticipates the major EGA [215]. The fast reactivation of such EGA genes in SCNT embryos generated with cohesin-depleted cells strongly suggests them as direct targets of cohesin [213].

9 Concluding Remarks and Perspectives

Cloning by SCNT was pivotal to the conceptual basis of how epigenetic mechanisms contribute to the establishment of cellular identity during development. Beyond its scientific impact, animal cloning exploitation contributed to improving livestock production, provided numerous recombinant bioproducts, assisted in making disease models, and offered the foundation to reprogramming human cells for therapeutic applications. Despite much inventiveness in employing SCNT for scientific and commercial endeavors, progress in understanding nuclear reprogramming (and improving cloning efficiency) proved much harder. Genome-scale studies underscored multiple epigenetic barriers to nuclear reprogramming. Such approaches provided contextual epigenetic constraints to nuclear reprogramming but also revealed few dysregulated *loci* had substantial contribution to common phenotypes found in clones and their placentas.

More recent developments in genomics may prompt additional discoveries in nuclear reprogramming. Single-cell genomics contemplates cellular heterogeneity which is found in early embryos. There are studies that applied single-cell technologies to SCNT embryos, but their epigenetic heterogeneity remains much unexplored. Further, single-cell multi-omics dissects the interplay of different regulatory layers (e.g., DNA methylation maps, transcriptome) and should disclose the interplay of epigenetic marks and chromatin states [216]. The combined use of single-cell multi-omics [216] with spatial transcriptomics [217] or molecular recording technologies [218] may assist in solving the cell-cell interactions and timely constraints of nuclear reprogramming, respectively.

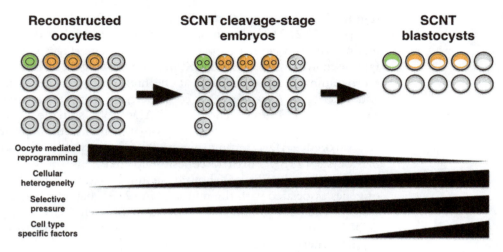

Fig. 3 Exploring somatic cell nuclear transfer (SCNT)-mediated reprogramming using single-cell multi-omics. Reconstructed oocytes and SCNT embryos having full-term developmental potential (green) are rare events in cloning experiments (considering an ~5% cloning efficiency). Even the most efficient mouse cloning protocols (~20% efficiency) hold relatively few events of faithful nuclear reprogramming (green and orange combined). As development progresses, faithful reprogramming events become more representative among batches of SCNT embryos but turn into less informative entities about initial oocyte-mediated reprogramming processes and accumulate confounding factors (cellular heterogeneity, selective pressure for surviving embryos, and cellular differentiation at the blastocyst stage). Only extensive single-cell multi-omics would circumvent these limitations and thus allow to unravel these rare events of faithful reprogramming. Schematic cloning experiment with 80% cleavage and 50% blastocyst rates

The other front needing progress is in scaling SCNT embryo production. Irrespective of the progress in exploiting SCNT embryos at genome scales, the number of viable clones (harboring faithful reprogramming events) remains very low(<5%), and cloning experiments usually yield fewer than 100–200 SCNT embryos. Even genomic technologies that require low inputs (50–1,000 cells) will contemplate very few SCNT embryos that would reach full-term development due to the low cloning efficiency (Fig. 3), thus making their contribution to the analysis readout quite diminutive. Since genomics currently relies on a majority of SCNT embryos that would not develop beyond implantation, it captures more easily potential mechanisms that act as roadblocks to nuclear reprogramming than the conditions that support full-term development [213]. Therefore, the systematic identification of rare events of faithful reprogramming will require exploring hundreds or thousands of SCNT embryos at the single-cell level, depending on their developmental stage (Fig. 3). An example of such a scenario was the identification of cellular heterogeneity among human ESCs accompanied by the pinpointing of a totipotent-like ESC subset (~1.6% of cultured cells). These totipotent-like ESCs hold transcriptomes strikingly like the 8-cell human embryo [219, 220] and could become an alternative model to unrevealing EGA, just like the totipotent 2C-like ESCs in mice [221]. Hence, these promising technological advances in genomics and SCNT technology may reinvigorate the excitement in SCNT-mediated reprogramming for both research and commercial applications.

References

1. Moura MT (2011) Pluripotency and cellular reprogramming. An Acad Pernamb Ciênc Agron 8:138–168

2. Handel MA, Schimenti JC (2010) Genetics of mammalian meiosis: regulation, dynamics and impact on fertility. Nat Rev Genet 11:124–136

3. Young C, Brink R (2021) The unique biology of germinal center B cells. Immunity 54:1652–1664

4. Ciccia A, Elledge SJ (2010) The DNA damage response: making it safe to play with knives. Mol Cell 40:179–204

5. Ďurovcová I, Kyzek S, Fabová J, Makuková J, Gálová E, Ševčovičová A (2022) Genotoxic potential of bisphenol A: a review. Environ Pollut 306:119346

6. Vaddavalli PL, Schumacher B (2022) The p53 network: cellular and systemic DNA damage responses in cancer and aging. Trends Genet 38:598–612

7. Soto-Palma C, Niedernhofer LJ, Faulk CD, Dong X (2022) Epigenetics, DNA damage, and aging. J Clin Invest 132:e158446

8. Rossant J (2018) Genetic control of early cell lineages in the mammalian embryo. Annu Rev Genet 52:185–201

9. Buckley CE, St Johnston D (2022) Apical-basal polarity and the control of epithelial form and function. Nat Rev Mol Cell Biol 23:559–577

10. Zernicka-Goetz M, Morris SA, Bruce AW (2009) Making a firm decision: multifaceted regulation of cell fate in the early mouse embryo. Nat Rev Genet 10:467–477

11. Cree LM, Samuels DC, de Sousa Lopes SC, Rajasimha HK, Wonnapinij P, Mann JR et al (2008) A reduction of mitochondrial DNA molecules during embryogenesis explains the rapid segregation of genotypes. Nat Genet 40:249–254

12. Kim YS, Bedzhov I (2022) Mechanisms of formation and functions of the early embryonic cavities. Semin Cell Dev Biol 131:110–116

13. Alizadeh Z, Kageyama S, Aoki F (2005) Degradation of maternal mRNA in mouse embryos: selective degradation of specific mRNAs after fertilization. Mol Reprod Dev 72:281–290

14. Schier AF (2007) The maternal-zygotic transition: death and birth of RNAs. Science 316:406–407

15. Moore NW, Adams CE, Rowson LE (1968) Developmental potential of single blastomeres of the rabbit egg. J Reprod Fertil 17:527–531

16. Maemura M, Taketsuru H, Nakajima Y, Shao R, Kakihara A, Nogami J et al (2021) Totipotency of mouse zygotes extends to single blastomeres of embryos at the four-cell stage. Sci Rep 11(1):11167

17. Nichols J, Zevnik B, Anastassiadis K, Niwa H, Klewe-Nebenius D, Chambers I et al (1998) Formation of pluripotent stem cells in the mammalian embryo depends on the POU transcription factor Oct4. Cell 95:379–391

18. Avilion AA, Nicolis SK, Pevny LH, Perez L, Vivian N, Lovell-Badge R (2003) Multipotent cell lineages in early mouse development depend on SOX2 function. Genes Dev 17:126–140

19. Rinkevich Y, Lindau P, Ueno H, Longaker MT, Weissman IL (2011) Germ-layer and lineage-restricted stem/progenitors regenerate the mouse digit tip. Nature 476:409–413

20. Levine F, Itkin-Ansari P (2008) Beta-cell regeneration: neogenesis, replication or both? J Mol Med 86:247–258

21. Gao C, Peng J (2021) All routes lead to Rome: multifaceted origin of hepatocytes during liver regeneration. Cell Regen 10:2

22. Ji Z, Lu M, Xie H, Yuan H, Chen Q (2022) β cell regeneration and novel strategies for treatment of diabetes (review). Biomed Rep 17(3):72

23. Mikkelsen TS, Xu Z, Zhang X, Wang L, Gimble JM, Lander ES et al (2010) Comparative epigenomic analysis of murine and human adipogenesis. Cell 143:156–169

24. Ziller MJ, Edri R, Yaffe Y, Donaghey J, Pop R, Mallard W et al (2015) Dissecting neural differentiation regulatory networks through epigenetic footprinting. Nature 518:355–359

25. Tsankov AM, Gu H, Akopian V, Ziller MJ, Donaghey J, Amit I et al (2015) Transcription factor binding dynamics during human ES cell differentiation. Nature 518:344–349

26. Li E (2002) Chromatin modification and epigenetic reprogramming in mammalian development. Nat Rev Genet 3:662–673

27. Smith ZD, Meissner A (2013) DNA methylation: roles in mammalian development. Nat Rev Genet 14:204–220

28. Millán-Zambrano G, Burton A, Bannister AJ, Schneider R (2022) Histone post-translational modifications – cause and consequence of genome function. Nat Rev Genet 23:563–580

29. Fu K, Bonora G, Pellegrini M (2020) Interactions between core histone marks and DNA methyltransferases predict DNA methylation patterns observed in human cells and tissues. Epigenetics 15:272–282

30. Oswald J, Engemann S, Lane N, Mayer W, Olek A, Fundele R et al (2000) Active demethylation of the paternal genome in the mouse zygote. Curr Biol 10:475–478

31. Gou LT, Lim DH, Ma W, Aubol BE, Hao Y, Wang X et al (2020) Initiation of parental genome reprogramming in fertilized oocyte by splicing kinase SRPK1-catalyzed protamine phosphorylation. Cell 180:1212–1227.e14

32. Smith ZD, Chan MM, Humm KC, Karnik R, Mekhoubad S, Regev A et al (2014) DNA methylation dynamics of the human preimplantation embryo. Nature 511:611–615

33. Wang C, Liu X, Gao Y, Yang L, Li C, Liu W et al (2018) Reprogramming of H3K9me3-dependent heterochromatin during mammalian embryo development. Nat Cell Biol 20: 620–631

34. Yamaguchi S, Hong K, Liu R, Inoue A, Shen L, Zhang K et al (2013) Dynamics of 5-methylcytosine and 5-hydroxymethylcytosine during germ cell reprogramming. Cell Res 23:329–339

35. Liu Y, Zhang Y, Yin J, Gao Y, Li Y, Bai D et al (2019) Distinct H3K9me3 and DNA methylation modifications during mouse spermatogenesis. J Biol Chem 294:18714–18725

36. Gu C, Liu S, Wu Q, Zhang L, Guo F (2019) Integrative single-cell analysis of transcriptome, DNA methylome and chromatin accessibility in mouse oocytes. Cell Res 29:110–123

37. Gurdon JB (2014) A view of amphibian embryology during the last century. Int J Dev Biol 58:723–725

38. Briggs R, King TJ (1952) Transplantation of living nuclei from blastula cells into enucleated frogs' eggs. Proc Natl Acad Sci U S A 38:455–463

39. Schultz J (1947) Nuclear differentiation and the origin of tumors. Cancer Res 7:41

40. King TJ (1966) Nuclear transplantation in amphibia. In: Prescott DM (ed) Methods in cell biology, vol 2. Academic Press, pp 1–36

41. Lorch IJ, Danielli JF (1950) Transplantation of nuclei from cell to cell. Nature 166:329–330

42. Fankhauser G (1934) Cytological studies on egg fragments of the salamander triton. IV. The cleavage of egg fragments without the egg nucleus. J Exp Zool 67:349–393

43. Briggs R, Green EU, King TJ (1951) An investigation of the capacity for cleavage and differentiation in Rana pipiens eggs lacking "functional" chromosomes. J Exp Zool 116: 455–499

44. Briggs R, King TJ (1953) Factors affecting the transplantability of nuclei of frog embryonic cells. J Exp Zool 122:485–505

45. Di Berardino MA, McKinnell RG (2004) The pathway to animal cloning and beyond--Robert Briggs (1911–1983) and Thomas J. King (1921–2000). J Exp Zool A Comp Exp Biol 301:275–279

46. Briggs R, King TJ (1957) Changes in the nuclei of differentiating endoderm cells as revealed by nuclear transplantation. J Morphol 100:269–312

47. Hochedlinger K, Jaenisch R (2002) Nuclear transplantation: lessons from frogs and mice. Curr Opin Cell Biol 14:741–748

48. Fischberg M, Gurdon JB, Elsdale TR (1958) Nuclear transplantation in Xenopus laevis. Nature 181:424

49. Gurdon JB, Elsdale TR, Fischberg M (1958) Sexually mature individuals of Xenopus laevis from the transplantation of single somatic nuclei. Nature 182:64–65

50. Elsdale TR, Gurdon JB, Fischberg M (1960) A description of the technique for nuclear transplantation in Xenopus laevis. J Embryol Exp Morphol 8:437–444

51. Gurdon JB (2006) From nuclear transfer to nuclear reprogramming: the reversal of cell differentiation. Annu Rev Cell Dev Biol 22: 1–22

52. Wabl MR, Brun RB, Du Pasquier L (1975) Lymphocytes of the toad Xenopus laevis have the gene set for promoting tadpole development. Science 190:1310–1312

53. Gurdon JB (1962) The developmental capacity of nuclei taken from intestinal epithelium cells of feeding tadpoles. J Embryol Exp Morphol 10:622–640

54. Gurdon JB, Laskey RA, Reeves OR (1975) The developmental capacity of nuclei transplanted from keratinized skin cells of adult frogs. J Embryol Exp Morphol 34:93–112

55. Gurdon JB (1962) The transplantation of nuclei between two species of Xenopus. Dev Biol 5:68–83

56. Laskey R (2005) Solving mysteries of DNA replication and frog cloning. Cell 123:760–762

57. Illmensee K (1968) Transplantation of embryonic nuclei into unfertilized eggs of Drosophila melanogaster. Nature 219:1268–1269

58. Gasaryan KG, Hung NM, Neyfakh AA, Ivanenkov VV (1979) Nuclear transplantation in teleost Misgurnus fossilis L. Nature 280:585–587

59. Schubiger M, Schneiderman HA (1971) Nuclear transplantation in Drosophila melanogaster. Nature 230:185–186

60. Illmensee K (1972) Developmental potencies of nuclei from cleavage, preblastoderm, and syncytial blastoderm transplanted into unfertilized eggs of Drosophila melanogaster. Wilhelm Roux Arch Entwickl Mech Org 170:267–298

61. Bromhall JD (1975) Nuclear transplantation in the rabbit egg. Nature 258:719–722

62. Modliński JA (1975) Haploid mouse embryos obtained by microsurgical removal of one pronucleus. J Embryol Exp Morphol 33:897–905

63. McGrath J, Solter D (1983) Nuclear transplantation in the mouse embryo by microsurgery and cell fusion. Science 220:1300–1302

64. McLaren A (1984) Mammalian development: methods and success of nuclear transplantation in mammals. Nature 309(5970):671–672

65. Baranska W, Koprowski H (1970) Fusion of unfertilized mouse eggs with somatic cells. J Exp Zool 174:1–14

66. Modliński JA (1978) Transfer of embryonic nuclei to fertilised mouse eggs and development of tetraploid blastocysts. Nature 273:466–467

67. Hoppe PC, Illmensee K (1977) Microsurgically produced homozygous-diploid uniparental mice. Proc Natl Acad Sci U S A 74:5657–5661

68. Modliński JA (1980) Preimplantation development of microsurgically obtained haploid and homozygous diploid mouse embryos and effects of pretreatment with Cytochalasin B on enucleated eggs. J Embryol Exp Morphol 60:153–161

69. Modliński JA (1981) The fate of inner cell mass and trophectoderm nuclei transplanted to fertilized mouse eggs. Nature 292:342–343

70. Illmensee K, Hoppe PC (1981) Nuclear transplantation in Mus musculus: developmental potential of nuclei from preimplantation embryos. Cell 23:9–18

71. Solter D (2000) Mammalian cloning: advances and limitations. Nat Rev Genet 1:199–207

72. McGrath J, Solter D (1984) Completion of mouse embryogenesis requires both the maternal and paternal genomes. Cell 37:179–183

73. Barton SC, Surani MA, Norris ML (1984) Role of paternal and maternal genomes in mouse development. Nature 311:374–376

74. Surani MA, Barton SC, Norris ML (1984) Development of reconstituted mouse eggs suggests imprinting of the genome during gametogenesis. Nature 308:548–550

75. McGrath J, Solter D (1984) Inability of mouse blastomere nuclei transferred to enucleated zygotes to support development in vitro. Science 226:1317–1319

76. Collas P, Robl JM (1991) Relationship between nuclear remodeling and development in nuclear transplant rabbit embryos. Biol Reprod 45:455–465

77. Willadsen SM (1986) Nuclear transplantation in sheep embryos. Nature 320:63–65

78. Campbell KH, Loi P, Otaegui PJ, Wilmut I (1996) Cell cycle co-ordination in embryo cloning by nuclear transfer. Rev Reprod 1:40–46

79. Sun FZ, Moor RM (1995) Nuclear transplantation in mammalian eggs and embryos. Curr Top Dev Biol 30:147–176

80. Smith LC, Wilmut I (1989) Influence of nuclear and cytoplasmic activity on the development in vivo of sheep embryos after nuclear transplantation. Biol Reprod 40:1027–1035

81. Collas P, Balise JJ, Robl JM (1992) Influence of cell cycle stage of the donor nucleus on development of nuclear transplant rabbit embryos. Biol Reprod 46:492–500

82. Collas P, Pinto-Correia C, Ponce de Leon FA, Robl JM (1992) Effect of donor cell cycle stage on chromatin and spindle morphology

in nuclear transplant rabbit embryos. Biol Reprod 46:501–511

83. Collas P, Barnes FL (1994) Nuclear transplantation by microinjection of inner cell mass and granulosa cell nuclei. Mol Reprod Dev 38: 264–267

84. Keefer CL, Stice SL, Matthews DL (1994) Bovine inner cell mass cells as donor nuclei in the production of nuclear transfer embryos and calves. Biol Reprod 50:935–939

85. Sims M, First NL (1994) Production of calves by transfer of nuclei from cultured inner cell mass cells. Proc Natl Acad Sci U S A 91:6143–6147

86. Campbell KH, McWhir J, Ritchie WA, Wilmut I (1996) Sheep cloned by nuclear transfer from a cultured cell line. Nature 380:64–66

87. Szöllösi D, Czołowska R, Szöllösi MS, Tarkowski AK (1988) Remodeling of mouse thymocyte nuclei depends on the time of their transfer into activated, homologous oocytes. J Cell Sci 91:603–613

88. Solter D (1996) Lambing by nuclear transfer. Nature 380(6569):24–25

89. Moura MT, Nascimento PS, Silva JCF, Deus PR, Oliveira MAL (2016) The evolving picture in obtaining genetically modified livestock. An Acad Pernamb Ciênc Agron 13: 145–169

90. Tarkowski AK, Bałakier H (1980) Nucleocytoplasmic interactions in cell hybrids between mouse oocytes, blastomeres and somatic cells. J Embryol Exp Morphol 55: 319–330

91. Czołowska R, Modliński JA, Tarkowski AK (1984) Behaviour of thymocyte nuclei in non-activated and activated mouse oocytes. J Cell Sci 69:19–34

92. Borsuk E, Szöllösi MS, Besomebes D, Debey P (1996) Fusion with activated mouse oocytes modulates the transcriptional activity of introduced somatic cell nuclei. Exp Cell Res 225: 93–101

93. Kono T, Ogawa M, Nakahara T (1993) Thymocyte transfer to enucleated oocytes in the mouse. J Reprod Dev 39:301–307

94. Wilmut I, Schnieke AE, McWhir J, Kind AJ, Campbell KH (1997) Viable offspring derived from fetal and adult mammalian cells. Nature 385:810–813

95. Rossant J (2002) A monoclonal mouse? Nature 415:967–969

96. Rodriguez-Osorio N, Urrego R, Cibelli JB, Eilertsen K, Memili E (2012) Reprogramming mammalian somatic cells. Theriogenology 78:1869–1886

97. Zhou Q, Renard JP, Le Friec G, Brochard V, Beaujean N, Cherifi Y et al (2003) Generation of fertile cloned rats by regulating oocyte activation. Science 302:1179

98. Chesné P, Adenot PG, Viglietta C, Baratte M, Boulanger L, Renard JP (2002) Cloned rabbits produced by nuclear transfer from adult somatic cells. Nat Biotechnol 20:366–369

99. Kues WA, Niemann H (2004) The contribution of farm animals to human health. Trends Biotechnol 22:286–294

100. Wells DN, Misica PM, Tervit HR, Vivanco WH (1998) Adult somatic cell nuclear transfer is used to preserve the last surviving cow of the Enderby Island cattle breed. Reprod Fertil Dev 10:369–378

101. Loi P, Ptak G, Barboni B, Fulka J Jr, Cappai P, Clinton M (2001) Genetic rescue of an endangered mammal by cross-species nuclear transfer using post-mortem somatic cells. Nat Biotechnol 19:962–964

102. Selokar NL, Sharma P, Saini M, Sheoran S, Rajendran R, Kumar D et al (2019) Successful cloning of a superior buffalo bull. Sci Rep 9: 11366

103. Lee BC, Kim MK, Jang G, Oh HJ, Yuda F, Kim HJ et al (2005) Dogs cloned from adult somatic cells. Nature 436:641

104. Wakayama S, Ohta H, Hikichi T, Mizutani E, Iwaki T, Kanagawa O et al (2008) Production of healthy cloned mice from bodies frozen at −20 degrees C for 16 years. Proc Natl Acad Sci U S A 105:17318–17322

105. Li J, Mombaerts P (2008) Nuclear transfer-mediated rescue of the nuclear genome of nonviable mouse cells frozen without cryoprotectant. Biol Reprod 79(4):588–593

106. Wakayama S, Ito D, Hayashi E, Ishiuchi T, Wakayama T (2022) Healthy cloned offspring derived from freeze-dried somatic cells. Nat Commun 13:3666

107. Eyestone WH (1999) Production and breeding of transgenic cattle using in vitro embryo production technology. Theriogenology 51: 509–517

108. Bordignon V, Keyston R, Lazaris A, Bilodeau AS, Pontes JH, Arnold D et al (2003) Transgene expression of green fluorescent protein and germ line transmission in cloned calves derived from in vitro-transfected somatic cells. Biol Reprod 68:2013–2023

109. Kuroiwa Y, Kasinathan P, Matsushita H, Sathiyaselan J, Sullivan EJ, Kakitani M et al (2004) Sequential targeting of the genes encoding immunoglobulin-mu and prion protein in cattle. Nat Genet 36:775–780

110. Suzuki S, Iwamoto M, Saito Y, Fuchimoto D, Sembon S, Suzuki M et al (2012) Il2rg gene-

targeted severe combined immunodeficiency pigs. Cell Stem Cell 10:753–758

111. Kuroiwa Y, Kasinathan P, Choi YJ, Naeem R, Tomizuka K, Sullivan EJ et al (2002) Cloned transchromosomic calves producing human immunoglobulin. Nat Biotechnol 20:889–894

112. Niu D, Wei HJ, Lin L, George H, Wang T, Lee IH et al (2017) Inactivation of porcine endogenous retrovirus in pigs using CRISPR-Cas9. Science 357:1303–1307

113. Dongre A, Weinberg RA (2019) New insights into the mechanisms of epithelial-mesenchymal transition and implications for cancer. Nat Rev Mol Cell Biol 20:69–84

114. Hochedlinger K, Blelloch R, Brennan C, Yamada Y, Kim M, Chin L et al (2004) Reprogramming of a melanoma genome by nuclear transplantation. Genes Dev 18:1875–1885

115. Mombaerts P (2004) Genes and ligands for odorant, vomeronasal and taste receptors. Nat Rev Neurosci 5:263–278

116. Li J, Ishii T, Feinstein P, Mombaerts P (2004) Odorant receptor gene choice is reset by nuclear transfer from mouse olfactory sensory neurons. Nature 428:393–399

117. Eggan K, Baldwin K, Tackett M, Osborne J, Gogos J, Chess A et al (2004) Mice cloned from olfactory sensory neurons. Nature 428:44–49

118. Inoue K, Wakao H, Ogonuki N, Miki H, Seino K, Nambu-Wakao R et al (2005) Generation of cloned mice by direct nuclear transfer from natural killer T cells. Curr Biol 15:1114–1118

119. Hochedlinger K, Jaenisch R (2002) Monoclonal mice generated by nuclear transfer from mature B and T donor cells. Nature 415:1035–1038

120. Thomson JA, Itskovitz-Eldor J, Shapiro SS, Waknitz MA, Swiergiel JJ, Marshall VS et al (1998) Embryonic stem cell lines derived from human blastocysts. Science 282:1145–1147

121. Colman A, Kind A (2000) Therapeutic cloning: concepts and practicalities. Trends Biotechnol 18:192–196

122. Rhind SM, Taylor JE, De Sousa PA, King TJ, McGarry M, Wilmut I (2003) Human cloning: can it be made safe? Nat Rev Genet 4:855–864

123. Cherry AB, Daley GQ (2013) Reprogrammed cells for disease modeling and regenerative medicine. Annu Rev Med 64:277–290

124. Kawase E, Yamazaki Y, Yagi T, Yanagimachi R, Pedersen RA (2000) Mouse embryonic stem (ES) cell lines established from neuronal cell-derived cloned blastocysts. Genesis 28:156–163

125. Munsie MJ, Michalska AE, O'Brien CM, Trounson AO, Pera MF, Mountford PS (2000) Isolation of pluripotent embryonic stem cells from reprogrammed adult mouse somatic cell nuclei. Curr Biol 10:989–992

126. Wakayama T, Tabar V, Rodriguez I, Perry AC, Studer L, Mombaerts P (2001) Differentiation of embryonic stem cell lines generated from adult somatic cells by nuclear transfer. Science 292:740–743

127. Brambrink T, Hochedlinger K, Bell G, Jaenisch R (2006) ES cells derived from cloned and fertilized blastocysts are transcriptionally and functionally indistinguishable. Proc Natl Acad Sci U S A 103:933–938

128. Wakayama S, Jakt ML, Suzuki M, Araki R, Hikichi T, Kishigami S et al (2006) Equivalency of nuclear transfer-derived embryonic stem cells to those derived from fertilized mouse blastocysts. Stem Cells 24:2023–2033

129. Rideout WM 3rd, Hochedlinger K, Kyba M, Daley GQ, Jaenisch R (2002) Correction of a genetic defect by nuclear transplantation and combined cell and gene therapy. Cell 109:17–27

130. Tabar V, Tomishima M, Panagiotakos G, Wakayama S, Menon J, Chan B et al (2008) Therapeutic cloning in individual parkinsonian mice. Nat Med 14:379–381

131. Byrne JA, Pedersen DA, Clepper LL, Nelson M, Sanger WG, Gokhale S et al (2007) Producing primate embryonic stem cells by somatic cell nuclear transfer. Nature 450:497–502

132. Tachibana M, Amato P, Sparman M, Gutierrez NM, Tippner-Hedges R, Ma H et al (2013) Human embryonic stem cells derived by somatic cell nuclear transfer. Cell 153:1228–1238

133. Egli D, Chen AE, Saphier G, Powers D, Alper M, Katz K et al (2011) Impracticality of egg donor recruitment in the absence of compensation. Cell Stem Cell 9:293–294

134. Hochedlinger K, Jaenisch R (2006) Nuclear reprogramming and pluripotency. Nature 441:1061–1067

135. Wakayama T, Tateno H, Mombaerts P, Yanagimachi R (2000) Nuclear transfer into mouse zygotes. Nat Genet 24:108–109

136. Egli D, Rosains J, Birkhoff G, Eggan K (2007) Developmental reprogramming after

137. Egli D, Birkhoff G, Eggan K (2008) Mediators of reprogramming: transcription factors and transitions through mitosis. Nat Rev Mol Cell Biol 9:505–516

138. Kang E, Wu G, Ma H, Li Y, Tippner-Hedges-R, Tachibana M et al (2014) Nuclear reprogramming by interphase cytoplasm of two-cell mouse embryos. Nature 509:101–104

139. Egli D, Chen AE, Saphier G, Ichida J, Fitzgerald C, Go KJ et al (2011) Reprogramming within hours following nuclear transfer into mouse but not human zygotes. Nat Commun 2:488

140. Takahashi K, Yamanaka S (2006) Induction of pluripotent stem cells from mouse embryonic and adult fibroblast cultures by defined factors. Cell 126:663–676

141. Takahashi K, Tanabe K, Ohnuki M, Narita M, Ichisaka T, Tomoda K et al (2007) Induction of pluripotent stem cells from adult human fibroblasts by defined factors. Cell 131:861–872

142. Shi Y, Inoue H, Wu JC, Yamanaka S (2017) Induced pluripotent stem cell technology: a decade of progress. Nat Rev Drug Discov 16: 115–130

143. Johannesson B, Sagi I, Gore A, Paull D, Yamada M, Golan-Lev T et al (2014) Comparable frequencies of coding mutations and loss of imprinting in human pluripotent cells derived by nuclear transfer and defined factors. Cell Stem Cell 15:634–642

144. Yamanaka S (2020) Pluripotent stem cell-based cell therapy-promise and challenges. Cell Stem Cell 27:523–531

145. Wagoner EJ, Rosenkrans CF Jr, Gliedt DW, Pierson JN, Munyon AL (1996) Functional enucleation of bovine oocytes: effects of centrifugation and ultraviolet light. Theriogenology 46:279–284

146. Moura MT, de Sousa RV, de Oliveira Leme L, Rumpf R (2008) Analysis of actinomycin D treated cattle oocytes and their use for somatic cell nuclear transfer. Anim Reprod Sci 109:40–49

147. Greenwood J, Costanzo V, Robertson K, Hensey C, Gautier J (2001) Responses to DNA damage in Xenopus: cell death or cell cycle arrest. Novartis Found Symp 237:221–230

148. Moura MT, Badaraco J, Sousa RV, Lucci CM, Rumpf R (2019) Improved functional oocyte enucleation by actinomycin D for bovine somatic cell nuclear transfer. Reprod Fertil Dev 31:1321–1329

149. Moura MT, Sousa RV, Lucci CM, Rumpf R (2019) Bovine somatic cell nuclear transfer using mitomycin C-mediated chemical oocyte enucleation. Zygote 27:137–142

150. Rinaldi VD, Hsieh K, Munroe R, Bolcun-Filas E, Schimenti JC (2017) Pharmacological inhibition of the DNA damage checkpoint prevents radiation-induced oocyte death. Genetics 206:1823–1828

151. Tsunoda Y, Shioda Y, Onodera M, Nakamura K, Uchida T (1988) Differential sensitivity of mouse pronuclei and zygote cytoplasm to Hoechst staining and ultraviolet irradiation. J Reprod Fertil 82:173–178

152. Dominko T, Chan A, Simerly C, Luetjens CM, Hewitson L, Martinovich C et al (2000) Dynamic imaging of the metaphase II spindle and maternal chromosomes in bovine oocytes: implications for enucleation efficiency verification, avoidance of parthenogenesis, and successful embryogenesis. Biol Reprod 62:150–154

153. Liu L, Oldenbourg R, Trimarchi JR, Keefe DL (2000) A reliable, noninvasive technique for spindle imaging and enucleation of mammalian oocytes. Nat Biotechnol 18:223–225

154. Wang MK, Liu JL, Li GP, Lian L, Chen DY (2001) Sucrose pretreatment for enucleation: an efficient and non-damage method for removing the spindle of the mouse MII oocyte. Mol Reprod Dev 58:432–436

155. Liu JL, Sung LY, Barber M, Yang X (2002) Hypertonic medium treatment for localization of nuclear material in bovine metaphase II oocytes. Biol Reprod 66:1342–1349

156. Kawakami M, Tani T, Yabuuchi A, Kobayashi T, Murakami H, Fujimura T et al (2003) Effect of demecolcine and nocodazole on the efficiency of chemically assisted removal of chromosomes and the developmental potential of nuclear transferred porcine oocytes. Cloning Stem Cells 5:379–387

157. Tani T, Shimada H, Kato Y, Tsunoda Y (2006) Demecolcine-assisted enucleation for bovine cloning. Cloning Stem Cells 8:61–66

158. Wakayama T, Perry AC, Zuccotti M, Johnson KR, Yanagimachi R (1998) Full-term development of mice from enucleated oocytes injected with cumulus cell nuclei. Nature 394:369–374

159. Galli C, Lagutina I, Vassiliev I, Duchi R, Lazzari G (2002) Comparison of microinjection (piezo-electric) and cell fusion for nuclear transfer success with different cell types in cattle. Cloning Stem Cells 4:189–196

160. Kawano K, Kato Y, Tsunoda Y (2004) Comparison of in vitro development of porcine nuclear-transferred oocytes receiving fetal somatic cells by injection and fusion methods. Cloning Stem Cells 6:67–72

161. Yu Y, Ding C, Wang E, Chen X, Li X, Zhao C et al (2007) Piezo-assisted nuclear transfer affects cloning efficiency and may cause apoptosis. Reproduction 133:947–954

162. Peura TT, Lewis IM, Trounson AO (1998) The effect of recipient oocyte volume on nuclear transfer in cattle. Mol Reprod Dev 50:185–191

163. Vajta G, Lewis IM, Hyttel P, Thouas GA, Trounson AO (2001) Somatic cell cloning without micromanipulators. Cloning 3:89–95

164. Verma G, Arora JS, Sethi RS, Mukhopadhyay CS, Verma R (2015) Handmade cloning: recent advances, potential and pitfalls. J Anim Sci Biotechnol 6:43

165. Dean W, Santos F, Stojkovic M, Zakhartchenko V, Walter J, Wolf E et al (2001) Conservation of methylation reprogramming in mammalian development: aberrant reprogramming in cloned embryos. Proc Natl Acad Sci U S A 98:13734–13738

166. Santos F, Zakhartchenko V, Stojkovic M, Peters A, Jenuwein T, Wolf E et al (2003) Epigenetic marking correlates with developmental potential in cloned bovine preimplantation embryos. Curr Biol 13:1116–1121

167. Bortvin A, Eggan K, Skaletsky H, Akutsu H, Berry DL, Yanagimachi R et al (2003) Incomplete reactivation of Oct4-related genes in mouse embryos cloned from somatic nuclei. Development 130:1673–1680

168. Ng RK, Gurdon JB (2005) Epigenetic memory of active gene transcription is inherited through somatic cell nuclear transfer. Proc Natl Acad Sci U S A 102:1957–1962

169. Xue F, Tian XC, Du F, Kubota C, Taneja M, Dinnyes A et al (2002) Aberrant patterns of X chromosome inactivation in bovine clones. Nat Genet 31:216–220

170. Zhang S, Kubota C, Yang L, Zhang Y, Page R, O'Neill M et al (2004) Genomic imprinting of H19 in naturally reproduced and cloned cattle. Biol Reprod 71:1540–1544

171. Gao S, Chung YG, Williams JW, Riley J, Moley K, Latham KE (2003) Somatic cell-like features of cloned mouse embryos prepared with cultured myoblast nuclei. Biol Reprod 69:48–56

172. Gjørret JO, Wengle J, Maddox-Hyttel P, King WA (2005) Chronological appearance of apoptosis in bovine embryos reconstructed by somatic cell nuclear transfer from quiescent granulosa cells. Reprod Domest Anim 40:210–216

173. Palmieri C, Loi P, Ptak G, Della Salda L (2008) Review paper: a review of the pathology of abnormal placentae of somatic cell nuclear transfer clone pregnancies in cattle, sheep, and mice. Vet Pathol 45:865–880

174. Arnold DR, Bordignon V, Lefebvre R, Murphy BD, Smith LC (2006) Somatic cell nuclear transfer alters peri-implantation trophoblast differentiation in bovine embryos. Reproduction 132:279–290

175. Biase FH, Rabel C, Guillomot M, Hue I, Andropolis K, Olmstead CA et al (2016) Massive dysregulation of genes involved in cell signaling and placental development in cloned cattle conceptus and maternal endometrium. Proc Natl Acad Sci U S A 113:14492–14501

176. Ogura A, Inoue K, Ogonuki N, Lee J, Kohda T, Ishino F (2002) Phenotypic effects of somatic cell cloning in the mouse. Cloning Stem Cells 4:397–405

177. Wilmut I, Beaujean N, de Sousa PA, Dinnyes A, King TJ, Paterson LA et al (2002) Somatic cell nuclear transfer. Nature 419:583–586

178. Young LE, Sinclair KD, Wilmut I (1998) Large offspring syndrome in cattle and sheep. Rev Reprod 3:155–163

179. Scott KA, Yamazaki Y, Yamamoto M, Lin Y, Melhorn SJ, Krause EG et al (2010) Glucose parameters are altered in mouse offspring produced by assisted reproductive technologies and somatic cell nuclear transfer. Biol Reprod 83:220–227

180. Bolondi A, Kretzmer H, Meissner A (2022) Single-cell technologies: a new lens into epigenetic regulation in development. Curr Opin Genet Dev 76:101947

181. Biniszkiewicz D, Gribnau J, Ramsahoye B, Gaudet F, Eggan K, Humpherys D et al (2002) Dnmt1 overexpression causes genomic hypermethylation, loss of imprinting, and embryonic lethality. Mol Cell Biol 22:2124–2135

182. Enright BP, Kubota C, Yang X, Tian XC (2003) Epigenetic characteristics and development of embryos cloned from donor cells treated by trichostatin A or 5-aza-2'-deoxycytidine. Biol Reprod 69:896–901

183. Kishigami S, Mizutani E, Ohta H, Hikichi T, Thuan NV, Wakayama S et al (2006) Significant improvement of mouse cloning technique by treatment with trichostatin A after

somatic nuclear transfer. Biochem Biophys Res Commun 340:183–189

184. Rybouchkin A, Kato Y, Tsunoda Y (2006) Role of histone acetylation in reprogramming of somatic nuclei following nuclear transfer. Biol Reprod 74:1083–1089

185. Bui HT, Wakayama S, Kishigami S, Park KK, Kim JH, Thuan NV et al (2010) Effect of trichostatin A on chromatin remodeling, histone modifications, DNA replication, and transcriptional activity in cloned mouse embryos. Biol Reprod 83:454–463

186. Inoue K, Oikawa M, Kamimura S, Ogonuki N, Nakamura T, Nakano T et al (2015) Trichostatin A specifically improves the aberrant expression of transcription factor genes in embryos produced by somatic cell nuclear transfer. Sci Rep 5:10127

187. Wakayama S, Kohda T, Obokata H, Tokoro M, Li C, Terashita Y et al (2013) Successful serial recloning in the mouse over multiple generations. Cell Stem Cell 12:293–297

188. Wakayama T, Shinkai Y, Tamashiro KL, Niida H, Blanchard DC, Blanchard RJ et al (2000) Cloning of mice to six generations. Nature 407:318–319

189. Matoba S, Liu Y, Lu F, Iwabuchi KA, Shen L, Inoue A et al (2014) Embryonic development following somatic cell nuclear transfer impeded by persisting histone methylation. Cell 159:884–895

190. Inoue K, Kohda T, Sugimoto M, Sado T, Ogonuki N, Matoba S et al (2010) Impeding Xist expression from the active X chromosome improves mouse somatic cell nuclear transfer. Science 330:496–499

191. Beyhan Z, Ross PJ, Iager AE, Kocabas AM, Cunniff K, Rosa GJ et al (2007) Transcriptional reprogramming of somatic cell nuclei during preimplantation development of cloned bovine embryos. Dev Biol 305:637–649

192. Matoba S, Inoue K, Kohda T, Sugimoto M, Mizutani E, Ogonuki N et al (2011) RNAi-mediated knockdown of Xist can rescue the impaired postimplantation development of cloned mouse embryos. Proc Natl Acad Sci U S A 108:20621–20626

193. Chan MM, Smith ZD, Egli D, Regev A, Meissner A (2012) Mouse ooplasm confers context-specific reprogramming capacity. Nat Genet 44:978–980

194. Gao R, Wang C, Gao Y, Xiu W, Chen J, Kou X et al (2018) Inhibition of aberrant DNA re-methylation improves post-implantation development of somatic cell nuclear transfer embryos. Cell Stem Cell 23:426–435.e5

195. Yang L, Liu X, Song L, Su G, Di A, Bai C et al (2019) Inhibiting repressive epigenetic modification promotes telomere rejuvenation in somatic cell reprogramming. FASEB J 3:3982–13997

196. Liu W, Liu X, Wang C, Gao Y, Gao R, Kou X et al (2016) Identification of key factors conquering developmental arrest of somatic cell cloned embryos by combining embryo biopsy and single-cell sequencing. Cell Discov 2:16010

197. Miyamoto K, Tajima Y, Yoshida K, Oikawa M, Azuma R, Allen GE et al (2017) Reprogramming towards totipotency is greatly facilitated by synergistic effects of small molecules. Biol Open 6:415–424

198. Yang G, Zhang L, Liu W, Qiao Z, Shen S, Zhu Q et al (2021) Dux-mediated corrections of aberrant H3K9ac during 2-cell genome activation optimize efficiency of somatic cell nuclear transfer. Cell Stem Cell 28:150–163.e5

199. Yang L, Liu X, Song L, Di A, Su G, Bai C et al (2020) Transient Dux expression facilitates nuclear transfer and induced pluripotent stem cell reprogramming. EMBO Rep 21:e50054

200. Ng RK, Gurdon JB (2008) Epigenetic memory of an active gene state depends on histone H3.3 incorporation into chromatin in the absence of transcription. Nat Cell Biol 10:102–109

201. Hörmanseder E, Simeone A, Allen GE, Bradshaw CR, Figlmüller M, Gurdon J et al (2017) H3K4 methylation-dependent memory of somatic cell identity inhibits reprogramming and development of nuclear transfer embryos. Cell Stem Cell 21:135–143.e6

202. Matoba S, Wang H, Jiang L, Lu F, Iwabuchi KA, Wu X et al (2018) Loss of H3K27me3 imprinting in somatic cell nuclear transfer embryos disrupts post-implantation development. Cell Stem Cell 23:343–354.e5

203. Wang LY, Li ZK, Wang LB, Liu C, Sun XH, Feng GH et al (2020) Overcoming intrinsic H3K27me3 imprinting barriers improves post-implantation development after somatic cell nuclear transfer. Cell Stem Cell 27:315–325.e5

204. Chen Z, Zhang Y (2020) Maternal H3K27me3-dependent autosomal and X chromosome imprinting. Nat Rev Genet 21:555–571

205. Xie Z, Zhang W, Zhang Y (2022) Loss of Slc38a4 imprinting is a major cause of mouse placenta hyperplasia in somatic cell nuclear transferred embryos at late gestation. Cell Rep 38:110407

206. Zhou C, Wang Y, Zhang J, Su J, An Q, Liu X et al (2019) H3K27me3 is an epigenetic barrier while KDM6A overexpression improves nuclear reprogramming efficiency. FASEB J 33:4638–4652

207. Liu X, Chen L, Wang T, Zhou J, Li Z, Bu G et al (2021) TDG is a pig-specific epigenetic regulator with insensitivity to H3K9 and H3K27 demethylation in nuclear transfer embryos. Stem Cell Rep 16:2674–2689

208. Djekidel MN, Inoue A, Matoba S, Suzuki T, Zhang C, Lu F et al (2018) Reprogramming of chromatin accessibility in somatic cell nuclear transfer is DNA replication independent. Cell Rep 23:1939–1947

209. Miyamoto K, Nguyen KT, Allen GE, Jullien J, Kumar D, Otani T et al (2018) Chromatin accessibility impacts transcriptional reprogramming in oocytes. Cell Rep 24:304–311

210. Tao C, Li J, Zhang X, Chen B, Chi D, Zeng Y et al (2017) Dynamic reorganization of nucleosome positioning in somatic cells after transfer into porcine enucleated oocytes. Stem Cell Rep 9:642–653

211. Yang L, Xu X, Xu R, Chen C, Zhang X, Chen M et al (2022) Aberrant nucleosome organization in mouse SCNT embryos revealed by ULI-MNase-seq. Stem Cell Rep 17:1730–1742

212. Chen M, Zhu Q, Li C, Kou X, Zhao Y, Li Y et al (2020) Chromatin architecture reorganization in murine somatic cell nuclear transfer embryos. Nat Commun 11:1813

213. Zhang K, Wu DY, Zheng H, Wang Y, Sun QR, Liu X et al (2020) Analysis of genome architecture during SCNT reveals a role of cohesin in impeding minor ZGA. Mol Cell 79:234–250.e9

214. Ciabrelli F, Cavalli G (2015) Chromatin-driven behavior of topologically associating domains. J Mol Biol 427:608–625

215. Abe K, Yamamoto R, Franke V, Cao M, Suzuki Y, Suzuki MG et al (2015) The first murine zygotic transcription is promiscuous and uncoupled from splicing and 3′ processing. EMBO J 34:1523–1537

216. Wang Y, Yuan P, Yan Z, Yang M, Huo Y, Nie Y et al (2021) Single-cell multiomics sequencing reveals the functional regulatory landscape of early embryos. Nat Commun 12:1247

217. Peng G, Cui G, Ke J, Jing N (2020) Using single-cell and spatial transcriptomes to understand stem cell lineage specification during early embryo development. Annu Rev Genomics Hum Genet 21:163–181

218. Schmidt F, Cherepkova MY, Platt RJ (2018) Transcriptional recording by CRISPR spacer acquisition from RNA. Nature 562:380–385

219. Taubenschmid-Stowers J, Rostovskaya M, Santos F, Ljung S, Argelaguet R, Krueger F et al (2022) 8C-like cells capture the human zygotic genome activation program in vitro. Cell Stem Cell 29:449–459.e6

220. Mazid MA, Ward C, Luo Z, Liu C, Li Y, Lai Y et al (2022) Rolling back human pluripotent stem cells to an eight-cell embryo-like stage. Nature 605:315–324

221. Macfarlan TS, Gifford WD, Driscoll S, Lettieri K, Rowe HM, Bonanomi D et al (2012) Embryonic stem cell potency fluctuates with endogenous retrovirus activity. Nature 487:57–63

Chapter 2

Epigenetic Reprogramming and Somatic Cell Nuclear Transfer

Luna N. Vargas, Márcia M. Silveira, and Maurício M. Franco

Abstract

Epigenetics is an area of genetics that studies the heritable modifications in gene expression and phenotype that are not controlled by the primary sequence of DNA. The main epigenetic mechanisms are DNA methylation, post-translational covalent modifications in histone tails, and non-coding RNAs. During mammalian development, there are two global waves of epigenetic reprogramming. The first one occurs during gametogenesis and the second one begins immediately after fertilization. Environmental factors such as exposure to pollutants, unbalanced nutrition, behavioral factors, stress, in vitro culture conditions can negatively affect epigenetic reprogramming events. In this review, we describe the main epigenetic mechanisms found during mammalian preimplantation development (e.g., genomic imprinting, X chromosome inactivation). Moreover, we discuss the detrimental effects of cloning by somatic cell nuclear transfer on the reprogramming of epigenetic patterns and some molecular alternatives to minimize these negative impacts.

Key words Epigenetics, Early embryo development, Nuclear transplantation, Reprogramming, Somatic cell nuclear transfer

---

1 Epigenetics

Epigenetics is an area of genetics that studies the heritable modifications in gene expression that are not controlled by the primary DNA sequence. Although all cells in an organism contain the same genetic information, there are different patterns of gene expression within the diversity of cell types—that are controlled by epigenetic marks. These marks are chemical modifications in the chromatin, which are inherited in daughter cells after a mitotic or meiotic division and affect gene activity. These modifications are reversible; environmental factors such as exposure to pollutants, unbalanced nutrition, behavioral factors, and stress can influence them. The main epigenetic mechanisms are DNA methylation, post-translational covalent modifications in histone tails, and non-coding RNAs.

Marcelo Tigre Moura (ed.), *Somatic Cell Nuclear Transfer Technology*, Methods in Molecular Biology, vol. 2647, https://doi.org/10.1007/978-1-0716-3064-8_2,
© The Author(s), under exclusive license to Springer Science+Business Media, LLC, part of Springer Nature 2023

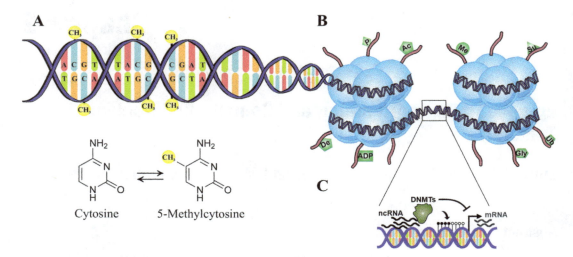

Fig. 1 Representation of the main epigenetic mechanisms. (**a**) DNA methylation is characterized by the addition of methyl group (CH$_3$) in the cytosine in the CpG context, creating the base 5-methylcytosine (5mC). (**b**) The DNA is compacted in a histones octamer forming the nucleosomes, which is characterized by the histones amino-terminal tails extend outside the nucleosome and are the target of post-translational covalent modifications such phosphorylation (P), acetylation (Ac), methylation (Me), sumoylation (Su), ubiquitination (Ub), glycosylation (Gly), deamination (De), and ADP ribosylation (ADP). These modifications are also responsible to control the state of heterochromatin and euchromatin. (**c**) The spaced nucleosomes, unmethylated DNA, and mRNA expression demonstrate a euchromatin state; however, the ncRNAs regulate the epigenetic machinery to control this transcriptional activity. The ncRNAs act recruiting chromatin remodelers, such as DNMTs, to specific sites in the genome; thus, these enzymes establish repressive marks of DNA methylation and block the transcriptional activity

1.1 DNA Methylation

DNA methylation involves the covalent addition of a methyl group from the methionine cycle to the fifth carbon of the cytosine base of DNA, creating a base called 5-methylcytosine (5mC) (Fig. 1a). In mammals, DNA methylation occurs more frequently in a CpG dinucleotides context. However, non-CpG (CpA, CpT, and CpC) methylation was found in human mitochondrial DNA, human pluripotent cells, and the mouse brain [1–5]. The CpG sites distributed throughout the genome are predominantly methylated (70–80%); on the other hand, regions with a high density of CpG, called CpG islands, are frequently unmethylated [6, 7]. DNA methylation may be involved in both repression and activation of gene transcription. Methylated CpG islands located at gene promoters prevent binding of transcription factors, thus leading to gene inactivation. Inversely, gene body methylation allows transcription factor binding on the promoter and gene transcription [8].

The enzymes responsible for DNA methylation are known as DNA methyltransferases (*DNMTs*). These enzymes methylate DNA by two different mechanisms: maintenance methylation and de novo methylation. The maintenance methylation, a process that occurs in mitotically active cells, is catalyzed by DNA methyltransferase 1 (*DNMT1*) [9]. During the semiconservative process of

DNA replication, this enzyme recognizes hemimethylated DNA and copies the preexisting methylation patterns to the new DNA strand. This mechanism maintains DNA methylation patterns in the differentiated cells (i.e., epigenetic memory). De novo methylation creates new DNA methylation patterns during gametogenesis and embryogenesis, a reaction catalyzed by DNA methyltransferase 3a (*DNMT3A*) and DNA methyltransferase 3a (*DNMT3B*) [10]. These enzymes rely on DNA methyltransferase 3L (*DNMT3L*) as a cofactor. *DNMT3L* lacks catalytic activity but stimulates *DNMT3A* and *DNMT3B* methyltransferase activities [11, 12]. There is an additional isoform of the DNMTs family, DNA methyltransferase 2 (*DNMT2*), which has a role in tRNA methylation [13, 14].

Inversely to DNA methylation, DNA demethylation is the process of removing the methyl group from cytosine, which can occur by passive or active mechanisms. Passive demethylation occurs when the DNA strands synthesized during replication are not methylated due to the loss of the maintenance methylation machinery, resulting in a dilution of 5mC marks in each cell division. On the other hand, active demethylation involves enzymatic reactions. The enzymes responsible for this process belong to the family of ten-eleven translocation (TET) enzymes, which include *TET1*, *TET2*, and *TET3*. These enzymes perform sequential oxidation reactions to convert 5mC into 5-hydroxymethylcytosine (5hmC), 5-formylcytosine (5fC), and 5-carboxylcytosine (5caC). Further, the base excision repair pathway (thymine DNA glycosylase—TDG) recognizes the modified bases 5fC and 5caC and replaces them with non-methylated cytosine.

1.2 Histone Code

The compaction of genomic DNA in eukaryotic cells involves the association of a histones octamer with 146 base pairs of DNA. The octamer is compost of four pairs of histones (H2A, H2B, H3, and H4), and its association with the DNA form the fundamental unit of the chromatin, the nucleosome. In the next level of chromatin packing, each nucleosome is connected to the adjacent through of linker DNA associate with the histone H1. The DNA wrapped in the nucleosome is normally inaccessible to binding proteins and transcription factors. However, histone proteins have amino-terminal tails that extend outside the nucleosome and may be a target of modifications that affect the global chromatin structure (Fig. 1b).

The main post-translational covalent modifications in histone tails include methylation, acetylation, phosphorylation, ubiquitination, glycosylation, sumoylation, deamination, ADP ribosylation, among others. The combinatorial potential of these modifications is called "histone code." These modifications control the recruitment of remodeling enzymes that affect the accessibility of DNA transcription machinery, thereby controlling the gene activation and repression.

The most studied modifications are histone methylation and acetylation. Histone methylation is a reaction that involves the transfer of a methyl group mainly to arginine (R) or lysine (K) residues, which are catalyzed by histone methyltransferases (HMTs). The effects of methylation depend on the modified residue and the degree of methylation; arginine can be mono- or dimethylated and lysine mono-, di-, and trimethylated (me1, me2, and me3, respectively). The most frequent methylation marks occur on the tail of histone H3 and lysine residues. Normally, active transcription marks are histone 3 lysine 4 methylation (H3K4me), histone 3 lysine 36 methylation (H3K36me), and histone 3 lysine 79 methylation (H3K79me), whereas the repressive marks are histone 3 lysine 9 methylation (H3K9me), histone 3 lysine 20 methylation (H3K20me), and histone H3 lysine 27 trimethylation (H3K27me) [15–17]. The transfer of an acetyl group to the lysine residues characterizes the histone acetylation, reaction catalyzed by histone acetyltransferases (HATs). The addition of the acetyl group influences the compaction state of chromatin by neutralize the positive charge of lysine and decrease the affinity between histones and DNA. Due to the potential of chromatin unpacking, acetylation is associated with active transcription, while deacetylation leads to transcriptional silencing.

1.3 Non-Coding RNA

Non-coding RNAs (ncRNAs) are RNAs that have regulatory roles in transcription but are not translated into proteins. The family of ncRNAs includes transfer RNAs (tRNA), ribosomal RNA (rRNA), small interfering RNAs (siRNAs), microRNAs (miRNAs), piwi-interacting RNAs (piRNAs), and long ncRNAs (lncRNAs). Among these, some are key regulators of gene expression. These regulatory ncRNAs act directing epigenetic machinery to specific sites in the genome, mainly altering DNA methylation and histone epigenetic patterns.

The piRNAs are a class of ncRNA with a length of 26–32 nucleotides, which are responsible for the control transposable elements in the germline and stem cells [18]. This ncRNA promotes the recruitment of DNA methylation enzymes and guides these proteins to transposon sequences to establish repressive marks, therefore modulating gene expression at the transcription level. This event is important to maintain genome stability during de novo DNA methylation [19].

The lncRNAs are transcripts with a length of >200 nucleotides, which play a role in several regulatory functions. They are involved in the recruitment of chromatin remodeling complexes that establish different epigenetic states, both activating and repressing gene expression [20]. Among the remodelers, the polycomb repressive complex 1 (PRC1) and polycomb repressive complex 2 (PRC2) have an essential role in chromatin regulation by lncRNAs [21]. Further, lncRNAs are involved in important epigenetic events, such as inactivation of the X chromosome and genomic imprinting [22, 23].

2 Epigenetic Reprogramming During Mammalian Preimplantation Development

There are two global waves of epigenetic reprogramming during mammalian development [24]. The first one occurs during gametogenesis. It initiates when primordial germ cells (PGC) migrate and colonize the gonads of the developing fetus, fact that initiates a genome-wide active DNA demethylation process [25]. PGCs lose their DNA methylation patterns, including imprinted and non-imprinted DNA methylation marks [25–27], as described in detail below (Fig. 2). Moreover, the inactive X chromosome in the female fetus is later reactivated [26]. Exceptions to this rule are the DNA methylation patterns of CpG-rich DNA sequences of young (retro)transposons (Fig. 2), which are partially protected of this global demethylation process [26, 28–30]. Concomitant with this global DNA demethylation, an accumulation of 5-hydroxymethylcytosine (5hmC) occurs, in agreement with this modified cytosine being part of the physiological process of cytosine DNA demethylation [31]. By the end of the DNA demethylation process, PGCs differentiate into oocyte and spermatozoa progenitors in female and male fetuses, respectively.

In the male fetus, the de novo DNA methylation process conducted especially by the *DNMT3A* and *DNMT3B* (Fig. 2). It starts during fetal development and males are born with prospermatogonias showing high DNA methylation levels [27, 28]. On the other hand, females are born with oogonias showing low DNA methylation levels [28]. These germ cells will experience genome-wide de novo DNA methylation during the oocyte growth phase (Fig. 2), while full-grown oocytes will complete their specific DNA

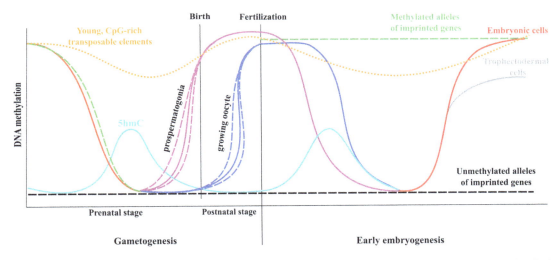

Fig. 2 DNA methylation reprogramming during the mammalian early development. All events are detailed described in the text. Curves do not represent the exact amount of methylation, and the axes are not represented in a scale

methylation pattern only after reaching puberty, when females may complete oogenesis [32]. Importantly, imprinted genes gain their new DNA methylation marks asynchronously during de novo DNA methylation (Fig. 2). For instance, the female fetus receives allele-specific DNA methylation in maternally imprinted genes before the paternally imprinted alleles [33]. The opposite occurs in the male fetus (Fig. 2). In sum, mature oocytes and spermatozoa are highly specialized cells showing highly methylated genomes.

The second wave of DNA demethylation begins immediately after fertilization. The paternal genome demethylates actively by *TET* enzymes (Fig. 2). Consequently, 5hmC marks accumulate [31]. On the other hand, the maternal genome is mainly demethylated by a passive mechanism dependent on DNA replication. Therefore, the maternal genome starts to be demethylated latter than the paternal one [24, 27, 34, 35], in a passive manner (Fig. 2). In cattle, around the 8-to-16-cell stage transition, embryos showed their lowest levels of DNA methylation [34]. From this embryonic stage onward (Fig. 2), the process of de novo methylation begins [34, 36], and the first cell differentiation occurs in the blastocyst, thus generating the trophectoderm and the inner cell mass (ICM). Similarly to what occurs during the first wave of DNA demethylation commented above, the DNA methylation patterns of (retro)-transposons are also partially protected from demethylation (Fig. 2). The maintenance of high DNA methylation levels during this entire developmental time window may be involved in sustaining these DNA elements silenced. There is evidence of microRNAs silencing in this phenomenon, in a similar way to piwiRNA-mediated transposon silencing during mammalian spermatogenesis [37].

After differential DNA methylation patterns are established at imprinted regions during gametogenesis, these marks remain refractory to demethylation during preimplantation development. This protection of epigenetic marks maintains imprinting in somatic tissues throughout the animal lifespan [27, 38] (Fig. 2). In sharp contrast, the trophectoderm shows lower methylation levels at non-imprinted regions compared to the ICM as development progresses [24].

Considering the entire lifespan of an animal, the early developmental time window, that encompasses early gametogenesis and preimplantation development, is one of the most critical periods regarding the relationship between the epigenome and the environment. This is the period in which the epigenome is most susceptible to the adverse effects stemming from the environment because it is the reprogramming events that take place. Among several harmful environmental effects that negatively affect the epigenome are the in vitro conditions in the context of assisted reproductive technologies (ARTs), such as the somatic cell nuclear transfer (SCNT) [39, 40]. This subject will be discussed in more detail later on this chapter.

2.1 Genomic Imprinting

Mammals are biparental diploid organisms that have two copies of each autosomal chromosome, being one copy maternal and the other of paternal origin. Mammalian genes were initially presumed to be expressed equally from both parental alleles, coherently with the fundamental rule of Mendelian genetics. This hypothesis was challenged in the mid-1980s through pronuclear transfer experiments carried out by Solter and Surani laboratories, which demonstrated the first evidence of functional non-equivalence of mammalian parental genomes [41, 42]. This subset of genes with such characteristics were termed imprinted genes [41–47].

Imprinted genes are expressed from a single parental allele, while the other is silenced. This monoallelic expression is controlled by DNA methylation, the histone code, and long non-coding RNAs [48]. Furthermore, the genomic imprinting frequently requires a more complex molecular machinery, such as enhancer or insulator activities and transcriptional silencing by an antisense gene [49–51]. Therefore, genomic imprinting is a highly regulated epigenetic process. The regulation of genes located at an imprinting cluster is controlled by differentially methylated regions (DMRs) at imprinting control regions (ICRs). During gametogenesis, gametes acquire parental-specific methylation patterns (Fig. 3). These differences in allelic methylation status are retained at fertilization and are maintained throughout development, including during the genome-wide demethylation by recruiting *DNMT1* through the recognition of a methylated sequence motif by Krüppel-associated box-containing (*KRAB*) zinc finger protein system [38, 52, 53].

DMRs acquired during gametogenesis are called primary or gDMRs. However, gDMRs can exist in a permanent or transient state after fertilization. The latter is a phenomenon termed transient imprinting [38]. Also, secondary or somatic DMRs (sDMRs) are regions that acquire differential DNA methylation during embryonic development [54, 55]. The first three characterized imprinted genes were (insulin-like growth factor 2 receptor, insulin-like growth factor 2, and H19 imprinted maternally expressed transcript) the mouse homologs of *Igf2r*, *Igf2*, and *H19* in 1991 [23, 56–58]. These three genes regulate normal embryo growth, with the paternally expressed gene *Igf2* being growth-promoting and the maternally expressed imprinted genes *Igf2r* and *H19* being negative growth regulators [59]. The mouse *Igf2* and *H19* genes form a well-characterized imprinted locus. Briefly, *Igf2* and *H19* genes are apart from each other by a genomic region containing an ICR, which has binding sites for *CCCTC-binding factor* (*CTCF*). Enhancers downstream *H19* regulate both genes. In the maternal allele, *CTCF* binds to the unmethylated ICR leading to its insulator function, and thereby not allowing the downstream enhancers to act on the *Igf2* promoter. Thus, *H19* expression is limited to the maternal allele. In the paternal allele,

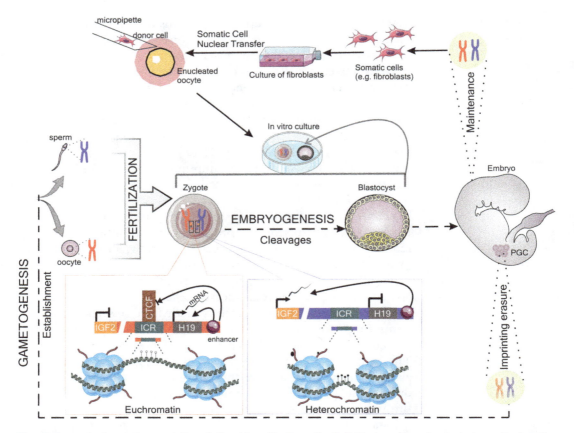

Fig. 3 Genomic imprinting and its relationship with Somatic Cell Nuclear Transfer technique. During the formation of the male (sperm) and female (oocyte) gametes, paternally imprinted (blue chromosome) and maternally imprinted (red chromosome) patterns are established. Each gamete acquires parental of origin-specific methylation patterns. Differences in allelic methylation statuses are retained after fertilization and maintained throughout embryogenesis. In Somatic Cell Nuclear Transfer (SCNT), a somatic cell from the donor is introduced into the perivitelline space of an enucleated oocyte. Imprinted marks need to be maintaining after the SCNT procedure. However, SCNT has been linked with loss of imprinting (LOI), thus resulting in imprinting disorders. LOI induces non-expression or biallelic expression of imprinted genes

ICR is methylated, so *CTCF* cannot bind to it, leading to loss of insulator function, and allows the enhancer to activate *Igf2* [60]. The main theory to explain the evolution of imprinting has been the parental conflict or kinship theory since the *Igf2/H19* cluster represents a parental tug-of-war between maternal and paternal genomes for the control of offspring growth during pregnancy [61]. Alternatively, the maternal-offspring co-adaptation theory argues that imprinted genes evolved to act co-adaptively to optimize both the offspring's development and maternal provisioning and nurturing [62].

Through the DNA methylation marks, the genomic imprinting involves the intergenerational transmission of epigenetic information from gametes to a newly formed embryo. These DNA methylation patterns have been shown as crucial to support fetal growth and placentation [63–65]. However, the loss of imprinting (LOI) induces non-expression or biallelic expression of imprinted genes, which may cause developmental abnormalities (Fig. 3). Many studies have linked ARTs, such as in vitro fertilization (IVF) and SCNT, with leads to LOI and associated disorders. For example, the congenital overgrowth condition in humans and ruminants namely Beckwith–Wiedemann syndrome and large offspring syndrome, respectively [66–69]. Studies demonstrate that placenta-specific maternally imprinted genes showed biallelic expression in SCNT, thus suggesting LOI [70, 71]. Furthermore, studies revealed complete loss of H3K27me3-dependent imprinting in SCNT blastocyst embryos, which likely are the cause of the observed developmental defects of preimplantation SCNT embryos [72, 73]. Additionally, a study revealed that the loss of H3K27me3-dependent imprinting is responsible for abnormal placental enlargement and low birth rates following SCNT, through upregulation of imprinted microRNAs [74].

ARTs (e.g., SCNT, intracytoplasmic sperm injection, IVF) overlap with the timing of extensive epigenetic reprogramming or parental-specific imprint acquisition (Fig. 3). Therefore, the expression of imprinted genes has been widely investigated in under ART conditions [49, 50, 75, 76]. Studies investigated the DNA methylation pattern in a CpG island of the *IGF2* imprinting gene, which is related to regulating fetal and placental growth, thus aiming to predict the impact of bovine ARTs. The spermatozoa showed a hypermethylated pattern [77]. In contrast, oocytes demonstrated a hypomethylated pattern [32]. However, this DNA methylation pattern varies in mature oocytes [78]. ARTs have been widely used in livestock production, albeit with a strong bias to cattle breeding (both beef and dairy), and in human-assisted reproduction. Therefore, the correct establishment of genomic imprinting during ARTs represents one enormous challenge for wider application of these technologies.

2.2 X Chromosome Inactivation

The X chromosome inactivation (XCI) is an evolutionary mechanism that evolved in animals to compensate for imbalanced gene dosage between sexes (dosage compensation). However, species rely on different strategies to accomplish XCI. In mammals, one X chromosome is inactivated in female embryos, thus leveraging gene transcription between male (XY) and female (XX) individuals. XCI was described for the first time by Mary F. Lyon in 1961 after studies of Barr & Bertram 1949 and S Ohno 1959, which showed that the nuclear spot present in female cells was an inactive X chromosome [79–81]. XCI is not complete in mammals and some genes escape inactivation [82]. Besides genes located in the pseudoautosomal region, genes from many other regions of the X chromosome also escape inactivation [83]. This incomplete X inactivation explains, at least in part, differences associated with sexual

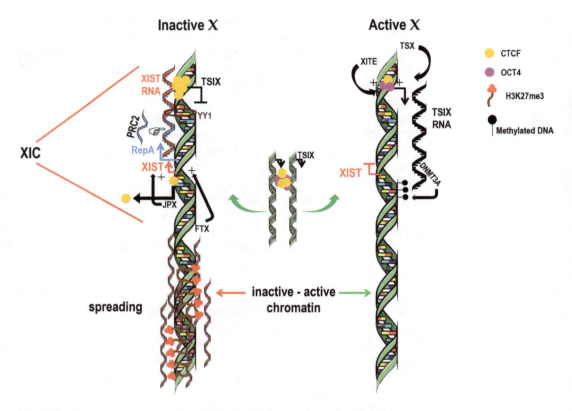

Fig. 4 Random X chromosome inactivation in mouse model. All events shown in the figure are detailed and described in the text

dimorphism and further supported by the phenotypes of X0 women (i.e., carrying only one X as sexual chromosomes) with Turner syndrome [84]. If XCI were complete, with no genes escaping inactivation, X0 women would not show altered phenotypes.

The XCI occurs usually during preimplantation development albeit with some differences among species. XCI takes place at around the blastocyst stage in the mouse, whereas in cattle it initiates in the elongated conceptus around implantation [85]. Mechanistically, XCI has an imprinted control in mice, where the paternal X chromosome (Xp) is chosen for inactivation in both ICM and trophectoderm cells [86–89]. From the blastocyst stage onward, trophectoderm cells remain with the Xp inactivated, whereas epiblast cells reactivate the Xp and immediately initiate XCI randomly [87] (Fig. 4). Once an X becomes inactive, this XCI pattern remains throughout all cell divisions of the lifespan of the animal [90]. During development, primordial germ cells (PGC) reactivate their inactive X in female fetuses [91] and both X chromosomes are in an active state in oocytes [87]. In humans and rabbits, XCI seems to happen randomly [88], whereas in cattle there is evidence of an imprinted control in the placenta [92, 93].

All epigenetic mechanisms (e.g., long non-coding RNAs—lncRNAs, DNA methylation, chromatin remodeling complexes, and post-translational histone) contribute to XCI. Some of them involved in the initiation of XCI, spreading of heterochromatic marks along the X, or involved in the maintenance of X inactivation throughout cell life. In embryonic stem cells, both X chromosomes show active chromatin states sustained by pluripotency factors [94]. During differentiation, it seems that an X pairing event is necessary to start XCI [95]. Pluripotency factors, such as *OCT4* (also known as *POU5F1*), accumulate on the X inactivation center (XIC) on both X chromosomes and recruit other factors as the (e.g., *CTCF*) for X pairing [94, 96]. Further, X chromosomes separate and XCI occurs randomly, albeit the X undergoing XCI remains enriched for bound *CTCF* [95]. To initiate XCI, the lncRNA X inactive-specific transcript (*XIST*), located in the X inactivation center (XIC), is transcribed only from the X chromosome that will become inactive [97]. An interesting fact is that although XCI begins much later in cattle, *XIST* mRNA accumulates early during preimplantation development [49]. On the other hand, before XCI is triggered, a *XIST* antisense lncRNA called (*TSIX*) is transcribed to block *XIST* expression [98].

The *XIST* gene has six conserved repeat sequences (A–F), which are essential for its role during XCI [99]. *XIST* is the major gene involved in the initiation of XCI and acts *in cis* on the X chromosome. The Ying Yang 1 (*YY1*) factor links to a motif into the repeat C of *XIST* and is essential to this *cis*-action during XCI [100]. Another repeat sequence, the repeat A (located into the exon one of *XIST*) transcribes a shorter lncRNA called repeat A (*RepA*) before *XIST* expression [101]. XIC is a region on the X chromosome that is enriched for lncRNAs and besides *XIST*, *RepA*, and *TSIX*, other lncRNAs located in the XIC also participate in the initiation of XCI [90, 102]. For instance, the *Expressed neighbor of Xist* (*JPX/ENOX*) and Five Prime to Xist (*FTX*) promote *XIST* transcription, while the X Inactivation Intergenic Transcription Elements (*XITE*) and testes-specific X-linked (*TSX*) promote *TSIX* transcription [90, 103]. To initiate *XIST* transcription, after X pairing is complete, on the future inactive X, *CTCF* is removed from the *XIST* promoter by *JPX*, thus considering that *CTCF* seems to have a binding preference for RNA instead of DNA [104]. Thus, RepA is transcribed, then recruiting *PRC2* to the future inactive X and promoting *XIST* expression and its accumulation in *cis* [101]. Then, *PRC2* promotes the establishment of repressive marks on the chromatin such as H3K27me3.

Other marks are deposited on the future inactive X such as the histone variant *MacroH2A*, a specific variant histone of the inactive X [94]. At the same time that *XIST* starts spreading on the future inactive X chromosome, those repressive marks, also including DNA methylation, start to accumulate from the unique XIC region present in the X chromosome. Repeat B of the XIST gene is essential to spread the XIST RNA along the inactive X chromosome and carrying PRC1 and PRC2 complexes to establish repressive marks [105]. At the same time, the higher *CTCF* enrichment near the *TSIX* promoter on the future inactive X prevents *TSIX* expression, thus allowing *XIST* expression. After this process, the inactive X rearranges itself in a specific tridimensional folding pattern near the nuclear membrane, with contributions of several nuclear factors (*CTCF*, *Cohesin*, Structural Maintenance Of Chromosomes Flexible Hinge Domain Containing 1—*SMCHD1*) among others [50, 106]. On the other hand, the presence of pluripotency factors that remained near *TSIX*, XITE, and TSX RNAs, along with the lower amount of CTCF induces TSIX expression on the future active X. At least in the mouse, where the XCI is well understood, *TSIX* recruits *DNMT3A* to the *XIST* promoter and for *XIST* transcription repression on the active X [94, 107, 108]. Further, it seems that *TSIX* also blocks the interaction between *RepA* and *PRC2*, thus leading to impaired *XIST* expression [94, 101]. In the absence of *XIST* accumulation, this X chromosome remains in an active state. The ability of *TSIX* to induce methylation at the *XIST* promoter does not seem to happen in cattle due to the lack of CpG dinucleotides in the *XIST* promoter [92]. Furthermore, despite *TSIX* expression being essential in mice to repress *XIST* expression, this mechanism remains unclear in cattle. *TSIX* is predicted as a pseudogene in several species, including cattle. Its original structure seems to have been lost along the course of evolution, and in consequence, its sequence is not conserved among mammalian species [92, 109].

The XCI is essentially controlled by epigenetic mechanisms. Therefore, it may be susceptible to harmful environmental conditions, such as under ARTs [93, 110, 111]. X inactivation is essential throughout the female lifespan and faithful XCI is required for full-term development [89, 112]. Therefore, it becomes paramount to better understand XCI at the molecular level in livestock and humans, thus aiming to improve the efficiency of ARTs [49].

3 Epigenetic Reprogramming After Somatic Cell Nuclear Transfer

In the context of cloning by SCNT, the cytoplasm of the enucleated oocyte needs to reprogram a highly differentiated somatic cell into an undifferentiated zygote-like state, consequently restoring totipotency in the donor cell. In domestic species, SCNT embryos have an in vitro developmental potential (i.e., up to the blastocyst stage) similar to IVF embryos [113]. However, post-implantation development of SCNT embryos is much lower than IVF counterparts, which fewer than 10% of transferred blastocysts develop into viable offspring, mostly due to the high prevalence of gestational losses and developmental abnormalities [114–116]. The low cloning efficiency is frequently associated with functional deficiencies occurring at the onset of placentation [117]. There is vast evidence that suggests that placentation deficiencies come from an erroneous epigenetic reprogramming after SCNT [118–121].

In mammals, the first cellular differentiation takes place at the blastocyst formation. Blastomeres segregate into two cell lineages resulting in the ICM and the trophectoderm. The ICM that will give rise to the embryo proper and the trophectoderm will form the chorion tissue and the embryonic part of the placenta to implant into the uterine wall [122] During cloning by SCNT, it appears to be an unbalance between embryonic and extra-embryonic tissues in developing cloned pregnancies. A study with bovine conceptuses derived by SCNT demonstrated a high incidence of uncoupling at embryonic and extra-embryonic differentiation (both morphologically and molecularly) independent of their cell of origin, which compromised further development [123]. This study also evaluated day 18 extra-embryonic tissue and linked differentially expressed genes (DEGs) with defects in microvilli formation or in the extracellular matrix composition [123]. Another study on cotyledon tissues revealed DEGs by transcriptome-wide analysis. This material was retrieved from cloned cattle at day 180 of gestation and showed that the DEGs were enriched for urea and ions transmembrane transport components, thus indicating that the disturbed maternal-fetal interactions in placentas [124]. Interestingly, clones display several placental disorders, such as a reduced number of placentomes with compensatory enlargement in ruminants, placental edema, abnormal vascular development of the placenta, and hydroallantois [39, 125–127].

The extra-embryonic tissue of the placental lineage seems to be more susceptible to problems related to SCNT cloning [128]. Two hypotheses were proposed by Yang et al. [129] to elucidate the possible reasons. The first hypothesis is that cell fate determination, achieved by activating or suppressing genes through epigenetic modifications, is affected by SCNT. In a physiological attempt to protect the cells that will give rise to the embryo, cells more prone to errors are preferentially incorporated into the extra-embryonic tissue rather than ICM. The second hypothesis, which is not mutually exclusive to the previous, is that the trophoblast cells may be affected by epigenetic deregulations of genomic imprinting [129]. Intriguingly, most imprinted genes are located in the placenta [63, 130, 131].

A study reveals that non-canonical imprints, characterized by genes mono-allelically expressed independent of inherited DNA methylation, are localized preferentially at active endogenous retrovirus-K (*ERVK*) long terminal repeats insertions, which act as imprinted promoters specifically in murine extra-embryonic lineages [132]. In eutherians, the expression of the retroviral envelope proteins of syncytin genes derived from an ERV is crucial for the formation of the multinucleated syncytiotrophoblast, playing an essential role in placental morphogenesis [133]. Analysis of syncytin-like genes, *FEMATRIN-1* and *SYNCYTIN-RUM1*, in SCNT bovine placental cotyledon showed hypermethylation at the *FEMATRIN-1 locus* in cloned stillborn calves. Furthermore, deregulation in the expression of syncytin-like genes in cloned calves compared to artificial insemination animals was found [40]. These evidences reveal a fascinating symbiosis relationship. The retroviruses are involved in a fundamental stage for the perpetuation of the host species [134–136]. Understanding this relationship will bring far-reaching implications for improving ARTs.

3.1 Molecular Strategies to Improve SCNT Efficiency

The erroneous epigenetic reprogramming in SCNT preimplantation embryos is one of the main causes of the low efficiency of this technology. Therefore, many studies explored different strategies aiming to improve reprogramming efficiency and overall full-term cloning efficiency.

The treatment of fibroblasts with demethylating agents such as procaine, S-adenosyl-l-homocysteine (SAH), and 5-aza-2′-deoxycytidine (5-Aza-dc) has shown a significant reduction in global and *locus*-specific DNA methylation levels [137, 138]. One study demonstrated that donor cells treated with SAH led to improved SCNT preimplantation embryonic development [139]. Similarly, cloned embryos treated with 5-Aza-dc also displayed enhanced preimplantation development [140]. The use of small interfering RNA (siRNA) to silence *DNMT1* in donor cells showed an improvement in SCNT blastocyst rates but not at full-term development [141, 142]. In turn, the injection of siRNA to both *DNMT3A* and *DNMT3B* and exogenous mRNAs of histone lysine demethylases (KDMs) for overexpression into enucleated oocytes decreased DNA methylation levels and improvement of cloning efficiency [143]. This strategy was successful because it led to a hypomethylated donor genome that ultimately facilitates epigenetic reprogramming after SCNT.

Another promising strategy is to treat donor cells with histone deacetylase inhibitors (HDACi), to increase chromatin accessibility of the donor cell to the oocyte reprogramming machinery [144]. One study demonstrated that the donor cell treated with the HDACi valproic acid (VPA) led to overexpression of pluripotent genes and, consequently, an increase in the SCNT blastocyst rates [145]. Another HDACi used widely is trichostatin A (TSA). Reconstructed oocytes are treated with TSA occurs for 8–12 h post-activation and it improves the blastocyst rates, but long-term TSA treatment is toxic [146]. In contrast, SCNT embryos exposed to the TSA for 20 h also improve embryo production [147]. In non-human primates, the strategy that improved blastocyst rates and allowed the production of healthy cloned monkeys was the injection of lysine demethylase 4D (*KDM4D*) mRNA at the one-cell and incubation with TSA for 10 h post-activation [148, 149].

An alternative to chromatin-modifying chemicals is to screen viable SCNT embryos by molecular analyses. Several studies have identified alterations at the transcriptional level in cloned embryos, which point out to important genes such as monoamine oxidase A (*MAO-A*), *IGF2*, and *H19* [110, 150]. A better understanding of these SCNT-associated molecular alterations would allow us to identify reliable molecular markers and further select SCNT embryos with less probability of abnormal development, with high full-term development rates, and increased offspring survival.

The abnormal development of SCNT embryos also occurs due to the alterations in genes involved in XCI, especially the ectopic expression of XIST. For this reason, strategies using gene editing to knockout and knockdown of *XIST* in cloned embryos showed significant increase in cloning efficiency [151, 152]. Although genetic editing has been used to improve the efficiency of SCNT cloning, the most promising tool is the epigenetic editing using Clustered Regularly Interspaced Short Palindromic Repeats/deactivated Cas9 (CRISPR/dCas9). This technology allows the recruitment of chromatin remodelers to specific target sites to establish an euchromatin permissive state to epigenetic reprogramming [153]. This CRISPR/dCas9 approach was already successful to reprogramming somatic cells into induced pluripotent cells [154]. Therefore, CRISPR/dCas9 could be useful to prevent epigenetic errors during epigenetic reprogramming, and consequently, to improve SCNT cloning efficiency.

References

1. Patil V, Cuenin C, Chung F, Aguilera JRR, Fernandez-Jimenez N, Romero-Garmendia I et al (2019) Human mitochondrial DNA is extensively methylated in a non-CpG context. Nucleic Acids Res 47:10072–10085

2. Ziller MJ, Müller F, Liao J, Zhang Y, Gu H, Bock C et al (2011) Genomic distribution and inter-sample variation of non-CpG methylation across human cell types. PLoS Genet 7: e1002389

3. Lister R, Pelizzola M, Dowen RH, Hawkins RD, Hon G, Tonti-Filippini J et al (2009) Human DNA methylomes at base resolution show widespread epigenomic differences. Nature 462:315–322

4. Keown CL, Berletch JB, Castanon R, Nery JR, Disteche CM, Ecker JR et al (2017) Allele-specific non-CG DNA methylation marks domains of active chromatin in female

mouse brain. Proc Natl Acad Sci U S A 114: 2882–2890

5. Hadad N, Unnikrishnan A, Jackson JA, Masser DR, Otalora L, Stanford DR et al (2018) Caloric restriction mitigates age-associated hippocampal differential CG and non-CG methylation. Neurobiol Aging 67:53–66

6. Larsen F, Gundersen G, Lopez R, Prydz H (1992) CpG islands as gene markers in the human genome. Genomics 13:1095–1107

7. Bird AP (1986) CpG-rich islands and the function of DNA methylation. Nature 321: 209–213

8. Hellman A, Chess A (2007) Gene body-specific methylation on the active X chromosome. Science 315:1141–1143

9. Shirane K, Toh H, Kobayashi H, Miura F, Chiba H, Ito T et al (2013) Mouse oocyte

methylomes at base resolution reveal genome-wide accumulation of non-CpG methylation and role of DNA methyltransferases. PLoS Genet 9:e1003439

10. Okano M, Bell DW, Haber DA, Li E (1999) DNA methyltransferases Dnmt3a and Dnmt3b are essential for De novo methylation and mammalian development. Cell 99: 247–257

11. Veland N, Lu Y, Hardikar S, Gaddis S, Zeng Y, Liu B et al (2019) DNMT3L facilitates DNA methylation partly by maintaining DNMT3A stability in mouse embryonic stem cells. Nucleic Acids Res 47:152–167

12. Jia D, Jurkowska RZ, Zhang X, Jeltsch A, Cheng X (2007) Structure of Dnmt3a bound to Dnmt3L suggests a model for de novo DNA methylation. Nature 449:248–251

13. Goll MG, Kirpekar F, Maggert KA, Yoder JA, Hsieh CL, Zhang X et al (2006) Methylation of tRNAAsp by the DNA methyltransferase homolog Dnmt2. Science 311:395–398

14. Kiani J, Grandjean V, Liebers R, Tuorto F, Ghanbarian H, Lyko F et al (2013) RNA–mediated epigenetic heredity requires the cytosine methyltransferase Dnmt2. PLoS Genet 9:e1003498

15. Bernstein BE, Humphrey EL, Erlich RL, Schneider R, Bouman P, Liu JS et al (2002) Methylation of histone H3 Lys 4 in coding regions of active genes. Proc Natl Acad Sci U S A 99:8695–8700

16. Bannister AJ, Schneider R, Myers FA, Thorne AW, Crane-Robinson C, Kouzarides T (2005) Spatial distribution of di-and tri-methyl lysine 36 of histone H3 at active genes. J Biol Chem 280:17732–17736

17. Barski A, Cuddapah S, Cui K, Roh T-Y, Schones DE, Wang Z et al (2007) High-resolution profiling of histone methylations in the human genome. Cell 129:823–837

18. Kuramochi-Miyagawa S, Watanabe T, Gotoh K, Totoki Y, Toyoda A, Ikawa M et al (2008) DNA methylation of retrotransposon genes is regulated by Piwi family members MILI and MIWI2 in murine fetal testes. Genes Dev 22:908–917

19. Huang XA, Yin H, Sweeney S, Raha D, Snyder M, Lin H (2013) A major epigenetic programming mechanism guided by piRNAs. Dev Cell 24:502–516

20. Rinn JL, Kertesz M, Wang JK, Squazzo SL, Xu X, Brugmann SA et al (2007) Functional demarcation of active and silent chromatin domains in human HOX loci by noncoding RNAs. Cell 129:1311–1323

21. Almeida M, Pintacuda G, Masui O, Koseki Y, Gdula M, Cerase A et al (2017) PCGF3/5–PRC1 initiates Polycomb recruitment in X chromosome inactivation. Science 356: 1081–1084

22. Maclary E, Buttigieg E, Hinten M, Gayen S, Harris C, Sarkar MK et al (2014) Differentiation-dependent requirement of Tsix long non-coding RNA in imprinted X-chromosome inactivation. Nat Commun 5:1–14

23. Bartolomei MS, Zemel S, Tilghman SM (1991) Parental imprinting of the mouse H19 gene. Nature 351:153–155

24. Reik W, Dean W, Walter J (2001) Epigenetic reprogramming in mammalian development. Science 293:1089

25. Hajkova P, Ancelin K, Waldmann T, Lacoste N, Lange UC, Cesari F et al (2008) Chromatin dynamics during epigenetic reprogramming in the mouse germ line. Nature 452:877–881

26. Seisenberger S, Andrews S, Krueger F, Arand J, Walter J, Santos F et al (2012) The dynamics of genome-wide DNA methylation reprogramming in mouse primordial germ cells. Mol Cell 48:849–862

27. MacDonald WA, Mann MRW (2014) Epigenetic regulation of genomic imprinting from germ line to preimplantation. Mol Reprod Dev 81:126–140

28. Kobayashi H, Sakurai T, Miura F, Imai M, Mochiduki K, Yanagisawa E et al (2013) High-resolution DNA methylome analysis of primordial germ cells identifies gender-specific reprogramming in mice. Genome Res 23:616–627

29. Cowley M, Oakey Rebecca J (2012) Resetting for the next generation. Mol Cell 48:819–821

30. Edwards JR, Yarychkivska O, Boulard M, Bestor TH (2017) DNA methylation and DNA methyltransferases. Epigenetics Chromatin 10:23–23

31. Hill PWS, Amouroux R, Hajkova P (2014) DNA demethylation, Tet proteins and 5-hydroxymethylcytosine in epigenetic reprogramming: an emerging complex story. Genomics 104:324–333

32. Fagundes NS, Michalczechen-Lacerda VA, Caixeta ES, Machado GM, Rodrigues FC, Melo EO et al (2011) Methylation status in the intragenic differentially methylated region of the IGF2 locus in Bos taurus indicus oocytes with different developmental competencies. Mol Hum Reprod 17:85–91

33. Morgan HD, Santos F, Green K, Dean W, Reik W (2005) Epigenetic reprogramming in mammals. Hum Mol Genet 14:47–58

34. Dean W, Santos F, Stojkovic M, Zakhartchenko V, Walter J, Wolf E et al (2001) Conservation of methylation reprogramming in mammalian development: aberrant reprogramming in cloned embryos. Proc Natl Acad Sci U S A 98:13734–13738

35. Sasaki H, Matsui Y (2008) Epigenetic events in mammalian germ-cell development: reprogramming and beyond. Nat Rev Genet 9: 129–140

36. Maalouf WE, Alberio R, Campbell KH (2008) Differential acetylation of histone H4 lysine during development of in vitro fertilized, cloned and parthenogenetically activated bovine embryos. Epigenetics 3:199–209

37. Aravin AA, Bourc'his D (2008) Small RNA guides for de novo DNA methylation in mammalian germ cells. Genes Dev 22:970–975

38. Proudhon C, Duffié R, Ajjan S, Cowley M, Iranzo J, Carbajosa G et al (2012) Protection against de novo methylation is instrumental in maintaining parent-of-origin methylation inherited from the gametes. Mol Cell 47: 909–920

39. Silveira MM, Salgado Bayão HX, dos Santos Mendonça A, Borges NA, Vargas LN, Caetano AR et al (2018) DNA methylation profile at a satellite region is associated with aberrant placentation in cloned calves. Placenta 70:25–33

40. Silveira MM, Vargas LN, Bayão HXS, Schumann NAB, Caetano AR, Rumpf R et al (2019) DNA methylation of the endogenous retrovirus Fematrin-1 in fetal placenta is associated with survival rate of cloned calves. Placenta 88:52–60

41. McGrath J, Solter D (1984) Completion of mouse embryogenesis requires both the maternal and paternal genomes. Cell 37: 179–183

42. Surani MA, Barton SC (1983) Development of gynogenetic eggs in the mouse: implications for parthenogenetic embryos. Science 222:1034

43. McGrath J, Solter D (1983) Nuclear transplantation in mouse embryos. J Exp Zool 228:355–362

44. McGrath J, Solter D (1983) Nuclear transplantation in the mouse embryo by microsurgery and cell fusion. Science 220:1300

45. Ferguson-Smith AC, Bourc'his D (2018) The discovery and importance of genomic imprinting. eLife 7:e42368. https://doi.org/10.7554/eLife.42368

46. Cattanach BM, Kirk M (1985) Differential activity of maternally and paternally derived chromosome regions in mice. Nature 315: 496–498

47. Surani MAH, Barton SC, Norris ML (1984) Development of reconstituted mouse eggs suggests imprinting of the genome during gametogenesis. Nature 308:548–550

48. Adalsteinsson BT, Ferguson-Smith AC (2014) Epigenetic control of the genome-lessons from genomic imprinting. Genes 5: 635–655

49. Mendonça ADS, Silveira MM, Rios ÁFL, Mangiavacchi PM, Caetano AR, Dode MAN et al (2019) DNA methylation and functional characterization of the XIST gene during in vitro early embryo development in cattle. Epigenetics 14:568–588

50. Franco MM, Prickett AR, Oakey RJ (2014) The role of CCCTC-binding factor (CTCF) in genomic imprinting, development, and reproduction. Biol Reprod 91:125

51. Leighton PA, Saam JR, Ingram RS, Tilghman SM (1996) Genomic imprinting in mice: its function and mechanism 1. Biol Reprod 54: 273–278

52. Barlow DP, Bartolomei MS (2014) Genomic imprinting in mammals. Cold Spring Harb Perspect Biol 6:a018382. https://doi.org/10.1101/cshperspect.a018382

53. Quenneville S, Verde G, Corsinotti A, Kapopoulou A, Jakobsson J, Offner S et al (2011) In embryonic stem cells, ZFP57/KAP1 recognize a methylated hexanucleotide to affect chromatin and DNA methylation of imprinting control regions. Mol Cell 44:361–372

54. Ferguson-Smith AC (2011) Genomic imprinting: the emergence of an epigenetic paradigm. Nat Rev Genet 12:565–575

55. Nechin J, Tunstall E, Raymond N, Hamagami N, Pathmanabhan C, Forestier S et al (2019) Hemimethylation of CpG dyads is characteristic of secondary DMRs associated with imprinted loci and correlates with 5-hydroxymethylcytosine at paternally methylated sequences. Epigenetics Chromatin 12: 64–64

56. Barlow DP, Stöger R, Herrmann BG, Saito K, Schweifer N (1991) The mouse insulin-like growth factor type-2 receptor is imprinted and closely linked to the Tme locus. Nature 349:84–87

57. DeChiara TM, Robertson EJ, Efstratiadis A (1991) Parental imprinting of the mouse

insulin-like growth factor II gene. Cell 64: 849–859

58. Ferguson-Smith AC, Cattanach BM, Barton SC, Beechey CV, Surani MA (1991) Embryological and molecular investigations of parental imprinting on mouse chromosome 7. Nature 351:667–670

59. Ideraabdullah FY, Vigneau S, Bartolomei MS (2008) Genomic imprinting mechanisms in mammals. Mutat Res 647:77–85

60. Nordin M, Bergman D, Halje M, Engström W, Ward A (2014) Epigenetic regulation of the Igf2/H19 gene cluster. Cell Prolif 47:189–199

61. Moore T, Haig D (1991) Genomic imprinting in mammalian development: a parental tug-of-war. Trends Genet 7:45–49

62. Wolf JB, Hager R (2006) A maternal–offspring coadaptation theory for the evolution of genomic imprinting. PLoS Biol 4:e380. https://doi.org/10.1371/journal.pbio.0040380

63. Hanna CW (2020) Placental imprinting: emerging mechanisms and functions. PLoS Genet 16:e1008709. https://doi.org/10.1371/journal.pgen.1008709

64. Cleaton MAM, Edwards CA, Ferguson-Smith AC (2014) Phenotypic outcomes of imprinted gene models in mice: elucidation of pre- and postnatal functions of imprinted genes. Annu Rev Genomics Hum Genet 15: 93–126

65. Plasschaert RN, Bartolomei MS (2014) Genomic imprinting in development, growth, behavior and stem cells. Development 141: 1805–1813

66. Li Y, Donnelly CG, Rivera RM (2019) Overgrowth syndrome. Vet Clin North Am Food Anim Pract 35:265–276

67. Chen Z (2013) Large offspring syndrome: a bovine model for the human loss-of-imprinting overgrowth syndrome Beckwith–Wiedemann. Epigenetics 8:591–601

68. Chen Z, Hagen DE, Elsik CG, Ji T, Morris CJ, Moon LE et al (2015) Characterization of global loss of imprinting in fetal overgrowth syndrome induced by assisted reproduction. Proc Natl Acad Sci U S A 112:4618–4623

69. Li Y, Hagen DE, Ji T, Bakhtiarizadeh MR, Frederic WM, Traxler EM et al (2019) Altered microRNA expression profiles in large offspring syndrome and Beckwith-Wiedemann syndrome. Epigenetics 14:850–876

70. Hirose M, Hada M, Kamimura S, Matoba S, Honda A, Motomura K et al (2018) Aberrant

imprinting in mouse trophoblast stem cells established from somatic cell nuclear transfer-derived embryos. Epigenetics 13: 693–703

71. Okae H, Matoba S, Nagashima T, Mizutani E, Inoue K, Ogonuki N et al (2013) RNA sequencing-based identification of aberrant imprinting in cloned mice. Hum Mol Genet 23:992–1001

72. Matoba S, Wang H, Jiang L, Lu F, Iwabuchi KA, Wu X et al (2018) Loss of H3K27me3 imprinting in somatic cell nuclear transfer embryos disrupts post-implantation development. Cell Stem Cell 23:343–354

73. Matoba S, Liu Y, Lu F, Iwabuchi KA, Shen L, Inoue A et al (2014) Embryonic development following somatic cell nuclear transfer impeded by persisting histone methylation. Cell 159:884–895

74. Inoue K, Ogonuki N, Kamimura S, Inoue H, Matoba S, Hirose M et al (2020) Loss of H3K27me3 imprinting in the Sfmbt2 miRNA cluster causes enlargement of cloned mouse placentas. Nat Commun 11:2150–2150

75. dos Santos Mendonça A, Franco MM, de Oliveira Carvalho J, Machado GM, Dode MAN (2019) DNA methylation of the insulin-like growth factor 2-imprinted gene in trophoblast cells of elongated bovine embryo: effects of the in vitro culture. Cell Reprogram 21: 260–269

76. Poirier M, Smith OE, Therrien J, Rigoglio NN, Miglino MA, Silva LA et al (2019) Resiliency of equid H19 imprint to somatic cell reprogramming by oocyte nuclear transfer and genetically induced pluripotency. Biol Reprod 102:211–219

77. Carvalho JO, Michalczechen-Lacerda VA, Sartori R, Rodrigues FC, Bravim O, Franco MM et al (2012) The methylation patterns of the IGF2 and IGF2R genes in bovine spermatozoa are not affected by flow-cytometric sex sorting. Mol Reprod Dev 79:77–84

78. Mendonça AS, Guimarães ALS, da Silva NMA, Caetano AR, Dode MAN, Franco MM (2015) Characterization of the IGF2 imprinted gene methylation status in bovine oocytes during folliculogenesis. PLoS One 10:e0142072. https://doi.org/10.1371/journal.pone.0142072

79. Barr ML, Bertram EG (1949) A morphological distinction between neurones of the male and female, and the behaviour of the nucleolar

satellite during accelerated nucleoprotein synthesis. Nature 163:676–677

80. Ohno S, Kaplan WD, Kinosita R (1959) Formation of the sex chromatin by a single X-chromosome in liver cells of Rattus norvegicus. Exp Cell Res 18:415–418

81. Lyon MF (1961) Gene action in the X-chromosome of the mouse (Mus musculus L.). Nature 190:372–373

82. Shapiro LJ, Mohandas T, Weiss R, Romeo G (1979) Non-inactivation of an X-chromosome locus in man. Science 204:1224

83. Brown CJ, Carrel L, Willard HF (1997) Expression of genes from the human active and inactive X chromosomes. Am J Hum Genet 60:1333–1343

84. Zinn AR, Page DC, Fisher EMC (1993) Turner syndrome: the case of the missing sex chromosome. Trends Genet 9:90–93

85. Bermejo-Alvarez P, Ramos-Ibeas P, Gutierrez-Adan A (2012) Solving the "X" in embryos and stem cells. Stem Cells Dev 21:1215–1224

86. Okamoto I, Heard E (2006) The dynamics of imprinted X inactivation during preimplantation development in mice. Cytogenet Genome Res 113:318–324

87. Payer B, Lee JT (2008) X chromosome dosage compensation: how mammals keep the balance. Annu Rev Genet 42:733–772

88. Okamoto I, Patrat C, Thépot D, Peynot N, Fauque P, Daniel N et al (2011) Eutherian mammals use diverse strategies to initiate X-chromosome inactivation during development. Nature 472:370–374

89. Gontan C, Mira-Bontenbal H, Magaraki A, Dupont C, Barakat TS, Rentmeester E et al (2018) REX1 is the critical target of RNF12 in imprinted X chromosome inactivation in mice. Nat Commun 9:4752

90. Maclary E, Hinten M, Harris C, Kalantry S (2013) Long noncoding RNAs in the X-inactivation center. Chromosom Res 21:601–614

91. Monk M (1992) The X chromosome in development in mouse and man. J Inherit Metab Dis 15:499–513

92. Dindot SV, Kent KC, Evers B, Loskutoff N, Womack J, Piedrahita JA (2004) Conservation of genomic imprinting at the XIST, IGF2, and GTL2 loci in the bovine. Mamm Genome 15:966–974

93. Xue F, Tian XC, Du F, Kubota C, Taneja M, Dinnyes A et al (2002) Aberrant patterns of X chromosome inactivation in bovine clones. Nat Genet 31:216–220

94. Jeon Y, Sarma K, Lee JT (2012) New and Xisting regulatory mechanisms of X chromosome inactivation. Curr Opin Genet Dev 22:62–71

95. Xu N, Tsai C-L, Lee JT (2006) Transient homologous chromosome pairing marks the onset of X inactivation. Science 311:1149

96. Donohoe ME, Silva SS, Pinter SF, Xu N, Lee JT (2009) The pluripotency factor Oct4 interacts with Ctcf and also controls X-chromosome pairing and counting. Nature 460:128–132

97. Brockdorff N, Ashworth A, Kay GF, McCabe VM, Norris DP, Cooper PJ, Swift S, Rastan S (1992) The product of the mouse Xist gene is a 15 kb inactive X-specific transcript containing no conserved ORF and located in the nucleus. Cell 71:515–526

98. Lee J, Davidow LS, Warshawsky D (1999) Tsix, a gene antisense to Xist at the X-inactivation centre. Nat Genet 21:400–404

99. Elisaphenko EA, Kolesnikov NN, Shevchenko AI, Rogozin IB, Nesterova TB, Brockdorff N et al (2008) A dual origin of the Xist gene from a protein-coding gene and a set of transposable elements. PLoS One 3:e2521. https://doi.org/10.1371/journal.pone.0002521

100. Jeon Y, Lee Jeannie T (2011) YY1 tethers Xist RNA to the inactive X nucleation center. Cell 146:119–133

101. Zhao J, Sun BK, Erwin JA, Song J-J, Lee JT (2008) Polycomb proteins targeted by a short repeat RNA to the mouse X chromosome. Science 322:750

102. Furlan G, Rougeulle C (2016) Function and evolution of the long noncoding RNA circuitry orchestrating X-chromosome inactivation in mammals. WIREs RNA 7:702–722

103. Anguera MC, Ma W, Clift D, Namekawa S, Kelleher RJ III, Lee JT (2011) Tsx produces a long noncoding RNA and has general functions in the germline, stem cells, and brain. PLoS Genet 7:e1002248. https://doi.org/10.1371/journal.pgen.1002248

104. Sun S, Del Rosario BC, Szanto A, Ogawa Y, Jeon Y, Lee JT (2013) Jpx RNA activates Xist by evicting CTCF. Cell 153:1537–1551

105. Colognori D, Sunwoo H, Kriz AJ, Wang C-Y, Lee JT (2019) Xist deletional analysis reveals an interdependency between Xist RNA and polycomb complexes for spreading along the inactive X. Mol Cell 74:101–117

106. Wang C-Y, Jégu T, Chu H-P, Oh HJ, Lee JT (2018) SMCHD1 merges chromosome compartments and assists formation of superstructures on the inactive X. Cell 174:406–421

107. Sado T, Hoki Y, Sasaki H (2005) Tsix silences Xist through modification of chromatin structure. Dev Cell 9:159–165

108. Ohhata T, Hoki Y, Sasaki H, Sado T (2008) Crucial role of antisense transcription across the Xist promoter in Tsix-mediated Xist chromatin modification. Development 135:227

109. Chureau C, Prissette M, Bourdet A, Barbe V, Cattolico L, Jones L et al (2002) Comparative sequence analysis of the X-inactivation center region in mouse, human, and bovine. Genome Res 12:894–908

110. Ferreira A, Aguiar Filho L, Sousa R, Sartori R, Franco M (2015) Characterization of allele-specific expression of the X-linked gene MAO-A in trophectoderm cells of bovine embryos produced by somatic cell nuclear transfer. Genet Mol Res 14:12128–12136

111. Urrego R, Rodriguez-Osorio N, Niemann H (2014) Epigenetic disorders and altered gene expression after use of Assisted Reproductive Technologies in domestic cattle. Epigenetics 9:803–815

112. Borensztein M, Syx L, Ancelin K, Diabangouaya P, Picard C, Liu T et al (2017) Xist-dependent imprinted X inactivation and the early developmental consequences of its failure. Nat Struct Mol Biol 24:226–233

113. Yvan H (2005) Nuclear transfer: a new tool for reproductive biotechnology in cattle. Reprod Nutr Dev 45:353–361

114. Wells DN, Misica PM, Tervit HR (1999) Production of cloned calves following nuclear transfer with cultured adult mural granulosa cells. Biol Reprod 60:996–1005

115. Kato Y, Tani T, Tsunoda Y (2000) Cloning of calves from various somatic cell types of male and female adult, newborn and fetal cows. J Reprod Fertil 120:231–237

116. Hirasawa R, Matoba S, Inoue K, Ogura A (2013) Somatic donor cell type correlates with embryonic, but not extra-embryonic, gene expression in postimplantation cloned embryos. PLoS One 8:e76422. https://doi.org/10.1371/journal.pone.0076422

117. Hill JR, Burghardt RC, Jones K, Long CR, Looney CR, Shin T et al (2000) Evidence for placental abnormality as the major cause of mortality in first-trimester somatic cell cloned bovine fetuses. Biol Reprod 63:1787–1794

118. Matoba S, Zhang Y (2018) Somatic cell nuclear transfer reprogramming: mechanisms and applications. Cell Stem Cell 23:471. https://doi.org/10.1016/j.stem.2018.06.018

119. Wang X, Qu J, Li J, He H, Liu Z, Huan Y (2020) Epigenetic reprogramming during somatic cell nuclear transfer: recent Progress and future directions. Front Genet 11:205–205

120. Long CR, Westhusin ME, Golding MC (2014) Reshaping the transcriptional frontier: epigenetics and somatic cell nuclear transfer. Mol Reprod Dev 81:183–193

121. Niemann H, Tian XC, King WA, Lee RSF (2008) Epigenetic reprogramming in embryonic and foetal development upon somatic cell nuclear transfer cloning. Reproduction 135:151–163

122. Mauch T, Schoenwolf G (2001) Developmental biology. Sixth Edition. By Scott F. Gilbert. Am J Med Genet 99(2):170–171

123. Degrelle SA, Jaffrezic F, Campion E, Lê Cao K-A, Le Bourhis D, Richard C et al (2012) Uncoupled embryonic and extra-embryonic tissues compromise blastocyst development after somatic cell nuclear transfer. PLoS One 7:e38309. https://doi.org/10.1371/journal.pone.0038309

124. Gao G, Wang S, Zhang J, Su G, Zheng Z, Bai C et al (2019) Transcriptome-wide analysis of the SCNT bovine abnormal placenta during mid-to late gestation. Sci Rep 9:20035

125. Miglino MA, Pereira FT, Visintin JA, Garcia JM, Meirelles FV, Rumpf R et al (2007) Placentation in cloned cattle: structure and microvascular architecture. Theriogenology 68:604–617

126. Batchelder CA, Bertolini M, Mason JB, Moyer AL, Hoffert KA, Petkov SG et al (2007) Perinatal physiology in cloned and normal calves: physical and clinical characteristics. Cloning Stem Cells 9:63–82

127. Palmieri C, Loi P, Ptak G, Della Salda L (2008) Review paper: a review of the pathology of abnormal placentae of somatic cell nuclear transfer clone pregnancies in cattle, sheep, and mice. Vet Pathol 45:865–880

128. Chavatte-Palmer P, Camous S, Jammes H, Le Cleac'h N, Guillomot M, Lee RSF (2012) Review: Placental perturbations induce the developmental abnormalities often observed

in bovine somatic cell nuclear transfer. Placenta 33:99–104

129. Yang X, Smith SL, Tian XC, Lewin HA, Renard JP, Wakayama T (2007) Nuclear reprogramming of cloned embryos and its implications for therapeutic cloning. Nat Genet 39:295–302

130. Babak T, DeVeale B, Tsang EK, Zhou Y, Li X, Smith KS et al (2015) Genetic conflict reflected in tissue-specific maps of genomic imprinting in human and mouse. Nat Genet 47:544–549

131. Andergassen D, Dotter CP, Wenzel D, Sigl V, Bammer PC, Muckenhuber M et al (2017) Mapping the mouse Allelome reveals tissue-specific regulation of allelic expression. eLife 6:e25125. https://doi.org/10.7554/eLife.25125

132. Hanna CW, Pérez-Palacios R, Gahurova L, Schubert M, Krueger F, Biggins L et al (2019) Endogenous retroviral insertions drive non-canonical imprinting in extra-embryonic tissues. Genome Biol 20:225

133. Denner J (2016) Expression and function of endogenous retroviruses in the placenta. APMIS 124:31–43

134. Weiss RA (2016) Human endogenous retroviruses: friend or foe? APMIS 124:4–10

135. Chuong EB (2018) The placenta goes viral: retroviruses control gene expression in pregnancy. PLoS Biol 16:e3000028. https://doi.org/10.1371/journal.pbio.3000028

136. Meyer TJ, Rosenkrantz JL, Carbone L, Chavez SL (2017) Endogenous retroviruses: with us and against us. Front Chem 5:23

137. Schumann NA, Mendonça AS, Silveira MM, Vargas LN, Leme LO, Sousa RV et al (2020) Procaine and S-adenosyl-l-homocysteine affect the expression of genes related to the epigenetic machinery and change the DNA methylation status of in vitro cultured bovine skin fibroblasts. DNA Cell Biol 39:37–49

138. Shen C-J, Lin C-C, Shen P-C, Cheng WT, Chen H-L, Chang T-C et al (2013) Imprinted genes and satellite loci are differentially methylated in bovine somatic cell nuclear transfer clones. Cell Reprogram 15:413–424

139. Jeon B-G, Coppola G, Perrault SD, Rho G-J, Betts DH, King WA (2008) S-adenosylhomocysteine treatment of adult female fibroblasts alters X-chromosome inactivation and improves in vitro embryo development after somatic cell nuclear transfer. Reproduction 135:815–828

140. Huan YJ, Zhu J, Xie BT, Wang JY, Liu SC, Zhou Y et al (2013) Treating cloned embryos, but not donor cells, with 5-aza-2′-deoxycytidine enhances the developmental competence of porcine cloned embryos. J Reprod Dev 59:442. https://doi.org/10.1262/jrd.2013-026

141. Yamanaka K-i, Sakatani M, Kubota K, Balboula AZ, Sawai K, Takahashi M (2011) Effects of downregulating DNA methyltransferase 1 transcript by RNA interference on DNA methylation status of the satellite I region and in vitro development of bovine somatic cell nuclear transfer embryos. J Reprod Dev 57:393. https://doi.org/10.1262/jrd.10-181A

142. Golding MC, Williamson GL, Stroud TK, Westhusin ME, Long CR (2011) Examination of DNA methyltransferase expression in cloned embryos reveals an essential role for Dnmt1 in bovine development. Mol Reprod Dev 78:306–317

143. Gao R, Wang C, Gao Y, Xiu W, Chen J, Kou X et al (2018) Inhibition of aberrant DNA re-methylation improves post-implantation development of somatic cell nuclear transfer embryos. Cell Stem Cell 23:426–435

144. Wang L-J, Zhang H, Wang Y-S, Xu W-B, Xiong X-R, Li Y-Y et al (2011) Scriptaid improves in vitro development and nuclear reprogramming of somatic cell nuclear transfer bovine embryos. Cell Reprogram 13:431–439

145. Li X, Ao X, Bai L, Li D, Liu X, Wei Z et al (2018) VPA selectively regulates pluripotency gene expression on donor cell and improve SCNT embryo development. In Vitro Cell Dev Biol Anim 54:496–504

146. Tsuji Y, Kato Y, Tsunoda Y (2009) The developmental potential of mouse somatic cell nuclear-transferred oocytes treated with trichostatin A and 5-aza-2 [variant prime]-deoxycytidine. Zygote 17:109

147. Silva CG, Martins CF, Bessler HC, da Fonseca Neto ÁM, Cardoso TC, Franco MM et al (2019) Use of trichostatin A alters the expression of HDAC3 and KAT2 and improves in vitro development of bovine embryos cloned using less methylated mesenchymal stem cells. Reprod Domest Anim 54:289–299

148. Cibelli JB, Gurdon JB (2018) Custom-made oocytes to clone non-human primates. Cell 172:647–649

149. Liu Z, Cai Y, Wang Y, Nie Y, Zhang C, Xu Y et al (2018) Cloning of macaque monkeys by somatic cell nuclear transfer. Cell 172:881–887

150. Ruan Z, Zhao X, Qin X, Luo C, Liu X, Deng Y et al (2018) DNA methylation and expression of imprinted genes are associated with the viability of different sexual cloned buffaloes. Reprod Domest Anim 53:203–212

151. Inoue K, Kohda T, Sugimoto M, Sado T, Ogonuki N, Matoba S et al (2010) Impeding Xist expression from the active X chromosome improves mouse somatic cell nuclear transfer. Science 330:496–499

152. Matoba S, Inoue K, Kohda T, Sugimoto M, Mizutani E, Ogonuki N et al (2011) RNAi-mediated knockdown of Xist can rescue the impaired postimplantation development of cloned mouse embryos. Proc Natl Acad Sci U S A 108:20621–20626

153. Braun SM, Kirkland JG, Chory EJ, Husmann D, Calarco JP, Crabtree GR (2017) Rapid and reversible epigenome editing by endogenous chromatin regulators. Nat Commun 8:1–8

154. Liu P, Chen M, Liu Y, Qi LS, Ding S (2018) CRISPR-based chromatin remodeling of the endogenous Oct4 or Sox2 locus enables reprogramming to pluripotency. Cell Stem Cell 22:252–261

Chapter 3

Early Cell Specification in Mammalian Fertilized and Somatic Cell Nuclear Transfer Embryos

Marcelo D. Goissis and Jose B. Cibelli

Abstract

Early cell specification in mammalian preimplantation embryos is an intricate cellular process that leads to coordinated spatial and temporal expression of specific genes. Proper segregation into the first two cell lineages, the inner cell mass (ICM) and the trophectoderm (TE), is imperative for developing the embryo proper and the placenta, respectively. Somatic cell nuclear transfer (SCNT) allows the formation of a blastocyst containing both ICM and TE from a differentiated cell nucleus, which means that this differentiated genome must be reprogrammed to a totipotent state. Although blastocysts can be generated efficiently through SCNT, the full-term development of SCNT embryos is impaired mostly due to placental defects. In this review, we examine the early cell fate decisions in fertilized embryos and compare them to observations in SCNT-derived embryos, in order to understand if these processes are affected by SCNT and could be responsible for the low success of reproductive cloning.

Key words Blastocyst, Epiblast, Inner cell mass, Primitive endoderm, Totipotency, Trophectoderm

---

1 Introduction

A multicellular organism that originates from a single cell must undergo a plethora of cellular differentiation events. The first events of mammalian cellular differentiation occur during the pre-implantation stages of embryo development soon after fertilization. The totipotent zygote will experience multiple rounds of cell division, passing through cleavage stages until reaching the morula stage, which is the first stage that different cell populations become noticeable as outer and inner cells. Subsequent formation of the blastocoel allows the clear distinction of the inner cell mass (ICM) from the trophectoderm (TE).

An embryo generated by somatic cell nuclear transfer (SCNT) must endure the same differentiation events before implantation as fertilized embryos. However, since the genetic material comes from a differentiated cell, it must be reprogrammed. It contains marks of

Marcelo Tigre Moura (ed.), *Somatic Cell Nuclear Transfer Technology*, Methods in Molecular Biology, vol. 2647,
https://doi.org/10.1007/978-1-0716-3064-8_3,
© The Author(s), under exclusive license to Springer Science+Business Media, LLC, part of Springer Nature 2023

epigenetic differentiation that allow or repress genes that relate to cell function not suitable for embryonic development. Successful nuclear reprogramming is necessary for the establishment of embryonic stem cell (ESC) lines [1], pregnancy, and live births [2].

Although live births have been reported for several species, the rate of full-term pregnancies from cloned embryos is much lower when compared to in vitro fertilized embryos [2, 3]. A common issue in different species is placental alteration [4], and these alterations could be linked to a failure of gene imprinting in the placenta of SCNT-derived embryos and X chromosome inactivation [5]. SCNT-derived blastocysts can be effectively obtained, but obtaining healthy offspring from them continues to be a challenge. This review aims to link early cell specification in the early embryo to the fate of SCNT embryos and overall cloning efficiency. First, we will review cell differentiation in the fertilized embryo. This would allow us to understand the significance of observations related to cell differentiation in the context of SCNT.

2 Early Cell Specification During Mammalian Preimplantation Development

2.1 Differentiation of the Trophectoderm and the Inner Cell Mass

Segregation of the trophectoderm (TE) and the inner cell mass (ICM) is the first cell differentiation event during mammalian embryonic development. The TE is an epithelial cell layer surrounding the blastocoel and will form the placenta, while the ICM will originate the embryo proper and other extra-embryonic tissues [6]. This differentiation is preceded by a state of totipotency, in which blastomeres have the potential to contribute to both TE and ICM [7–9].

Blastomere isolation at the 2-cell [10], 4-cell [11], and 8-cell [12] stage was reported to yield live births in mice, cattle, and sheep, respectively, suggesting that some cells of the embryo retain totipotency at earlier stages of development. This totipotency potential reduces as development progresses [13]. In mice, at the 8-cell stage, there is embryo compaction leading to morula formation. At this stage, some blastomeres attain apicobasal polarity, and further cell divisions lead to outside and inside cell populations [14].

Most outer cells become polarized from this developmental stage onwards, while inner cells remain apolar. Beyond the 16-cell stage in mouse embryos, outer cells will become the TE and inner cells will form the ICM. Thus, there are two proposed cell fate specification models in the early embryo, the polarity model and the inside-outside model [15]. Mounting evidence throughout the years suggests that cell polarity precedes cell positioning in defining the cell fate of the blastomeres.

In mouse embryos, inhibition of polarization by reducing the level of activity of polarization-related proteins Par-3 family cell polarity regulator (PARD3) and protein kinase C alpha (PRKCA) leads to the internalization of outside cells [16]. Outer apolar cells are frequently internalized [17], and reduction of cell contractility can be influenced by polarization, leading to internalization of cells [18]. It was also shown that the presence of an apical domain leads to TE differentiation and that transplantation of an apical domain to an apolar cell induces TE features [19].

Polarization and cell contractility influence cellular localization of Yes1 associated transcriptional regulator (YAP1) [18, 20], a co-factor for transcription of genes and part of the HIPPO signaling pathway. HIPPO signaling was implicated in the segregation of the TE after the observation that TEA domain transcription factor 4 (*Tead4*) knockout mice were infertile due to the absence of blastocyst formation [21]. TEAD4 is a transcription factor present in outside and inside cells [22]; however, TEAD4-mediated transcription occurs when YAP1 is nuclear, as large tumor suppressor kinase 2 (LATS2) kinase phosphorylates YAP1 in inside cells, keeping YAP1 in the cytoplasm of mouse embryos [23]. It was recently shown that YAP1 activity could be modulated by glucose metabolism, which influences TE cell fate in the mouse [24] (Fig. 1). On the other hand, the reduction of *TEAD4* in bovine embryos did not impair blastocyst formation and did not change caudal type homeobox 2 (*CDX2*) expression [25, 26].

The nuclear localization of YAP1 and TEAD4 leads to transcription of *Cdx2* [23], which is a transcription factor associated with TE specification in the mouse [27]. Thus, *Cdx2* expression is downstream of cell polarization, which in turn influences HIPPO signaling. Although CDX2 is not required for segregation into TE, it is needed for the maintenance of epithelial integrity in mouse [28, 29] and bovine embryos [30] and also required for expression of TE genes, such as heart and neural crest derivatives expressed 1 (*Hand1*) [28] and interferon-tau (*IFNT*) [31]. Curiously, *CDX2* reduction in pigs impaired apical domain formation, while knockdown of *PRKCA* did not affect *CDX2* gene expression [32].

HIPPO signaling is required for silencing (directly or indirectly) pluripotency-associated genes in the ICM. POU Class 5 Homeobox 1 (OCT4), SRY-Box Transcription Factor 2 (SOX2), and Nanog Homeobox (NANOG) are transcription factors expressed in the ICM of mouse embryos [33–35] and known as core members of a pluripotency network in ESCs [36–38]. *Cdx2*—a direct target of YAP1 *and* TEAD4—specifically silences *Oct4* in mouse TE cells [28, 29, 39], leading to suppression of pluripotency [39]. In humans [40], and cattle [41], OCT4 is expressed in the TE despite CDX2 presence [42, 43]. It was demonstrated that the bovine genome does not contain the promoter regulatory element that allows CDX2 binding to silence *OCT4*, and this difference at the genome sequence is shared with

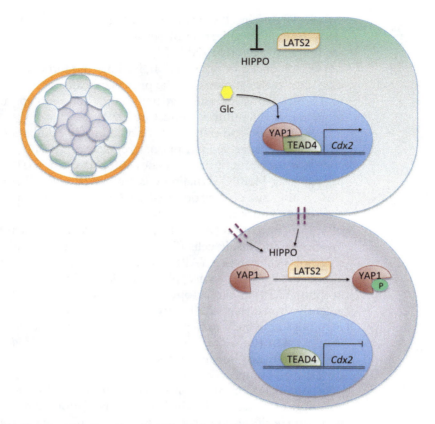

Fig. 1 Schematic representation of molecular mechanisms involved in the first lineage segregation between the ICM and the TE in the mouse preimplantation embryo. Glc – glucose

other species such as humans, chimp, rhesus, rat, horses, dogs, and rabbits [44].

SOX2 is the first marker restricted to inside cells in mouse embryos [45, 46], and its absence in outside cells is due to repression caused by members of the HIPPO signaling pathway [46, 47]. However, in bovine embryos, SOX2 is present in all cells of the morula and then restricted to the ICM [48]. SOX2 is not required for blastocyst formation as shown by its gene deletion in mice [46] or knockdown experiments in bovine [48]. Interestingly, it is proposed that heterogeneities in OCT4 and SOX2 action, location, or responses may bias cell fate to ICM as early as the 4-cell stage [49–51]. However, others found no lineage bias in early embryos in the mouse [52, 53] or cattle [54].

NANOG expression was also shown to be controlled by the HIPPO pathway in human ESCs, as loss of YAP1 and TEAD4 led to an increase in *NANOG* [55], but this has yet to be shown in embryos. NANOG is associated with restricted expression in the ICM of mouse embryos [35, 56]; however, *Nanog* transcripts can be detected in the TE [45]. The deletion of *Nanog* leads to abnormal development after implantation in the mouse [56], which is related to the role of NANOG in the second cell lineage specification between the epiblast and the primitive endoderm (PE) within the preimplantation embryo.

2.2 Differentiation of the Epiblast and the Primitive Endoderm

After the segregation between the TE and the ICM, the second event of differentiation occurs within the blastocyst, in which ICM cells will either commit to the epiblast or PE cell fates. The epiblast cells will differentiate into the three germ cell layers and the germline, whereas PE will form the visceral and parietal yolk sacs [57, 58].

The PE emerges as a cell layer lining on top of the ICM and facing the blastocoel cavity [57, 58]. Similar to the first segregation between TE and ICM, there was a positional model, in which ICM cells exposed to the blastocoel would become the PE [6]. However, cells in the early mouse ICM are a mosaic of cells that have the potential to become epiblast or PE [59]. After differentiation, PE cells are sorted to the surface of the ICM by cell movement and apoptosis [60].

Cells within the early mouse ICM express both transcription factors NANOG and GATA binding protein 6 (GATA6) [60, 61], which are markers of epiblast and PE, respectively. As development progresses, ICM cells will become mutually restricted to NANOG or GATA6 expression in a mixed pattern known as "salt-and-pepper" [59]. Initial studies suggested that fibroblast growth factor (FGF) signaling could participate in PE formation [62]. Further studies reinforced that FGF signaling via mitogen-activated protein kinase kinase 1/mitogen-activated protein kinase 1 (MEK/ERK) was involved in PE specification, more specifically through FGF receptor activation of growth factor receptor bound protein 2 (GRB2) and subsequent activation of the MEK/ERK pathway (Fig. 2). *Grb2* knockout in mouse embryos resulted in the absence of PE, while the ICM was restricted to NANOG-positive cells [59], similar to treatment of embryos with MEK inhibitors [63]. In turn, embryos treated with fibroblast growth factor 4 (FGF4) displayed an ICM restricted to GATA6-positive cells [63].

The interaction among NANOG, GATA6, and FGF4 has been dissected in the early mouse embryo. Deletion of *Nanog* led to the expression of GATA6 in all cells of the ICM; however, it impaired differentiation of the PE due to lack of FGF4 and failure to express downstream PE markers SRY-Box transcription factor 17 (SOX17) and GATA binding protein 4 (GATA4) that are downstream transcription factors involved with PE differentiation [64]. *Fgf4* knockout corroborated its need for PE specification as only NANOG was expressed in ICM cells, and no PE differentiation occurred [61]. As expected, deletion of *GATA6* in mouse embryos caused all ICM cells to express NANOG. More interestingly, treatment of *Gata6*-null embryos with FGF4 did not restore PE differentiation, thus suggesting additional GATA6 regulated processes that are essential for this event [65, 66].

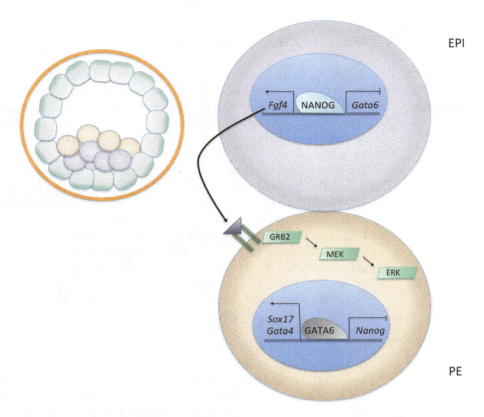

Fig. 2 Schematic representation of molecular mechanisms involved in the segregation of the epiblast (EPI) and primitive endoderm (PE) in the preimplantation embryo

In bovine embryos, FGF signaling reduces NANOG-positive cells [67]. Further, MEK inhibition did not change the number of NANOG-positive cells but abolished SOX17-positive cells [68]. The deletion of *NANOG* using CRISPR/Cas9 resulted in a reduced *GATA6* expression level with no apparent increase in the number of GATA6-positive cells [69], different than what is observed in the mouse.

2.3 Epigenetic Dynamics Related to Early Cell Specification

Early cell lineage specification in the developing embryo requires coordinated spatial and temporal gene expression, as discussed above. Expression of these genes is not only controlled by cellular processes, signaling pathways, and transcription factor networks, but it is also under the control of both transcriptional permissive or repressive epigenetic mechanisms. The most understood epigenetic modifications are DNA methylation and histone post-translational modifications. Gain or loss of such epigenetic marks supports the commitment to a differentiated state, preventing the expression of genes related to other cell lineages and facilitating the expression of other genes for further specialization [70].

There are important events that change the epigenetic landscape after fertilization. These include protamine to histone exchange in the paternal chromatin [71] and global erasure of DNA methylation marks [72, 73]. Although re-establishment of DNA methylation is more extensive in the ICM than in the TE [74], it is possible that these patterns occur after cell specification in the early embryo as no DNA methylation was observed in *Oct4* or E74-like ETS transcription factor 5 (*Elf5*) in both ICM or TE cells [75]. Thus, we will focus on histone post-translational modifications linked to cell differentiation events in the embryo.

Histone post-translational modifications include methylation, acetylation, ubiquitination of lysine residues, or phosphorylation of serine or threonine residues. Methylation of histone residues is can be associated with repressive chromatin states, such as histone 3 lysine 9 trimethylation (H3K9me3) or histone 3 lysine 27 trimethylation (H3K27me3), while histone 3 lysine 9 trimethylation (H3K4me3) correlates with active gene expression. On the other hand, acetylation of histones 3 and 4 tails is normally associated with a permissive chromatin state. Enzymes, such as histone acetyltransferases, histone deacetylases, histone methylases, and histone demethylases, catalyze these post-translational modifications [76, 77].

After fertilization and exchange of paternal protamines with oocyte-derived histones, an asymmetry of histone modifications is observed between the paternal and maternal genomes. For example, H3K9me3 was shown to be highly present in the maternal pronucleus while undetected in the paternal pronucleus, but after genome activation symmetry is restored, the paternal genome acquires the same level of H3K9me3 as the maternal [78], similarly to what is observed for H3K4me3 [79]. On the other hand, histone 4 lysine 20 trimethylation (H4K20me3) [80] and histone 3 lysine 64 trimethylation (H3K64me3) [81], both epigenetic marks associated with heterochromatin, are removed from the maternal genome at the pronuclear stage.

An absence of H4K20me was observed throughout preimplantation development, despite the presence of another heterochromatin marker—H3K9me3—suggesting immature heterochromatin present in preimplantation embryos [82]. Indeed, the chromatin of early embryos was observed as uncompacted fibers, which was also observed in the ICM, but not in the TE of mouse blastocysts [83]. Also, corroborating the idea that a more permissive chromatin state occurs in embryo development, chromatin immunoprecipitation of developing embryos revealed that H3K4me3 accumulated more rapidly and broadly than H3K27me3; however, a large gain of H3K27me3 was observed in ICM and TE [84].

Interestingly, mouse embryos revealed intense staining for H3K27me3 in the ICM, while the staining was restricted to the X chromosome in the TE [80, 83]; whereas no clear differences were observed in blastomeres of earlier stages [82]. Also, ICM and TE show differential enrichment in promoter regions of developmentally important genes. In the TE, H3K27me3 is enriched in *Sox2* [86], and repressive marker H3K9me2 was enriched in the promoter region of *Nanog* and *Oct4*. In the ICM, the *Cdx2* promoter region is enriched with H3K9 dimethylation (H3K9me2) [87].

Although differences in some of the most studied epigenetic markers seem to appear at the blastocyst stage as described above, there are data suggesting that cell fate is biased toward ICM or TE is predetermined at earlier stages [49–51, 86]. Differences in histone 3 arginine 26 methylation (H3R26me) were observed in blastomeres of 4-cell embryos. The cells with higher levels of H3R26me and coactivator-associated arginine methyltransferase 1 (CARM1) would be biased to contribute to the ICM of mouse embryos [88]. The knockdown of *CARM1* in mouse zygotes reduces H3R26me levels and increased the number of PE cells [50], while injection of *CARM1* mRNA led to increased OCT4 and SOX2 expression [50, 87]. Another methyltransferase, PR/SET domain 14 (PRDM14), was also implicated in the early bias of 4-cell blastomeres as injection of *PRDM14* mRNA led to an increase in H3R26me, leading to a preferential contribution of daughter cells to the ICM of mouse embryos [89].

2.4 X Chromosome Inactivation and Cell Lineage Specification

As mentioned above, X chromosome inactivation is noticeable in TE cells by H3K27me3 staining [85]. This is in agreement with the notion that XX cells must silence one X chromosome for dosage compensation of gene expression [90]. Briefly, X chromosome inactivation (XCI) involves the expression of a long non-coding RNA (lncRNA) named X inactive specific transcript (*Xist*) that coats the X chromosome in *cis*. It recruits both polycomb repressive complexes 1 and 2 (PRC1 and PRC2) that will lead to histone H2A lysine 119 ubiquitination (H2AK119ub) and H3K27me3, respectively [91].

During early development of female embryos, the paternal X chromosome is silenced in all cells by the establishment of H3K27me3 marks, but this silencing is reversed, and random inactivation of the paternal or maternal X chromosome occurs in the late ICM [90, 91]. Interestingly, cells that expressed NANOG lack *Xist* expression and XCI [92]. NANOG-positive cells also lack focal localization of embryonic ectoderm development protein (EED), which is part of the histone methylation complex PRC2. On the other hand, PE cells positive for GATA4 also display *EED* foci [93]. These findings indicate that paternal imprinting of the X chromosome occurs during development in the extra-embryonic tissues, at least in the mouse.

Human embryos seem to express *XIST* throughout early development without silencing one copy of the X chromosome, maintaining a biallelic expression with a gradual dosage compensation [94]. This is likely promoted by concomitant expression of *XIST* and another lncRNA named *XACT* [94, 95]. In the rabbit embryo, one X chromosome is inactive from the 4-cell stage until early blastocyst stage, when ICM and TE cells may have even two inactive X chromosomes, while late blastocyst cells will only present one inactive X chromosome [96]. In bovine embryos, it is proposed that the paternal X chromosome is inactive at the morula stage and reactivated at the blastocyst stage [97] as *XIST* is detected from the morula stage onwards but very few H3K27me3 spots were observed in the ICM when compared to the TE, indicating that both copies are active in the ICM [98].

3 Early Cell Lineage Specification in SCNT Embryos

Transfer of a nucleus to a recipient egg cytoplasm was initially performed to test if the same nuclear content is present in cells after differentiation. The classic experiment from Briggs and King revealed that differentiated embryonic cells were able to generate other embryos after nuclear transplantation [99]. Subsequent nuclear transfer experiments in the 1960s with amphibians found that fully differentiated cells from the intestinal epithelium were able to generate adult animals [100]. Only in the 1990s, the birth of a mammalian species after nuclear transfer from adult cells was obtained with Dolly the sheep [101]. Several species were cloned by SCNT since then [102].

After this breakthrough, SCNT emerged as a powerful tool for generating identical copies of an animal—reproductive cloning—and deriving embryonic stem cells (ESCs) from a preimplantation cloned embryo—therapeutic cloning. Derivation of pluripotent ESCs does not seem to be an issue [103, 104], as we will discuss ahead; however, the percentage of live births from cloned embryos is reduced when compared to in vitro fertilized (IVF) equivalents [2]. There are many reports of placental alterations in various species [105–109]. Interestingly, aggregation of SCNT-derived ICM with tetraploid TE fertilized embryos improved full-term development in mice [110], corroborating the idea that defects in the TE interfere with SCNT success. Although not completely elucidated, faulty epigenetic reprogramming is key to some of these failures [5] and is the subject of other reviews [102, 111]. Here, we will focus on aspects related to early cell lineage specification, comparing SCNT—with fertilized—embryonic development.

3.1 Differences in ICM or TE Cell Numbers in SCNT Embryos

A higher number of cells are often associated with improved embryo quality [112]. Cell allocation to the ICM and TE is also essential. Fewer cells allocated to the ICM can impair subsequent development [113]. Different media during in vitro culture can influence ICM cell numbers [114]. In vivo-derived embryos presented more ICM cells than their in vitro fertilized mouse, rat [115], bovine [116], and porcine [117] embryos.

As there is a significant influence of culture conditions in ICM cell number of in vitro fertilized embryos, SCNT-derived embryos could be impacted as they are exposed to in vitro culture conditions, and their initial genome must undergo extensive reprogramming. Indeed, studies designed to verify cell number in SCNT embryos revealed that the total cell number is reduced compared to in vivo and IVF embryos [118, 119].

Specifically, the cell number of SCNT embryos was reduced throughout all days of embryonic development when compared to in vivo fertilized embryos and reduced at days 3, 5, 7, and 8 when compared to IVF embryos [119]. With regard to cell allocation, IVF embryos presented fewer TE cells than in vivo ones, while SCNT presented fewer cells than both IVF and in vivo embryos, which resulted in a lower ICM/TE cell ratio [118]. Many other SCNT studies that evaluated ICM and TE cell allocation, although not as a primary goal, found that the total cell number is reduced in SCNT embryos compared to IVF equivalents, but ICM and TE cell numbers are variable among studies (Table 1). It was reported that aggregating mouse embryos improved the reproductive success of SCNT [120]. However, aggregation of bovine SCNT embryos did not improve efficiency [121, 122].

3.2 Expression of Differentiation-Associated Genes in SCNT Embryos

Since there are changes in the allocation of ICM and TE cells in most SCNT studies that compared these embryos with fertilized ones, it is plausible to consider that expression of genes related to differentiation is altered. It was shown that most SCNT-derived mouse embryos had incomplete reactivation of OCT4 [123]. Induced pluripotent stem cells (iPSC), which are reprogrammed in vitro to pluripotency by expression of transcription factors [124], allowed stepwise studies of gene expression changes during reprogramming and reinforced that activation of endogenous pluripotency network genes must occur [125]. Expression of pluripotency factors such as *Nanog* and *Oct4* is dependent on maternal histone variant 3.3 (H3.3) [126], relating successful epigenetic reprogramming to the expression of pluripotency genes after SCNT.

Interestingly, the use of different bovine cell lines yielded different expression of pluripotency genes *OCT4*, *SOX2*, and *NANOG* among SCNT embryos, which had either higher or similar expression levels than IVF counterparts [127]. These findings are in agreement with the idea that global gene expression can be

Table 1
Relative total cell numbers and cell type allocation in somatic cell nuclear transfer (SCNT) blastocysts compared to embryo counterparts produced by in vitro fertilization (IVF)

Species	Cell numbers				Reference
	Total	ICM	TE	ICM/TE	
Bovine (*Bos taurus*)	–	–	Down	N/A	[118]
Bovine (*Bos taurus*)	Down	–	Down	–	[156]
Bovine (*Bos taurus*)	–	–	–	Down	[157]
Bovine (*Bos taurus*)	–	Up	–	Up	[121]
Bovine (*Bos taurus*)	Down	N/A	N/A	N/A	[119]
Bovine (*Bos taurus*)	–	–	–	Up	[158]
Bovine (*Bos taurus*)	–	–	Down	N/A	[159]
Bovine (*Bos taurus*)	–	–	N/A	–	[160]
Bovine (*Bos taurus*)	Down	–	–	–	[161]
Bovine (*Bos taurus*)	–	Down	–	Down	[162]
Bovine (*Bos taurus*)	Down	Down	Down	N/A	[163]
Bovine (*Bos taurus*)	Down	–	–	Down	[164]
Bovine (*Bos taurus*)	–	–	–	N/A	[138]
Bovine (*Bos taurus*)	Down	Down	–	Down	[165]
Bovine (*Bos taurus*)	Down	–	N/A	Down	[166]
Buffalo (*Bubalus bubalis*)	Up	–	–	Down	[167]
Mouse (*Mus musculus*)	Down	Down	Down	N/A	[168]
Mouse (*Mus musculus*)	N/A	Down	Down	N/A	[169]
Pig (*Sus scrofa*)	N/A	N/A	N/A	Down	[170]
Rhesus monkey (*Macaca mulatta*)	–	N/A	N/A	–	[171]

This compilation was restricted to studies presenting direct comparisons between SCNT and IVF embryos
ICM – inner cell mass, *TE* – trophectoderm, *ICM/TE* – *ratio* number of ICM cells divided by the number of TE cells, *N/A* – not analyzed, (–) similar

different in SCNT embryos generated from cell lines shown to have different efficiencies at generating preimplantation embryos [128]. However, in this latter case, no differences in *OCT4, SOX2*, and *NANOG* were observed. The expression of these pluripotency genes and *CDX2* in SCNT compared to IVF blastocysts is observed throughout the literature (Table 2). Most studies showed no significant changes in the gene expression of these transcriptional regulators, consistent with studies that performed transcriptome analysis [128–132]. Nonetheless, the use of mouse *Oct4-*

70   Marcelo D. Goissis and Jose B. Cibelli

Table 2
Relative gene expression by RT-qPCR of early cell lineage specifying transcription factors in somatic cell nuclear transfer (SCNT) blastocysts relative to embryo counterparts produced by in vitro fertilization (IVF)

Species	Transcription factor				Reference
	OCT4	NANOG	SOX2	CDX2	
Bovine (*Bos taurus*)	–	Up	–	–	[172]
Bovine (*Bos taurus*)	–	–	–	–	[158]
Bovine (*Bos taurus*)	–	N/A	–	–	[160]
Bovine (*Bos taurus*)	Down	–	–	–	[163]
Bovine (*Bos taurus*)	Down	–	–	–	[164]
Bovine (*Bos taurus*)	–	–	–	Down	[138]
Bovine (*Bos taurus*)	–	–	–	–	[166]
Bovine (*Bos taurus*)	Down	Down	Down	–	[173]
Buffalo (*Bubalus bubalis*)	Down	Up	–	–	[167]
Pig (*Sus scrofa*)	–	Down	N/A	–	[174]

This compilation was restricted to studies presenting direct comparisons between SCNT and IVF embryos
CDX2 – caudal type homeobox 2, *NANOG* – Nanog homeobox, *OCT4* – (also known as *POU5F1*) Octamer of transcription 4, *SOX2* – SRY-Box transcription factor 2, *N/A* – not analyzed, (–) similar

green fluorescent protein (*GFP*) promoter-reporter in bovine SCNT embryos revealed heterogeneities in the number of cells which would express the *GFP* reporter [133].

Even though *CDX2* expression is often unchanged in SCNT embryos, downstream genes such as *HAND1* and *IFNT* are altered at later stages of embryonic development [134]. This could be due to the fact that an enhancer region for *Cdx2* shows different activity at preimplantation and post-implantation mouse embryos [135], implying that different mechanisms other than HIPPO signaling and notch [136] control *CDX2* expression at later stages. Unfortunately, there are no reports on HIPPO signaling pathway or establishment of cell polarity in SCNT-derived embryos. However, some studies provide glimpses on *TEAD4* expression in bovine SCNT embryos. In one study, *TEAD4* was reduced in the TE of SCNT embryos compared to in vivo fertilized ones [137]. However, *TEAD4* expression was not different between SCNT and IVF embryos, although it was highly increased after *CDX2* knockdown [138].

3.3 Derivation of ESC Lines from SCNT Embryos

The efficiency of reproductive cloning by SCNT at generating healthy offspring is low [2]. However, the derivation of ESC lines from cloned embryos (i.e., therapeutic cloning) has similar efficiencies than using IVF embryos. Shortly after the announcement of full reprogramming of mammalian somatic cells using SCNT

[101], ESC lines from mouse SCNT embryos (NT-ESC) were obtained [103, 139]. It was reported that the efficiency of obtaining NT-ESC lines per embryo was similar to conventional ESC lines from fertilized embryos, even when using donor cells from different mouse strains [140]. These NT-ESC cells were able to differentiate into all three germ layers and the germline [103, 139, 140], indicating proper reprogramming to pluripotency.

Two decades after the announcement of Dolly, and the derivation of human ESCs from fertilized embryos [141], pluripotent human NT-ESCs were obtained [104]. The delay was mostly related to technical hurdles to obtain cloned blastocysts in humans. Once cloned blastocysts were generated, the efficiency of NT-ESC derivation was as high as that obtained using IVF-derived blastocysts [104]. Interestingly, the X chromosome in female cell lines was less methylated in both NT-ESC cells and IVF-ESCs than iPSCs [1], suggesting that NT-ESC cells could have achieved better or broader reprogramming than iPSCs.

In order to understand issues with placental development, trophoblast stem cells (TSC) were also obtained from mouse SCNT embryos (NT-TSC). Interestingly, mouse NT-TSC were obtained more efficiently from SCNT embryos than fertilized ones. These NT-TSC proliferate more rapidly and relied less on self-renewal conditions using *FGF4* and activin to maintain their TSC phenotype [142]. Based on the activity of an enhancer regulated by HIPPO and NOTCH pathways, it was shown that TSC resemble TE cells from post-implantation embryos [135]. Nonetheless, when injected into host fertilized embryos and transferred into a recipient female, NT-TSC contributed to the placenta of clone fetuses that reached full-term development [143].

3.4 Aberrant X Chromosome Inactivation in SCNT Embryos

As mentioned previously, XCI is paternally imprinted in the TE while random in the ICM [90, 91]. Interestingly, it was reported that the placenta of deceased bovine clones presented random XCI, in contrast with a paternal XCI imprinting in the placenta of a fertilized fetus [144]. In previous studies, a random XCI was associated with NANOG expression in epiblast cells [90, 92], suggesting that *XIST* may be repressed by NANOG [145]. NANOG was shown to be present in the TE of bovine embryos [146]; however, in SCNT embryos, *NANOG* expression level appears normal compared to IVF and NANOG is absent in the TE of SCNT blastocysts [147], possibly excluding a role for NANOG in random placental XCI in cattle.

In bovine- and mouse-cloned fetuses, the expression of *XIST* was increased in the placenta [143, 144]. In the mouse, such increased expression led to a reduction in the expression of X-linked genes [148]. Similar findings were reported in pig-cloned fetuses [149]. In the mouse, the deletion or reduction of expression of *Xist* increased live births of cloned embryos

[145, 147]. Diminishing *Xist* in SCNT embryos normalized global gene expression [148–150]. The PE marker *GATA6* was among these normalized genes, as its expression increased in pig SCNT embryos devoid of *XIST* [149]. Interestingly, severe alterations in the yolk-sac were observed in bovine SCNT-derived pregnancies [151].

4 Perspectives and Conclusions

During early embryo development, cell differentiation requires coordinated expression of genes, which are controlled by integrated cellular processes such as cell polarization, microtubule contractility, carbohydrate metabolism, and cell signaling pathways. SCNT embryos have a proper ICM and TE, indicating that in a short period of time the oocyte's cytoplasm is capable to reprogram differentiated cells to carry these early events of differentiation.

On the other hand, it is well known that SCNT embryos' full-term development after transfer into surrogate mothers is severely impaired. Most of the abnormalities encountered are related to placental malformations. Epigenetic perturbations appear to be responsible for these events, including lack of paternal imprinting of the X chromosome. However, despite perturbations in *XIST* gene expression, differentiation into ICM and TE cells is not affected in SCNT embryos; although differentiation of the PE could warrant further investigation.

In contrast to the low rates in full-term development, the derivation of NT-ESC is feasible, and it is as efficient as that obtained using fertilized embryos. For the derivation of NT-TSC, SCNT embryos appear to be more effective. Albeit in vitro pluripotency may capture a snapshot of pluripotency in vivo [152, 153], the ability to derive NT-ESC and NT-TSC from SCNT embryos indicates that early differentiation occurs similarly to fertilized embryos.

The possibility of using CRISPR/Cas9 technology to study differentiation in early embryos is particularly interesting for mammalian species where the developmental window until reaching the blastocyst stage is longer, such as bovine, pigs, and primates. Interestingly, the use of IVF [154] or SCNT [147] to delete *OCT4* in bovine embryos pointed that the effects on *NANOG* and *CDX2* expression are closer to human [155] than to mouse [33] embryos lacking *Oct4*. One clear advantage of using SCNT over embryo microinjection, to study the effect of genetic deletions in the early embryo, is the ability to screen cell lines for the desired deletion, guaranteeing that all embryos studied will carry the same modification.

Cellular reprogramming through SCNT can reprogram a somatic cell to become totipotent again and to differentiate into a viable blastocyst. It is important to remember that when using in vitro matured oocytes, the efficiency of blastocyst formation by SCNT is not different from IVF [2]. Thus, evidence in the literature suggests that the intricate mechanisms of cell differentiation during the preimplantation period occur correctly in SCNT embryos, tempting us to postulate that these mechanisms related to cell polarization and cell contractility will push differentiation regardless of nuclear reprogramming status; however, only a correctly reprogrammed nucleus will be capable of developing properly beyond the blastocyst stage.

Acknowledgments

The Sao Paulo State Research Foundation (FAPESP 2017/09576-3) and the National Council of Science and Technology (CNPq 408634/2018-9) currently fund Marcelo D. Goissis.

References

1. Ma H, Morey R, O'Neil RC, He Y, Daughtry B, Schultz MD et al (2014) Abnormalities in human pluripotent cells due to reprogramming mechanisms. Nature 511:177–183

2. Yang X, Smith SL, Tian XC, Lewin H, Renard J-P, Wakayama T (2007) Nuclear reprogramming of cloned embryos and its implications for therapeutic cloning. Nat Genet 39:295–302

3. Heyman Y, Chavatte-Palmer P, LeBourhis D, Camous S, Vignon X, Renard JP (2002) Frequency and occurrence of late-gestation losses from cattle cloned embryos. Biol Reprod 66:6–13

4. Palmieri C, Loi P, Ptak G, Della Salda L (2008) Review Paper: A review of the pathology of abnormal placentae of somatic cell nuclear transfer clone pregnancies in cattle, sheep, and mice. Vet Pathol 45:865–880

5. Matoba S, Wang H, Jiang L, Lu F, Iwabuchi KA, Wu X et al (2018) Loss of H3K27me3 imprinting in somatic cell nuclear transfer embryos disrupts post-implantation development. Cell Stem Cell 23:343–354.e5

6. Cockburn K, Rossant J (2010) Making the blastocyst: lessons from the mouse. J Clin Invest 120:995–1003

7. Hillman N, Sherman MI, Graham C (1972) The effect of spatial arrangement on cell determination during mouse development. J Embryol Exp Morphol 28:263–278

8. Garner W, McLaren A (1974) Cell distribution in chimaeric mouse embryos before implantation. J Embryol Exp Morphol 32:495–503

9. Tabansky I, Lenarcic A, Draft RW, Loulier K, Keskin DB, Rosains J et al (2013) Developmental bias in cleavage-stage mouse blastomeres. Curr Biol 23:21–31

10. Casser E, Israel S, Witten A, Schulte K, Schlatt S, Nordhoff V et al (2017) Totipotency segregates between the sister blastomeres of two-cell stage mouse embryos. Sci Rep 7:1–15

11. Johnson WH, Loskutoff NM, Plante Y, Betteridge KJ (1995) Production of four identical calves by the separation of blastomeres from an in vitro derived four-cell embryo. Vet Rec 137:15–16

12. Willadsen SM (1980) The viability of early cleavage stages containing half the normal number of blastomeres in the sheep. J Reprod Fertil 59:357–362

13. Boiani M, Casser E, Fuellen G, Christians ES (2019) Totipotency continuity from zygote to early blastomeres: a model under revision. Reproduction 158:R49–R65

14. Stephenson RO, Rossant J, Tam PPL (2012) Intercellular interactions, position, and polarity in establishing blastocyst cell lineages and embryonic axes. Cold Spring Harb Perspect Biol 4:1–15

15. Sasaki H (2010) Mechanisms of trophectoderm fate specification in preimplantation

mouse development. Develop Growth Differ 52:263–273

16. Plusa B, Frankenberg S, Chalmers A, Hadjan-tonakis A-K, Moore CA, Papalopulu N et al (2005) Downregulation of Par3 and aPKC function directs cells towards the ICM in the preimplantation mouse embryo. J Cell Sci 118:505–515

17. Anani S, Bhat S, Honma-Yamanaka N, Krawchuk D, Yamanaka Y (2014) Initiation of Hippo signaling is linked to polarity rather than to cell position in the pre-implantation mouse embryo. Development 141:2813–2824

18. Maître J-L, Turlier H, Illukkumbura R, Eismann B, Niwayama R, Nédélec F et al (2016) Asymmetric division of contractile domains couples cell positioning and fate specification. Nature 536:344–348

19. Korotkevich E, Niwayama R, Courtois A, Friese S, Berger N, Buchholz F et al (2017) The apical domain is required and sufficient for the first lineage segregation in the mouse embryo. Dev Cell 40:235–247.e7

20. Hirate Y, Hirahara S, Inoue KI, Suzuki A, Alarcon VB, Akimoto K et al (2013) Polarity-dependent distribution of angiomotin localizes hippo signaling in preimplantation embryos. Curr Biol 23:1181–1194

21. Yagi R, Kohn MJ, Karavanova I, Kaneko KJ, Vullhorst D, DePamphilis ML et al (2007) Transcription factor TEAD4 specifies the trophectoderm lineage at the beginning of mammalian development. Development 134:3827–3836

22. Nishioka N, Yamamoto S, Kiyonari H, Sato H, Sawada A, Ota M et al (2008) Tead4 is required for specification of trophectoderm in pre-implantation mouse embryos. Mech Dev 125:270–283

23. Nishioka N, Inoue KI, Adachi K, Kiyonari H, Ota M, Ralston A et al (2009) The Hippo signaling pathway components Lats and Yap pattern Tead4 activity to distinguish mouse trophectoderm from inner cell mass. Dev Cell 16:398–410

24. Chi F, Sharpley MS, Nagaraj R, Roy SS, Banerjee U (2020) Glycolysis-independent glucose metabolism distinguishes TE from ICM fate during mammalian embryogenesis. Dev Cell 53:9–26.e4

25. Sakurai N, Takahashi K, Emura N, Hashizume T, Sawai K (2017) Effects of downregulating TEAD4 transcripts by RNA interference on early development of bovine embryos. J Reprod Dev 63:135–142

26. Akizawa H, Kobayashi K, Bai H, Takahashi M, Kagawa S, Nagatomo H et al (2018) Reciprocal regulation of TEAD4 and CCN2 for the trophectoderm development of the bovine blastocyst. Reproduction 155:563–571

27. Beck F, Erler T, Russell A, James R (1995) Expression of Cdx-2 in the mouse embryo and placenta: possible role in patterning of the extra-embryonic membranes. Dev Dyn 204:219–227

28. Strumpf D, Mao CA, Yamanaka Y, Ralston A, Chawengsaksophak K, Beck F et al (2005) Cdx2 is required for correct cell fate specification and differentiation of trophectoderm in the mouse blastocyst. Development 132:2093–2102

29. Wu G, Gentile L, Fuchikami T, Sutter J, Psathaki K, Esteves TC et al (2010) Initiation of trophectoderm lineage specification in mouse embryos is independent of Cdx2. Development 137:4159–4169

30. Goissis MD, Cibelli JB (2014) Functional characterization of CDX2 during bovine preimplantation development in vitro. Mol Reprod Dev 81:962–970

31. Sakurai T, Bai H, Konno T, Ideta A, Aoyagi Y, Godkin JD et al (2010) Function of a transcription factor CDX2 beyond its trophectoderm lineage specification. Endocrinology 151:5873–5881

32. Bou G, Liu S, Sun M, Zhu J, Xue B, Guo J et al (2017) CDX2 is essential for cell proliferation and polarity in porcine blastocysts. Development 144:1296–1306

33. Nichols J, Zevnik B, Anastassiadis K, Niwa H, Klewe-Nebenius D, Chambers I et al (1998) Formation of pluripotent stem cells in the mammalian embryo depends on the POU transcription factor Oct4. Cell 95:379–391

34. Avilion AA, Nicolis SK, Pevny LH, Perez L, Vivian N, Lovell-Badge R (2003) Multipotent cell lineages in early mouse development depend on SOX2 function. Genes Dev 17:126–140

35. Chambers I, Colby D, Robertson M, Nichols J, Lee S, Tweedie S et al (2003) Functional expression cloning of Nanog, a pluripotency sustaining factor in embryonic stem cells. Cell 113:643–655

36. Loh Y-H, Wu Q, Chew J-L, Vega VB, Zhang W, Chen X et al (2006) The Oct4 and Nanog transcription network regulates pluripotency in mouse embryonic stem cells. Nat Genet 38:431–440

37. Chew J, Loh Y, Zhang W, Chen X, Tam W, Yeap L et al (2006) Reciprocal transcriptional

regulation of complex in embryonic stem cells reciprocal transcriptional regulation of Pou5f1 and Sox2 via the Oct4/Sox2 complex in embryonic stem cells. Mol Cell Biol 25: 6031–6046

38. Chen X, Xu H, Yuan P, Fang F, Huss M, Vega VB et al (2008) Integration of external signaling pathways with the core transcriptional network in embryonic stem cells. Cell 133: 1106–1117

39. Niwa H, Toyooka Y, Shimosato D, Strumpf D, Takahashi K, Yagi R et al (2005) Interaction between Oct3/4 and Cdx2 determines trophectoderm differentiation. Cell 123:917–929

40. Chen AE, Egli D, Niakan K, Deng J, Akutsu H, Yamaki M et al (2009) Optimal timing of inner cell mass isolation increases the efficiency of human embryonic stem cell derivation and allows generation of sibling cell lines. Cell Stem Cell 4:103–106

41. Cao S, Wang F, Chen Z, Liu Z, Mei C, Wu H et al (2009) Isolation and culture of primary bovine embryonic stem cell colonies by a novel method. J Exp Zool Part A Ecol Genet Physiol 311:368–376

42. Adjaye J, Huntriss J, Herwig R, BenKahla A, Brink TC, Wierling C et al (2005) Primary differentiation in the human blastocyst: comparative molecular portraits of inner cell mass and trophectoderm cells. Stem Cells 23: 1514–1525

43. Kuijk EW, Du Puy L, Van Tol HT, Oei CHY, Haagsman HP, Colenbrander B et al (2008) Differences in early lineage segregation between mammals. Dev Dyn 237:918–927

44. Berg DK, Smith CS, Pearton DJ, Wells DN, Broadhurst R, Donnison M et al (2011) Trophectoderm lineage determination in cattle. Dev Cell 20:244–255

45. Guo G, Huss M, Tong GQ, Wang C, Li Sun L, Clarke ND et al (2010) Resolution of cell fate decisions revealed by single-cell gene expression analysis from zygote to blastocyst. Dev Cell 18:675–685

46. Wicklow E, Blij S, Frum T, Hirate Y, Lang R, Sasaki H et al (2014) HIPPO pathway members restrict SOX2 to the inner cell mass where it promotes ICM fates in the mouse blastocyst. PLoS Genet 10:e1004618

47. Frum T, Watts JL, Ralston A (2019) TEAD4, YAP1 and WWTR1 prevent the premature onset of pluripotency prior to the 16-cell stage. Development 146:dev179861. https://doi.org/10.1242/dev.179861

48. Goissis MD, Cibelli JB (2014) Functional characterization of SOX2 in bovine preimplantation embryos. Biol Reprod 90:30–30

49. Plachta N, Bollenbach T, Pease S, Fraser SE, Pantazis P (2011) Oct4 kinetics predict cell lineage patterning in the early mammalian embryo. Nat Cell Biol 13:117–123

50. Goolam M, Scialdone A, Graham SJL, Voet T, Marioni JC, Zernicka-goetz M et al (2016) Heterogeneity in Oct4 and Sox2 targets biases cell fate in 4-cell mouse embryos article heterogeneity in Oct4 and Sox2 targets biases cell fate in 4-cell mouse embryos. Cell 165: 61–74

51. White MD, Angiolini JF, Alvarez YD, Kaur G, Zhao ZW, Mocskos E et al (2016) Long-lived binding of Sox2 to DNA predicts cell fate in the four-cell mouse embryo. Cell 165:75–87

52. Waksmundzka M, Wiśniewska A, Maleszewski M (2006) Allocation of cells in mouse blastocyst is not determined by the order of cleavage of the first two blastomeres. Biol Reprod 75: 582–587

53. Kurotaki Y, Hatta K, Nakao K, Nabeshima YI, Fujimori T (2007) Blastocyst axis is specified independently of early cell lineage but aligns with the ZP shape. Science 316:719–723

54. Sepulveda-Rincon LP, Dube D, Adenot P, Laffont L, Ruffini S, Gall L et al (2016) Random allocation of blastomere descendants to the trophectoderm and ICM of the bovine blastocyst. Biol Reprod 95:123–123

55. Beyer TA, Weiss A, Khomchuk Y, Huang K, Ogunjimi AA, Varelas X et al (2013) Switch enhancers interpret TGF-β and hippo signaling to control cell fate in human embryonic stem cells. Cell Rep 5:1611–1624

56. Mitsui K, Tokuzawa Y, Itoh H, Segawa K, Murakami M, Takahashi K et al (2003) The homeoprotein nanog is required for maintenance of pluripotency in mouse epiblast and ES cells. Cell 113:631–642

57. Gardner RL, Rossant J (1979) Investigation of the fate of 4–5 day post-coitum mouse inner cell mass cells by blastocyst injection. J Embryol Exp Morphol 52:141–152

58. Gardner RL (1982) Investigation of cell lineage and differentiation in the extraembryonic endoderm of the mouse embryo. J Embryol Exp Morphol 68:175–198

59. Chazaud C, Yamanaka Y, Pawson T, Rossant J (2006) Early lineage segregation between epiblast and primitive endoderm in mouse blastocysts through the Grb2-MAPK pathway. Dev Cell 10:615–624

60. Plusa B, Piliszek A, Frankenberg S, Artus J, Hadjantonakis A-K (2008) Distinct

sequential cell behaviours direct primitive endoderm formation in the mouse blastocyst. Development 135:3081–3091

61. Kang M, Piliszek A, Artus J, Hadjantonakis A-K (2013) FGF4 is required for lineage restriction and salt-and-pepper distribution of primitive endoderm factors but not their initial expression in the mouse. Development 140:267–279

62. Arman E, Haffner-Krausz R, Chen Y, Heath JK, Lonai P (1998) Targeted disruption of fibroblast growth factor (FGF) receptor 2 suggests a role for FGF signaling in pregastrulation mammalian development. Proc Natl Acad Sci U S A 95:5082–5087

63. Yamanaka Y, Lanner F, Rossant J (2010) FGF signal-dependent segregation of primitive endoderm and epiblast in the mouse blastocyst. Development 137:715–724

64. Frankenberg S, Gerbe F, Bessonnard S, Belville C, Pouchin P, Bardot O et al (2011) Primitive endoderm differentiates via a three-step mechanism involving Nanog and RTK signaling. Dev Cell 21:1005–1013

65. Schrode N, Saiz N, Di Talia S, Hadjantonakis AK (2014) GATA6 levels modulate primitive endoderm cell fate choice and timing in the mouse blastocyst. Dev Cell 29:454–467. https://doi.org/10.1016/j.devcel.2014.04.011

66. Bessonnard S, De Mot L, Gonze D, Barriol M, Dennis C, Goldbeter A et al (2014) Gata6, Nanog and Erk signaling control cell fate in the inner cell mass through a tristable regulatory network. Development 141:3637–3648

67. Kuijk EW, van Tol LTA, van de Velde H, Wubbolts R, Welling M, Geijsen N et al (2012) The roles of FGF and MAP kinase signaling in the segregation of the epiblast and hypoblast cell lineages in bovine and human embryos. Development 139:871–882

68. Canizo JR, Rivolta AEY, Echegaray CV, Suvá M, Alberio V, Aller JF et al (2019) A dose-dependent response to MEK inhibition determines hypoblast fate in bovine embryos. BMC Dev Biol 9:1–13

69. Ortega MS, Kelleher AM, O'Neil E, Benne J, Cecil R, Spencer TE (2020) NANOG is required to form the epiblast and maintain pluripotency in the bovine embryo. Mol Reprod Dev 87:152–160

70. Atlasi Y, Stunnenberg HG (2017) The interplay of epigenetic marks during stem cell differentiation and development. Nat Rev Genet 18:643–658

71. McLay DW, Clarke HJ (2003) Remodelling the paternal chromatin at fertilization in mammals. Reproduction 125:625–633

72. Santos F, Dean W (2004) Epigenetic reprogramming during early development in mammals. Reproduction 127:643–651

73. Saitou M, Kagiwada S, Kurimoto K (2012) Epigenetic reprogramming in mouse pre-implantation development and primordial germ cells. Development 139:15–31

74. Santos F, Hendrich B, Reik W, Dean W (2002) Dynamic reprogramming of DNA methylation in the early mouse embryo. Dev Biol 241:172–182

75. Nakanishi MO, Hayakawa K, Nakabayashi K, Hata K, Shiota K, Tanaka S (2012) Trophoblast-specific DNA methylation occurs after the segregation of the trophectoderm and inner cell mass in the mouse peri-implantation embryo. Epigenetics 7:173–182

76. Gaspar-Maia A, Alajem A, Meshorer E, Ramalho-Santos M (2011) Open chromatin in pluripotency and reprogramming. Nat Rev Mol Cell Biol 12:36–47

77. Beaujean N (2014) Histone post-translational modifications in preimplantation mouse embryos and their role in nuclear architecture. Mol Reprod Dev 81:100–112

78. Liu H, Kim JM, Aoki F (2004) Regulation of histone H3 lysine 9 methylation in oocytes and early pre-implantation embryos. Development 131:2269–2280

79. Lepikhov K, Walter J (2004) Differential dynamics of histone H3 methylation at positions K4 and K9 in the mouse zygote. BMC Dev Biol 4:2–6

80. Kourmouli N, Jeppesen P, Mahadevhaiah S, Burgoyne P, Wu R, Gilbert DM et al (2004) Heterochromatin and tri-methylated lysine 20 of histone H4 in animals. J Cell Sci 117:2491–2501

81. Daujat S, Weiss T, Mohn F, Lange UC, Ziegler-Birling C, Zeissler U et al (2009) H3K64 trimethylation marks heterochromatin and is dynamically remodeled during developmental reprogramming. Nat Struct Mol Biol 16:777–781

82. Wongtawan T, Taylor JE, Lawson KA, Wilmut I, Pennings S (2011) Histone H4K20me3 and HP1α are late heterochromatin markers in development, but present in undifferentiated embryonic stem cells. J Cell Sci 124:1878–1890

83. Ahmed K, Dehghani H, Rugg-Gunn P, Fussner E, Rossant J, Bazett-Jones DP (2010) Global chromatin architecture reflects

pluripotency and lineage commitment in the early mouse embryo. PLoS One 5:0010531

84. Liu X, Wang C, Liu W, Li J, Li C, Kou X et al (2016) Distinct features of H3K4me3 and H3K27me3 chromatin domains in pre-implantation embryos. Nature 537:558–562

85. Erhardt S, Su IH, Schneider R, Barton S, Bannister AJ, Perez-Burgos L et al (2003) Consequences of the depletion of zygotic and embryonic enhancer of zeste 2 during preimplantation mouse development. Development 130:4235–4248

86. Dahl JA, Reiner AH, Klungland A, Wakayama T, Collas P (2010) Histone H3 lysine 27 methylation asymmetry on developmentally-regulated promoters distinguish the first two lineages in mouse preimplantation embryos. PLoS One 5:0009150

87. VerMilyea MD, O'Neill LP, Turner BM (2009) Transcription-independent heritability of induced histone modifications in the mouse preimplantation embryo. PLoS One 4:e6086

88. Torres-Padilla ME, Parfitt DE, Kouzarides T, Zernicka-Goetz M (2007) Histone arginine methylation regulates pluripotency in the early mouse embryo. Nature 445:214–218

89. Burton A, Muller J, Tu S, Padilla-Longoria P, Guccione E, Torres-Padilla ME (2013) Single-cell profiling of epigenetic modifiers identifies PRDM14 as an inducer of cell fate in the mammalian embryo. Cell Rep 5:687–701

90. Payer B, Lee JT (2008) X chromosome dosage compensation: how mammals keep the balance. Annu Rev Genet 42:733–772

91. Wutz A (2011) Gene silencing in X-chromosome inactivation: advances in understanding facultative heterochromatin formation. Nat Rev Genet 12:542–553

92. Mak W, Nesterova TB, De Napoles M, Appanah R, Yamanaka S, Otte AP et al (2004) Reactivation of the paternal X chromosome in early mouse embryos. Science 303:666–669

93. Silva J, Nichols J, Theunissen TW, Guo G, van Oosten AL, Barrandon O et al (2009) Nanog is the gateway to the pluripotent ground state. Cell 138:722–737

94. Petropoulos S, Edsgärd D, Reinius B, Deng Q, Panula SP, Codeluppi S et al (2016) Single-cell RNA-Seq reveals lineage and X chromosome dynamics in human pre-implantation embryos. Cell 165:1012–1026

95. Vallot C, Patrat C, Collier AJ, Huret C, Casanova M, Liyakat Ali TM et al (2017) XACT noncoding RNA competes with XIST in the control of X chromosome activity during human early development. Cell Stem Cell 20:102–111

96. Okamoto I, Patrat C, Thépot D, Peynot N, Fauque P, Daniel N et al (2011) Eutherian mammals use diverse strategies to initiate X-chromosome inactivation during development. Nature 472:370–374

97. Ferreira AR, Machado GM, Diesel TO, Carvalho JO, Rumpf R, Melo EO et al (2010) Allele-specific expression of the MAOA gene and X chromosome inactivation in in vitro produced bovine embryos. Mol Reprod Dev 77:615–621

98. Yu B, van Tol HTA, Stout TAE, Roelen BAJ (2020) Initiation of X chromosome inactivation during bovine embryo development. Cell 9:1016

99. Briggs R, King TJ (1952) Transplantation of living nuclei from blastula cells into enucleated frogs' eggs. Proc Natl Acad Sci 38:455–463

100. Gurdon JB, Melton DA (2008) Nuclear reprogramming in cells. Science 322:1811–1815

101. Wilmut I, Schnieke AE, McWhir J, Kind AJ, Campbell KHS (1997) Viable offspring derived from fetal and adult mammalian cells. Nature 385:810–813

102. Matoba S, Zhang Y (2018) Somatic cell nuclear transfer reprogramming: mechanisms and applications. Cell Stem Cell 23:471–485

103. Wakayama T, Tabar V, Rodriguez I, Perry ACF, Studer L, Mombaerts P (2001) Differentiation of embryonic stem cell lines generated from adult somatic cells by nuclear transfer. Science 292:740–743

104. Tachibana M, Amato P, Sparman M, Gutierrez NM, Tippner-Hedges R, Ma H et al (2013) Human embryonic stem cells derived by somatic cell nuclear transfer. Cell 153:1228–1238

105. Everts RE, Chavatte-Palmer P, Razzak A, Hue I, Green C, Oliveira R et al (2008) Aberrant gene expression patterns in placentomes are associated with phenotypically normal and abnormal cattle cloned by somatic cell nuclear transfer. Physiol Genomics 33:65–77

106. Chavatte-Palmer P, Camous S, Jammes H, Le Cleac'h N, Guillomot M, RSF L (2012) Review: Placental perturbations induce the developmental abnormalities often observed in bovine somatic cell nuclear transfer. Placenta 33:S99–S104

107. Ao Z, Liu D, Zhao C, Yue Z, Shi J, Zhou R et al (2017) Birth weight, umbilical and

placental traits in relation to neonatal loss in cloned pigs. Placenta 57:94–101

108. Loi P, Clinton M, Vackova I, Fulka J, Feil R, Palmieri C et al (2006) Placental abnormalities associated with post-natal mortality in sheep somatic cell clones. Theriogenology 65:1110–1121

109. Tanaka S, Oda M, Toyoshima Y, Wakayama T, Tanaka M, Hattori N et al (2001) Placentomegaly in cloned mouse concepti caused by expansion of the spongiotrophoblast layer. Biol Reprod 65:1813–1821

110. Lin J, Shi L, Zhang M, Yang H, Qin Y, Zhang J et al (2011) Defects in trophoblast cell lineage account for the impaired in vivo development of cloned embryos generated by somatic nuclear transfer. Cell Stem Cell 8:371–375

111. Wang X, Qu J, Li J, He H, Liu Z, Huan Y (2020) Epigenetic reprogramming during somatic cell nuclear transfer: recent progress and future directions. Front Genet 11:1–13

112. Van Soom A, Ysebaert MT, De Kruif A (1997) Relationship between timing of development, morula morphology, and cell allocation to inner cell mass and trophectoderm in in vitro-produced bovine embryos. Mol Reprod Dev 47:47–56

113. Leese HJ, Donnay I, Thompson JG (1998) Human assisted conception: a cautionary tale. Lessons from domestic animals. Hum Reprod 13:184–202

114. Van Soom A, Boerjan M, Ysebaert MT, De Kruif A (1996) Cell allocation to the inner cell mass and the trophectoderm in bovine embryos cultured in two different media. Mol Reprod Dev 45:171–182

115. Spielmann H, Jacob-Mueller U, Beckord W (1980) Immunosurgical studies on inner cell mass development in rat and mouse blastocysts before and during implantation in vitro. J Embryol Exp Morphol 60:255–269

116. Iwasaki S, Yoshiba N, Ushijima H, Watanabe S, Nakahara T (1990) Morphology and proportion of inner cell mass of bovine blastocysts fertilized in vitro and in vivo. J Reprod Fertil 90:279–284

117. Papaioannou VE, Ebert KM (1988) The preimplantation pig embryo: cell number and allocation to trophectoderm and inner cell mass of the blastocyst in vivo and in vitro. Development 102:793–803

118. Koo D-B, Kang Y-K, Choi Y-H, Park JS, Kim H-N, Oh KB et al (2002) Aberrant allocations of inner cell mass and trophectoderm cells in bovine nuclear transfer blastocysts. Biol Reprod 67:487–492

119. Ushijima H, Akiyama K, Tajima T (2008) Transition of cell numbers in bovine preimplantation embryos: in vivo collected and in vitro produced embryos. J Reprod Dev 54:239–243

120. Boiani M, Eckardt S, Leu NA, Schöler HR, McLaughlin KJ (2003) Pluripotency deficit in clones overcome by clone-clone aggregation: epigenetic complementation? EMBO J 22: 5304–5312

121. Misica-Turner PM, Oback FC, Eichenlaub M, Wells DN, Oback B (2007) Aggregating embryonic but not somatic nuclear transfer embryos increases cloning efficiency in cattle. Biol Reprod 76:268–278

122. Akagi S, Yamaguchi D, Matsukawa K, Mizutani E, Hosoe M, Adachi N et al (2011) Developmental ability of somatic cell nuclear transferred embryos aggregated at the 8-cell stage or 16- to 32-cell stage in cattle. J Reprod Dev 57:500–506

123. Bortvin A, Eggan K, Skaletsky H, Akutsu H, Berry DL, Yanagimachi R et al (2003) Incomplete reactivation of Oct4-related genes in mouse embryos cloned from somatic nuclei. Development 130:1673–1680

124. Takahashi K, Yamanaka S (2006) Induction of pluripotent stem cells from mouse embryonic and adult fibroblast cultures by defined factors. Cell 126:663–676

125. Buganim Y, Faddah DA, Jaenisch R (2013) Mechanisms and models of somatic cell reprogramming. Nat Rev Genet 14:427–439

126. Wen D, Banaszynski LA, Liu Y, Geng F, Noh KM, Xiang J et al (2014) Histone variant H3.3 is an essential maternal factor for oocyte reprogramming. Proc Natl Acad Sci U S A 111:7325–7330

127. Rodríguez-Alvarez L, Manriquez J, Velasquez A, Castro FO (2013) Constitutive expression of the embryonic stem cell marker OCT4 in bovine somatic donor cells influences blastocysts rate and quality after nucleus transfer. Vitr Cell Dev Biol Anim 49:657–667

128. Beyhan Z, Ross PJ, Iager AE, Kocabas AM, Cunniff K, Rosa GJ et al (2007) Transcriptional reprogramming of somatic cell nuclei during preimplantation development of cloned bovine embryos. Dev Biol 305:637–649

129. Pfister-Genskow M, Myers C, Childs LA, Lacson JC, Patterson T, Betthauser JM et al (2005) Identification of differentially expressed genes in individual bovine preimplantation embryos produced by nuclear transfer: improper reprogramming of genes

required for development. Biol Reprod 72:
546–555

130. Somers J, Smith C, Donnison M, Wells DN,
Henderson H, McLeay L et al (2006) Gene
expression profiling of individual bovine
nuclear transfer blastocysts. Reproduction
131:1073–1084

131. Min B, Cho S, Park JS, Lee YG, Kim N, Kang
YK (2015) Transcriptomic features of bovine
blastocysts derived by somatic cell nuclear
transfer. G3 Genes, Genomes, Genet 5:
2527–2538

132. Sood TJ, Lagah SV, Mukesh M, Singla SK,
Chauhan MS, Manik RS et al (2019) RNA
sequencing and transcriptome analysis of buffalo (Bubalus bubalis) blastocysts produced
by somatic cell nuclear transfer and in vitro
fertilization. Mol Reprod Dev 86:1149–1167

133. Wuensch A, Habermann FA, Kurosaka S,
Klose R, Zakhartchenko V, Reichenbach
H-D et al (2007) Quantitative monitoring
of pluripotency gene activation after somatic
cloning in cattle. Biol Reprod 76:983–991

134. Arnold DR, Bordignon V, Lefebvre R, Murphy BD, Smith LC (2006) Somatic cell
nuclear transfer alters peri-implantation trophoblast differentiation in bovine embryos.
Reproduction 132:279–290

135. Rayon T, Menchero S, Rollán I, Ors I,
Helness A, Crespo M et al (2016) Distinct
mechanisms regulate Cdx2 expression in the
blastocyst and in trophoblast stem cells. Sci
Rep 6:1–10

136. Rayon T, Menchero S, Nieto A,
Xenopoulos P, Crespo M, Cockburn K et al
(2014) Notch and Hippo converge on Cdx2
to specify the trophectoderm lineage in the
mouse blastocyst. Dev Cell 30:410–422

137. Fujii T, Moriyasu S, Hirayama H,
Hashizume T, Sawai K (2010) Aberrant
expression patterns of genes involved in segregation of inner cell mass and trophectoderm
lineages in bovine embryos derived from
somatic cell nuclear transfer. Cell Reprogram
12:617–625

138. Wu X, Song M, Yang X, Liu X, Liu K, Jiao C
et al (2016) Establishment of bovine embryonic stem cells after knockdown of CDX2. Sci
Rep 6:1–12

139. Munsie MJ, Michalska AE, O'Brien CM,
Trounson AO, Pera MF, Mountford PS
(2000) Isolation of pluripotent embryonic
stem cells from reprogrammed adult mouse
somatic cell nuclei. Curr Biol 10:989–992

140. Wakayama S, Ohta H, Kishigami S, Van
Thuan N, Hikichi T, Mizutani E et al (2005)
Establishment of male and female nuclear

transfer embryonic stem cell lines from different mouse strains and tissues. Biol Reprod 72:
932–936

141. Thomson JA, Itskovitz-Eldor J, Shapiro SS,
Waknitz MA, Swiergiel JJ, Marshall VSS et al
(1998) Embryonic stem cell lines derived
from human blastocysts. Science 282:1145–
1147

142. Rielland M, Brochard V, Lacroix MC, Renard
JP, Jouneau A (2009) Early alteration of the
self-renewal/differentiation threshold in trophoblast stem cells derived from mouse
embryos after nuclear transfer. Dev Biol 334:
325–334

143. Oda M, Tanaka S, Yamazaki Y, Ohta H,
Iwatani M, Suzuki M et al (2009) Establishment of trophoblast stem cell lines from
somatic cell nuclear-transferred embryos.
Proc Natl Acad Sci U S A 106:16293–16297

144. Xue F, Tian XC, Du F, Kubota C, Taneja M,
Dinnyes A et al (2002) Aberrant patterns of X
chromosome inactivation in bovine clones.
Nat Genet 31:216–220

145. Williams LH, Kalantry S, Starmer J, Magnu-
son T (2011) Transcription precedes loss of
Xist coating and depletion of H3K27me3
during X-chromosome reprogramming in
the mouse inner cell mass. Development
138:2049–2057

146. Muñoz M, Rodríguez A, De Frutos C, Caa-
maño JN, Díez C, Facal N et al (2008) Conventional pluripotency markers are unspecific
for bovine embryonic-derived cell-lines.
Theriogenology 69:1159–1164

147. Simmet K, Zakhartchenko V, Philippou-
Massier J, Blum H, Klymiuk N, Wolf E
(2018) OCT4/POU5F1 is required for
NANOG expression in bovine blastocysts.
Proc Natl Acad Sci 115:2770–2775

148. Inoue K, Kohda T, Sugimoto M, Sado T,
Ogonuki N, Matoba S et al (2010) Impeding
Xist expression from the active X chromosome improves mouse somatic cell nuclear
transfer. Science 330:496–499

149. Ruan D, Peng J, Wang X, Ouyang Z, Zou Q,
Yang Y et al (2018) XIST derepression in
active X chromosome hinders pig somatic
cell nuclear transfer. Stem Cell Rep 10:494–
508

150. Matoba S, Inoue K, Kohda T, Sugimoto M,
Mizutani E, Ogonuki N et al (2011) RNAi-
mediated knockdown of Xist can rescue the
impaired postimplantation development of
cloned mouse embryos. Proc Natl Acad Sci
108:20621–20626

151. Alberto ML, Meirelles FV, Perecin F, Ambró-
sio CE, Favaron PO, Franciolli ALR et al

(2012) Development of bovine embryos derived from reproductive techniques. Reprod Fertil Dev 25:907

152. Posfai E, Tam OH, Rossant J (2014) Mechanisms of pluripotency in vivo and in vitro, 1st edn. Elsevier Inc

153. Morgani S, Nichols J, Hadjantonakis AK (2017) The many faces of pluripotency: in vitro adaptations of a continuum of in vivo states. BMC Dev Biol 17:10–12

154. Daigneault BW, Rajput S, Smith GW, Ross PJ (2018) Embryonic POU5F1 is required for expanded bovine blastocyst formation. Sci Rep 8:7753

155. Fogarty NME, McCarthy A, Snijders KE, Powell BE, Kubikova N, Blakeley P et al (2017) Genome editing reveals a role for OCT4 in human embryogenesis. Nature 550:67–73

156. Tecirlioglu RT, Cooney MA, Lewis IM, Korfiatis NA, Hodgson R, Ruddock NT et al (2005) Comparison of two approaches to nuclear transfer in the bovine: hand-made cloning with modifications and the conventional nuclear transfer technique. Reprod Fertil Dev 17:573–585

157. Li Y, Li S, Dai Y, Du W, Zhao C, Wang L et al (2007) Nuclear reprogramming in embryos generated by the transfer of yak (Bos grunniens) nuclei into bovine oocytes and comparison with bovine-bovine SCNT and bovine IVF embryos. Theriogenology 67:1331–1338

158. Ross PJ, Rodriguez RM, Iager AE, Beyhan Z, Wang K, Ragina NP et al (2009) Activation of bovine somatic cell nuclear transfer embryos by PLCZ cRNA injection. Reproduction 137:427–437

159. Song BS, Kim JS, Kim CH, Han YM, Lee DS, Lee KK et al (2009) Prostacyclin stimulates embryonic development via regulation of the cAMP response element-binding protein-cyclo-oxygenase-2 signalling pathway in cattle. Reprod Fertil Dev 21:400–407

160. Cui X-S, Xu Y-N, Shen X-H, Zhang L-Q, Zhang J-B, Kim N-H (2011) Trichostatin A modulates apoptotic-related gene expression and improves embryo viability in cloned bovine embryos. Cell Reprogram 13:179–189

161. Hua S, Zhang H, Su JM, Zhang T, Quan FS, Liu J et al (2011) Effects of the removal of cytoplasm on the development of early cloned bovine embryos. Anim Reprod Sci 126:37–44

162. Su J, Wang Y, Li Y, Li R, Li Q, Wu Y et al (2011) Oxamflatin significantly improves

nuclear reprogramming, blastocyst quality, and in vitro development of bovine SCNT embryos. PLoS One 6:0023805

163. Goissis MD, Suhr ST, Cibelli JB (2013) Effects of donor fibroblasts expressing OCT4 on bovine embryos generated by somatic cell nuclear transfer. Cell Reprogram 15:24–34

164. Chen H, Zhang L, Guo Z, Wang Y, He R, Qin Y et al (2015) Improving the development of early bovine somatic-cell nuclear transfer embryos by treating adult donor cells with vitamin C. Mol Reprod Dev 82:867–879

165. An Q, Peng W, Cheng Y, Lu Z, Zhou C, Zhang Y et al (2019) Melatonin supplementation during in vitro maturation of oocyte enhances subsequent development of bovine cloned embryos. J Cell Physiol 234:17370–17381

166. Chang HY, Xie RX, Zhang L, Fu LZ, Zhang CT, Chen HH et al (2019) Overexpression of miR-101-2 in donor cells improves the early development of Holstein cow somatic cell nuclear transfer embryos. J Dairy Sci 102:4662–4673

167. Sah S, Sharma AK, Singla SK, Singh MK, Chauhan MS, Manik RS et al (2020) Effects of treatment with a microRNA mimic or inhibitor on the developmental competence, quality, epigenetic status and gene expression of buffalo (Bubalus bubalis) somatic cell nuclear transfer embryos. Reprod Fertil Dev 32:508–521

168. Hai T, Hao J, Wang L, Jouneau A, Zhou Q (2011) Pluripotency maintenance in mouse somatic cell nuclear transfer embryos and its improvement by treatment with the histone deacetylase inhibitor TSA. Cell Reprogram 13:47–56

169. Mizutani E, Torikai K, Wakayama S, Nagatomo H, Ohinata Y, Kishigami S et al (2016) Generation of cloned mice and nuclear transfer embryonic stem cell lines from urine-derived cells. Sci Rep 6:1–8

170. Kim G, Roy PK, Fang X, Hassan BMS, Cho J (2019) Improved preimplantation development of porcine somatic cell nuclear transfer embryos by caffeine treatment. J Vet Sci 20:1–12

171. Zhou Q, Yang SH, Ding CH, He XC, Xie YH, Hildebrandt TB et al (2006) A comparative approach to somatic cell nuclear transfer in the rhesus monkey. Hum Reprod 21:2564–2571

172. Iager AE, Ragina NP, Ross PJ, Beyhan Z, Cunniff K, Rodriguez RM et al (2008)

Trichostatin A improves histone acetylation in bovine somatic cell nuclear transfer early embryos. Cloning Stem Cells 10:371–379

173. Zhang J, Hao L, Wei Q, Zhang S, Cheng H, Zhai Y et al (2020) TET3 overexpression facilitates DNA reprogramming and early development of bovine SCNT embryos. Reproduction 160:379–391

174. Cervera RP, Martí-Gutiérrez N, Escorihuela E, Moreno R, Stojkovic M (2009) Trichostatin A affects histone acetylation and gene expression in porcine somatic cell nucleus transfer embryos. Theriogenology 72:1097–1110

Chapter 4

Mitochondrial Inheritance Following Nuclear Transfer: From Cloned Animals to Patients with Mitochondrial Disease

Jörg P. Burgstaller and Marcos R. Chiaratti

Abstract

Mitochondria are indispensable power plants of eukaryotic cells that also act as a major biochemical hub. As such, mitochondrial dysfunction, which can originate from mutations in the mitochondrial genome (mtDNA), may impair organism fitness and lead to severe diseases in humans. MtDNA is a multi-copy, highly polymorphic genome that is uniparentally transmitted through the maternal line. Several mechanisms act in the germline to counteract heteroplasmy (i.e., coexistence of two or more mtDNA variants) and prevent expansion of mtDNA mutations. However, reproductive biotechnologies such as cloning by nuclear transfer can disrupt mtDNA inheritance, resulting in new genetic combinations that may be unstable and have physiological consequences. Here, we review the current understanding of mitochondrial inheritance, with emphasis on its pattern in animals and human embryos generated by nuclear transfer.

Key words Cloning, Embryo, Heteroplasmy, Mitochondria, mtDNA, MRT, Nuclear transplantation, Oocyte, SCNT

---

1 Introduction

Mitochondria are double-membrane organelles that encompass multiple copies of a 16.6 kb circular genome (mtDNA; Fig. 1). In mammals, mtDNA is exclusively transmitted through the maternal lineage as sperm mitochondria are destroyed soon after fertilization through the action of mitophagy (mitochondria-targeted autophagy) [1–3]. Given the high polymorphic rate of mtDNA [4–8], such uniparental mode of inheritance contributes to the maintenance of homoplasmy (i.e., presence of a single mtDNA type within an individual). On the other hand, reproductive biotechnologies such as nuclear transfer can disrupt mitochondrial transmission, potentially resulting in the admixture of mtDNA variants (known as heteroplasmy). As such, nuclear transfer can lead to new nuclear-to-cytoplasmic genetic combinations, which can be unstable and have detrimental physiological consequences [9–14]. In this book

Marcelo Tigre Moura (ed.), *Somatic Cell Nuclear Transfer Technology*, Methods in Molecular Biology, vol. 2647,
https://doi.org/10.1007/978-1-0716-3064-8_4,
© The Author(s), under exclusive license to Springer Science+Business Media, LLC, part of Springer Nature 2023

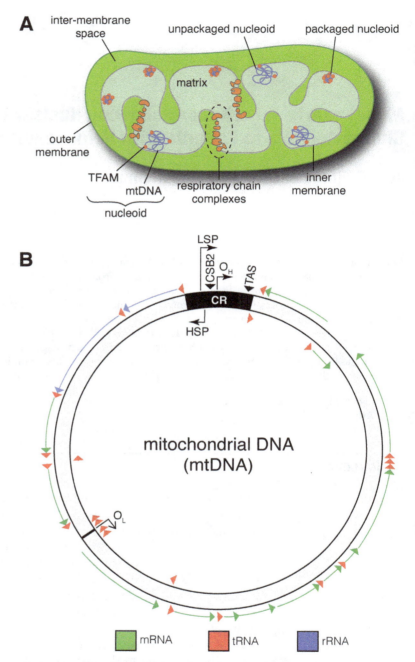

Fig. 1 Mitochondrial DNA (mtDNA) Localization, Structure, and Coding Capacity. (a) Mitochondria are composed of inner and outer membranes which delimitate two distinct compartments, the inter-membrane space and the matrix. The inner membrane folds toward the mitochondrial lumen to increase its length and host a higher number of the respiratory chain complexes, which are required for adenosine triphosphate (ATP) synthesis [15]. Within the matrix and attached to the inner membrane are one to several nucleoids, which can vary in their packaging degree and are composed of ~1.4 mtDNA molecules coated by (TFAM) and other proteins [50, 51]. (b) The mammalian mitochondrial genome

chapter, we outline basic aspects of mitochondrial inheritance, highlighting how it is altered in cloned animals. In addition, we discuss how this biotechnology has been translated to the clinical setting seeking to prevent transmission of mtDNA-encoded diseases in humans.

2 Mitochondrial Function and Dynamics

Mitochondria are organelles specialized in energy production (i.e., adenosine triphosphate—ATP) and thought to have originated from the engulfment of an ancestor bacterium by a primitive host cell [15]. Mitochondria retain several vestiges of their endosymbiotic origin, including their own genome, a double membrane, and the use of a proton (H^+) gradient to generate ATP. Four enzymatic complexes and the ATP synthase, which are all embedded in the inner mitochondrial membrane, are responsible for converting the energy derived from the oxidation of energetic substrates into ATP [15]. This process, known as oxidative phosphorylation (OXPHOS), is far more efficient in ATP generation than anaerobic glycolysis alone. In addition, mitochondria play important roles in phospholipid synthesis, reactive oxygen species (ROS) signaling, calcium handling, innate immunity, and programmed cell death through apoptosis. Collectively, these facts suggest that the symbiosis with mitochondria was a key event in eukaryotic evolution [16].

Although initially described as static and isolated, mitochondria can be remarkably dynamic and interact with other organelles including neighboring mitochondria, endoplasmic reticulum, lipid droplets, lysosomes, and the nucleus [17–20]. These aspects of mitochondria are determined by the balance between cyclic events of fusion and fission, which establish mitochondrial networks. Hence, under enhanced fusion mitochondria form long interconnected tubules, whereas enhanced fission leads to fragmentation of the mitochondrial network [20]. Mitochondrial dynamics are important to adapt the organelle to specific metabolic demands

Fig. 1 (continued) is a compact (~16.6 kb), circular molecule composed of two strands differing in their guanine content, the light and the heavy strands. Most of mtDNA sequence encompasses protein-coding genes, including messenger RNAs (mRNA), transfer RNAs (tRNA), and ribosomal RNAs (rRNA) [41]. However, there is also a non-coding region required for the control of mtDNA replication and transcription. This region, known as the control region (CR), harbors regulatory elements such as the light (LSP) and heavy (HSP)-strand promoters, the heavy-strand origin of replication (O_H), the conserved sequence block 2 (CSB2) and the termination associated sequence (TAS); the light-strand origin of replication (O_L) is located outside the CR [50, 55]

and nutrient availability, apart from controlling mitochondrial content exchange, transport, inheritance, and selective degradation [21]. The importance of mitochondrial dynamics is revealed by the range of diseases originating from defects in this organelle machinery. Furthermore, experimental disruption of either fusion or fission leads to impaired bioenergetics, mtDNA instability, poor cell growth, and, ultimately, apoptosis [18].

Amongst all cell types in mammals, oocytes own the largest mitochondrial count [22, 23], indicative of a high energy requirement. Contrary to this, the mitochondrial network is broadly fragmented in the female gamete. Oocyte mitochondria are characterized by their small size, spheric shape, and underdeveloped cristae structure [22, 24–29]. Collectively, these are indicative of low bioenergetic activity, which is corroborated by studies showing that deficient mitochondrial function has minimal impact on oocyte growth [28, 30–32]. In fact, mitochondrial fragmentation, decreased ROS production, and low respiration are intrinsic characteristics of most germ cells such that cellular reprogramming into induced pluripotent stem cells (iPSCs) leads to a metabolic reprogramming characterized by increased dependence on glycolysis [33–38]. Therefore, rather than a key role in energy supply, the high number of mitochondria in oocytes is likely required for assuring equitable organelle partitioning to embryonic cells until activation of mitochondrial biogenesis [39–41]. In addition, as discussed below, extensive fragmentation minimizes mitochondrial complementation, facilitating the action of selection against deleterious mtDNA mutations [29, 42, 43].

3 Mitochondrial DNA Structure and Maintenance

The mammalian mtDNA is a compact, circular genome encoding for 37 RNA molecules necessary for intramitochondrial translation of 13 polypeptides (Fig. 1). These polypeptides are essential subunits of the OXPHOS system and defects in their expression are associated with numerous mitochondrial diseases in humans [44]. Recently, new mitochondrial small open reading frames were discovered, coding for mitochondrial-derived peptides (MDPs) that are involved in (mitochondrial) metabolism, ROS production, apoptosis, and more. Both the number of known MDPs and their functions are currently under investigation [45]. Also, the existence of small mitochondrial RNAs is currently discussed, further indicating that mtDNA replication and transcription could be more complex than previously thought [46].

Mitochondria however rely on an intricated crosstalk with the nuclear compartment as the large majority of the mitochondrial proteome (~1100 peptides) is encoded in the nucleus, translated in the cytosol, and imported into mitochondria [13, 47, 48]. Among these are all of the proteins controlling mtDNA maintenance and expression, processes which are mainly regulated at a non-coding control region (CR) of mtDNA containing the transcription promoter of the light (LSP) and heavy (HSP) strands, as well as the origin of replication of the heavy strand (O_H; Fig. 1) [49–52]. Transcription initiated at both LSP and HSP generates two polycistronic molecules, which are then extensively processed to translation [52–54]. Replication of mtDNA, however, relies on the premature termination of the light-strand transcription near the O_H. According to the strand displacement model, the nascent RNA strand is used as a primer for unidirectional replication of the heavy strand; replication of the light strand only initiates when the replication fork in the leading strand reaches the replication origin of the light stand (O_L; Fig. 1) [50, 55]. Therefore, replication and transcription of mtDNA have been proposed to be mutually exclusive processes, with multiple levels of control [50, 52].

Replication of mtDNA and copy number control are regulated at least at four different levels: (i) by controlling transcription initiation from the LSP, which requires the mitochondrial RNA polymerase (POLRMT), the mitochondrial transcription factor A (TFAM), and the mitochondrial transcription factor B2 (TFB2M); (ii) by controlling termination/processing of the LSP transcript at the conserved sequence block-2 (CSB2), a process in which the mitochondrial transcription elongation factor (TEFM) has been negatively implicated; (iii) at the termination-associated sequence (TAS), a region located ~650 nucleotides downstream of the O_H that associates with the premature termination of mtDNA synthesis; and (iv) by regulating the level of mtDNA packaging, which is mainly determined by the action of TFAM (Fig. 1) [50]. Importantly, about two-thirds of all transcription events initiated from the LSP are prematurely terminated at the CSB2, which has been implicated in replicative advantage of certain mtDNA haplogroups with CSB2 polymorphisms [52, 56, 57]. Likewise, a number of polymorphisms have been mapped to the TAS, which can also interfere with mtDNA replication as the large majority of mtDNA replicative events are aborted at this region [50, 56–59]. Provided that the nascent replicative strand (known as 7S DNA, ranging from the O_H to the TSA region) remains bound to its template, forming a triple-stranded displacement loop (D-loop) structure, replication reinitiated from the 3′ end of the 7S DNA can enable a rapid increase in mtDNA copy number [50, 58, 60].

Due to its relative short half-life in mitotic (~10 days) and post-mitotic (~20 days) cells [61, 62], mtDNA is replicated at a correspondingly high rate. Thus, in spite of the action of proof-reading mechanisms, mtDNA is prone to single nucleotide variations (mtSNVs) and rearrangements arising from replicative errors [41, 63]. Replicative errors arise both in somatic tissues, where the resulting variations are discussed as an aging factor [64], but also in the germline, giving rise to various mtDNA haplotypes. For instance, differences of 20–80 mtSNVs are common between humans, even within the same haplogroup [10, 14, 65]. The mtDNA composition within cells is usually thought to be homoplasmic due to strict uniparental (maternal) inheritance. Massive parallel sequencing using next-generation sequencing (NGS) methods has shown that in humans intra-individual polymorphisms of mtDNA are nevertheless universal due to its high mutation rate, although at low (i.e., 0.5–1.5%) heteroplasmic levels [6, 7, 41, 66]. These mutations do not only arise de novo but are also transmitted to the next generation when present in the germline. These polymorphisms are mostly located in the CR, but can map to any part of the mtDNA, including protein-coding genes. Several safeguarding mechanisms reduce the impact of potential pathogenic mechanisms (*see* below), but generally the cell's energy demand can be supplied unless the mutation load reaches a threshold of typically >60% for deletions and >80% for mtSNVs [41].

4 Mitochondrial DNA Inheritance

Selective forces have long been shaping mtDNA evolution, resulting in several haplogroups within mammalian species, including cattle, pigs, sheep, goats, mice, and humans [10, 14, 62, 65, 67–76]. In principle, all mutations, even homoplasmic ones, originate from a single mtDNA molecule which expands through the action of mechanisms such as relaxed replication (i.e., different mtDNA molecules are copied at uneven rates in a cell cycle-independent fashion) and vegetative segregation (i.e., mtDNA variants are unevenly partitioned among daughter cells at cytokinesis) [41, 77–83]. Additionally, there is evidence that certain mtDNA variants can be biased replicated or partitioned, driving selection either for or against them [9, 29, 42, 66, 84–92].

Both relaxed replication and vegetative segregation act in the female germline through a mechanism known as the mtDNA genetic bottleneck (Fig. 2). Following fertilization, ~200,000 mtDNA molecules are progressively partitioned among cells of the early embryo while mtDNA replication remains downregulated; in embryonic cells that give rise to primordial germ cells (PGCs) mtDNA replication reinitiates only after implantation, thus forcing the stochastic segregation of mtDNA variants. Furthermore, only a few cells in the early embryo give rise to PGCs and the next generation, generating an important sampling effect [23, 41, 82, 90, 93–96]. Extensive studies in several species, including mice, cattle, non-human primates, and humans support that low-level heteroplasmic mtDNA variants can benefit from the mtDNA genetic bottleneck to expand and even become fixed in the population after few generations [12, 62, 66, 81, 89, 95, 97–100]. This is

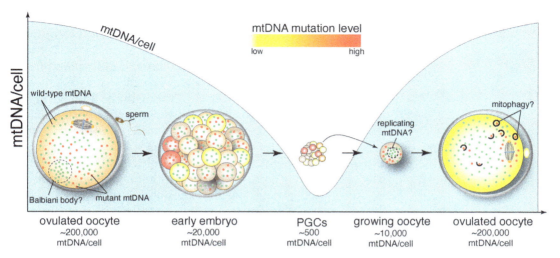

Fig. 2 The Mitochondrial DNA (mtDNA) Genetic Bottleneck and the Action of Selection in Germline Shape mtDNA Composition. Replication of mtDNA is downregulated during early embryogenesis, which extends up to post-implantation in cells originating primordial germ cells (PGCs). Hence, mtDNA copy number drops from ~200,000 in ovulated oocytes to ~500 in PGCs, thus forcing segregation of mtDNA variants amongst embryonic cells. Furthermore, only few cells out of hundreds in the developing embryo give rise to PGCs, resulting in dramatic heteroplasmic shifts in future oocytes and the next generation [23, 41, 82, 90, 93–96]. This mechanism, known as the mtDNA genetic bottleneck, can further be complemented by the action of selection either for or against deleterious mtDNA mutations [9, 29, 42, 66, 84–92]. Selection might take place through mitochondrial clusterization (Balbiani body) in oocytes, leading to biased organelle inheritance by PGCs [92]. Alternatively, extensive mitochondrial fragmentation during oogenesis seems to promote mtDNA segregation into individual organelles, enabling selection based on the variant effect on mitochondrial fitness. In the latter case, selection has been proposed to occur through either replication of a subgroup of mtDNA or mitochondria-targeted autophagy (mitophagy) [29, 42, 43, 106, 117]

corroborated by in vitro models of iPSC derivation, suggesting that mtDNA segregation takes advantage of intrinsic aspects of germ cells such as mitochondrial fragmentation and decreased OXPHOS dependence [38, 87, 101, 102]. Therefore, the mtDNA genetic bottleneck has been shown to be a key determinant of homoplasmy, while also enabling new mtDNA variants to be put to test at the organism level [41].

According to Muller's ratchet, the high mutation rate and the lack of germline recombination would make mtDNA susceptible to mutational meltdown. However, the mtDNA genetic bottleneck along with purifying selection probably counteract the Muller's ratchet by preventing intergenerational transmission of deleterious mtDNA mutations [103]. This hypothesis is in keeping with a number of studies showing that deleterious mutations on mtDNA genes coding for proteins, tRNAs, and rRNAs are more likely to decrease in level through generations [9, 66, 84, 88, 104–109]. In contrast, there is also evidence of positive (selfish) selection for deleterious mtDNA variants [79, 110–112]. In both cases, the mechanism(s) underpinning selection is unclear, albeit a few studies have provided some clues (Fig. 2). For instance, mutations in the CR may lead one mtDNA variant to outcompete another through an effect on mtDNA replication [57, 58, 113–115]. In this context, the sharp increase in mtDNA content from ~500 molecules in PGCs to ~200,000 in ovulated oocytes provides an excellent window of opportunity [23, 82, 83, 87]. Alternatively, findings in flies suggest that selection is driven by the variant effect on mitochondrial fitness, which has been linked to deficient import of factors required for mtDNA replication in oocytes [85, 86, 107, 116]. In accordance with these reports, extensive mitochondrial fragmentation in oocytes likely reduces complementation, enhancing the link between mtDNA variant and mitochondrial fitness, and enabling selection against deleterious mutations based on deficient mtDNA replication and/or mitophagy [29, 42, 43, 106, 117]. Additional mechanisms might take place at other stages of development (i.e., early embryos and PGCs), depending on nuclear-mitochondrial interactions and may take advantage of transient changes in metabolism [87, 104, 106, 108, 118–120].

5 Mitochondrial Inheritance in Cloned Animals

Livestock cloning by nuclear transfer is of great interest as a useful tool for producing genetically identical animals carrying desired inheritable traits, for assisting animal transgenesis, and for conservation of endangered species. In addition, cloning has been employed with success in humans for isolation of embryonic stem cells (ESCs), which can be used to model in vitro the development of diseases, besides a range of potential therapeutical applications [38, 121, 122]. Interestingly, techniques based on the cloning technology are now applied in human medicine to create healthy embryos of mothers with an inherited mtDNA-derived disease (*see* below) [123].

Cloned embryos can be generated by nuclear transfer through reconstruction of the recipient enucleated oocyte (i.e., at the metaphase II stage) with a donor cell, which can derive from either an early embryo or a somatic tissue (i.e., fibroblast). Depending on the amount of cytoplasm/mitochondria that is transported along with the nuclear material of the donor cell during oocyte reconstruction, nuclear transfer can generate embryos with an admixture of mitochondria of different sources. In the case of somatic cell nuclear transfer (SCNT), mtDNA contribution by the recipient oocyte is at least two orders of magnitude higher than that of nuclear-donor cells [124–128]. Such pattern of mitochondrial inheritance nevertheless contrasts with that of natural fertilization, resulting in animals with different mtDNA composition compared with the nuclear donor. In addition, due to poorly characterized mechanisms, nuclear-donor mtDNA can increase to high levels in offspring and be transmitted to the next generations (Fig. 3).

Given that polymorphic mtDNA variations are common within animals, cloning frequently results in heteroplasmic embryos and offspring [11, 73, 99, 124, 125, 129–132]. Most mtDNA polymorphisms are not deleterious, though their performance may vary for different nuclear-mitochondrial combinations [5, 9, 10, 14, 48, 65, 106, 133–137]. Such differences may become evident when polymorphic mtDNA variants are forced to coexist in cloned embryos, resulting in variant competition as heteroplasmy can be genetically unstable and cause adverse mitochondrial effects [9, 106, 133]. Accordingly, studies with cloned animals reveal that the levels of nuclear-donor mtDNA remain stable throughout development but, in a few cases, undergo stringent genetic drift [99, 124–126, 129–131, 138–140]. For instance, reports with cattle, pigs, sheep, and mice have shown that the nuclear-donor mtDNA can increase from <1% in reconstructed oocytes to >40% in offspring (Fig. 3) [11, 73, 130–132]. Additionally, nuclear-donor mtDNA levels can vary largely across tissues of the offspring, including the germline [11, 99, 124, 125, 130–132]. Although the mtDNA genetic bottleneck certainly contributes to these heteroplasmy shifts, mtDNA sequence analysis also indicates the action of other poorly characterized mechanisms (i.e., replicative advantage), as discussed above. If present, mtDNA heteroplasmy can be quantified (i.e., quantitative PCR) as long as both haplotypes are known and display genetic variability. Alternatively, if the mtDNA composition is unknown, heteroplasmy levels of ~1% and even much lower could be detected with NGS methods [141, 142], which might be used to ascertain whether an animal (or animal product) derives from a cloned animal. In opposition to this, cloning techniques that lead to homoplasmy have been developed [143, 144].

The overall success rate of SCNT is currently far from satisfactory, an outcome that is, in part, attributed to the abnormal pattern of mitochondrial inheritance in clones. Interspecies SCNT, for instance, has only been successfully applied to closely related animals such as wolf-canine, wildcat-cat, mouflon-ovine, and gaur-bovine [145–150]; evolutionary-driven genetic distances represent therefore a major barrier for the success of SCNT [151]. Recipient oocyte factors required for nuclear reprogramming, which in case of mismatch may result in aberrant gene expression, certainly play a fundamental role in this context [151]. However, provided that the nuclear and mitochondrial genomes coevolved to support their crosstalk and mitochondrial function [4, 5, 10, 14, 48, 50, 65, 137, 152], it is also plausible to consider that the donor nucleus from one species may not properly match recipient mitochondria from another, with implications to the overall SCNT cloning efficiency. In agreement with this, work with cybrid cells and mice has provided evidence of intersubspecies/interspecies (human-ape and

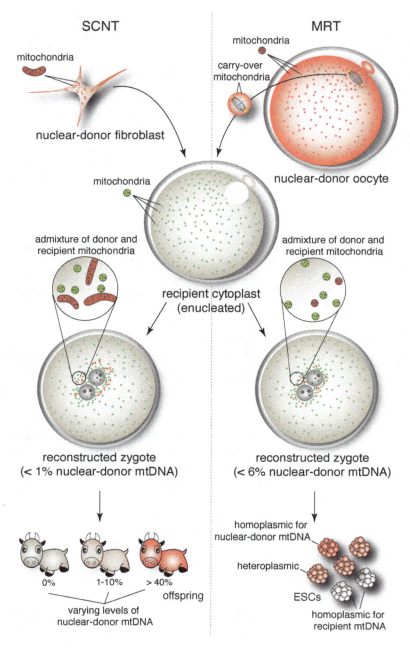

Fig. 3 Nuclear Transfer Results in Unpredicted Reversal back to Nuclear-Donor Mitochondrial DNA (mtDNA). In somatic cell nuclear transfer (SCNT), a somatic cell such as a fibroblast or a granulosa cell is transferred, including all its cytoplasmic structures, into an enucleated recipient oocyte [124–128]. The resulting embryo, therefore, contains an admixture of mitochondria [124–128], which diverge in several aspects (i.e., morphology and activity) due to their somatic and embryonic origins [36, 189–192]. Although the levels of nuclear-donor mtDNA in SCNT-derived embryos can be minimal (i.e., <1%), these can change dramatically during development to term, resulting in cloned offspring with variable heteroplasmy [11, 73, 130–132]. In contrast, only the nuclear

mouse-rat) nuclear-mitochondrial incompatibility, which was associated with mtDNA instability and mitochondrial dysfunction [137, 152–154]. Likewise, results from interspecies cloning support a similar trend, even though the data are not so clear as SCNT success relies on multiple factors [155–167]. On the other hand, evolutionary distance alone seems not to be a reliable indicator of nuclear-mitochondrial interaction potential since functional incompatibilities can be species-specific [153]. Also, if the donor nucleus fails to interact with recipient mitochondria, the embryo may attempt to rely on nuclear-donor mitochondria, likely explaining the high levels of heteroplasmy (and even homoplasmy) in certain fetuses and offspring (Fig. 3) [150, 157, 164, 166].

6 Manipulation of Mitochondrial Inheritance in Humans

Mitochondrial diseases caused by pathogenic mtDNA mutations are among the most common groups of neurological disorders, with an estimated prevalence of 1 in 5,000 births [8]. Manifestation of these diseases relies on a threshold effect, which can vary from <10% to 100% mutant mtDNA according to the specific mutation and affected tissue [168]. As a result, asymptomatic individuals carrying low-level pathogenic mtDNA mutations are even more frequent in the population, estimated to be 1 in 200 people [7]. This high incidence, along with the lack of effective treatments, supports the need for preventing mother-to-child transmission of mtDNA mutations [169]. However, genetic counseling is challenging for women with low to medium levels of pathogenic mtDNA and impossible when pathogenic mtDNA levels are high or even homoplasmic [44]. Fortunately, recent techniques based on nuclear transfer were developed to overcome these challenges.

Nuclear transfer enables to combine the nuclear genome from one individual with donor mtDNA from a recipient oocyte therefore preventing intergenerational transmission of pathogenic mtDNA mutations [169, 170]. Through this strategy, known as mitochondrial replacement therapy (MRT) or mitochondrial

Fig. 3 (continued) chromosomes and related structures are transferred into the recipient oocyte in the mitochondrial replacement therapy (MRT) or mitochondrial donation [169, 170]. In spite of this, carry-over mitochondria are most commonly transferred along the nuclear material, also resulting in heteroplasmic embryos with <6% nuclear-donor mtDNA [56, 171, 172]. As seen in cloned animals, these low-level heteroplasmies may increase during development to high levels or even reach homoplasmy, as reported for human embryonic stem cells (ESCs) generated following MRT [56, 171, 172]. This unexpected reversal back to nuclear-donor mtDNA highlights the need to eliminate carry-over mtDNA in order to assure offspring is homoplasmic for the recipient mtDNA [178]

donation, an oocyte harboring pathogenic mtDNA mutations has its nuclear genome transferred to a recipient (enucleated) oocyte, preferentially containing mtDNA of the same haplogroup, but free of pathogenic mtDNA mutations (Fig. 3). MRT can be carried out either before (using metaphase II-spindle oocytes) or after (using pronuclear zygotes) fertilization. Alternatively, either first or second polar bodies can be used as nuclear donors [170]. Importantly, these approaches have been shown to be compatible with development of human embryos to the blastocyst stage, which in turn were used to establish ESC lines [56, 171–175]. In addition, a clinical trial is currently under way to assess the outcome of MRT on the first 75 children born in the United Kingdom, and two babies were reported to be born following MRT in Mexico and Ukraine [44, 169, 176].

So far, proof-of-concept studies with humans have shown that MRT results in embryos in which the level of mtDNA derived from nuclear donors (i.e., patient's oocytes harboring pathogenic mtDNA mutations) ranges from 0% to <6% [56, 171, 172]. In principle, these residual levels are far below the mutation threshold reported for most mitochondrial diseases. However, extensive culture of ESCs obtained from the MRT-derived embryos led to a progressive reversal to the nuclear donor-derived mtDNA in ~15% of cases (Fig. 3) [56, 171, 172], a result also seen when ESCs were derived from human embryos following SCNT [56]. Given the bimodal pattern of mtDNA segregation reported for cultured cells [38, 101], one may argue that these findings do not correspond to that of embryos that are allowed to develop to term. Nonetheless, previous evidence from animal and human studies indicates that the mutation level can indeed change dramatically from early embryos to birth [95, 100, 177], a trend also seen when MRT-derived ESCs were allowed to differentiate [56], thus highlighting cell type-specific effects. In addition, these data are consistent with those from cloned animals discussed above (Fig. 3), arguing for long-term medical monitoring of the resulting children [169, 170]. This is even more important as the technique seems to get applied to overcome fertility problems, which might dramatically increase its use [178].

Kang and colleagues [56] have proposed two mechanisms to explain the rapid increase in nuclear-donor mtDNA following MRT: preferential mtDNA replication and cell growth advantage. The authors have provided experimental evidence in support of both mechanisms, which were associated to certain mtDNA haplotypes and seemed to be independent of an effect on mitochondrial fitness [56]. In addition, Wolf et al. [57] have suggested that the nuclear-donor mtDNA might be tagged for replication (i.e., epigenetic marks [179–187], enabling it to outcompete recipient mtDNA. In agreement, it was previously demonstrated in mice that karyoplast-derived, but not cytoplast-derived, mtDNA increases in preimplantation embryos following transfer to recipient zygotes [188]. Taking into account that mitochondria are physically connected with the nucleus [19], and only a fraction of mtDNA seems to be replicated in oocytes [83], easy access of perinuclear mitochondria to replication factors may facilitate mtDNA replication [80, 116, 188]. In summary, these findings highlight the need to fully prevent transfer of any carry-over mtDNA from the nuclear-donor cell to assure that children born following MRT will be free of the mitochondrial disease.

7 Conclusions and Future Prospects

Although our understanding of mitochondrial inheritance following nuclear transfer has experienced a great progress during the past two decades, there are many unknowns yet. For instance, the evolving field of mitochondrial dynamics suggests that admixture of morphologically and functionally divergent mitochondria may be critical for the success of SCNT, with potential consequences to mitochondrial inheritance [36, 189–192]. Cell cycle synchronization through starvation of nuclear donor cells, which is routinely applied in SCNT, can further enhance these differences once it leads to mitochondrial hyperfusion [193–195]. Moreover, mounting evidence supports the existence of epigenetic marks on mtDNA, which rely on cell type and environmental changes, and have been shown to regulate mtDNA replication and expression [179–187]. Therefore, apart from its potential use for generation of different nuclear-mitochondrial combinations, nuclear transfer should be employed with caution to avoid unintended consequences. This is of particular importance in humans, as MRT demands further optimization to fully prevent transmission of pathogenic mtDNA mutations and assure that the children will be free of the mitochondrial disease.

References

1. Rojansky R, Cha M-Y, Chan DC (2016) Elimination of paternal mitochondria in mouse embryos occurs through autophagic degradation dependent on PARKIN and MUL1. eLife 5:e17896. https://doi.org/10.7554/eLife.17896

2. Wei W, Pagnamenta AT, Gleadall N et al (2020) Nuclear-mitochondrial DNA segments resemble paternally inherited mitochondrial DNA in humans. Nat Commun 11:1740. https://doi.org/10.1038/s41467-020-15336-3

3. Sutovsky P, Moreno RD, Ramalho-Santos J et al (1999) Ubiquitin tag for sperm mitochondria. Nature 402:371–372. https://doi.org/10.1038/46466

4. Allio R, Donega S, Galtier N, Nabholz B (2017) Large variation in the ratio of mitochondrial to nuclear mutation rate across animals: implications for genetic diversity and the use of mitochondrial DNA as a molecular marker. Mol Biol Evol 34:2762–2772. https://doi.org/10.1093/molbev/msx197

5. James JE, Piganeau G, Eyre-Walker A (2016) The rate of adaptive evolution in animal mitochondria. Mol Ecol 25:67–78. https://doi.org/10.1111/mec.13475

6. Payne BAI, Wilson IJ, Yu-Wai-Man P et al (2013) Universal heteroplasmy of human

mitochondrial DNA. Hum Mol Genet 22:
384–390. https://doi.org/10.1093/hmg/
dds435

7. Elliott HR, Samuels DC, Eden JA et al (2008)
Pathogenic mitochondrial DNA mutations
are common in the general population. Am J
Hum Genet 83:254–260. https://doi.org/
10.1016/j.ajhg.2008.07.004

8. Gorman GS, Schaefer AM, Ng Y et al (2015)
Prevalence of nuclear and mitochondrial
DNA mutations related to adult mitochondrial disease. Ann Neurol 77:753–759.
https://doi.org/10.1002/ana.24362

9. Sharpley MS, Marciniak C, Eckel-Mahan K
et al (2012) Heteroplasmy of mouse
mtDNA is genetically unstable and results in
altered behavior and cognition. Cell 151:
333–343. https://doi.org/10.1016/j.cell.
2012.09.004

10. Røyrvik EC, Burgstaller JP, Johnston IG
(2016) mtDNA diversity in human populations highlights the merit of haplotype match-
ing in gene therapies. Mol Hum Reprod 22:
809–817. https://doi.org/10.1093/
molehr/gaw062

11. Burgstaller JP, Schinogl P, Dinnyes A et al
(2007) Mitochondrial DNA heteroplasmy in
ovine fetuses and sheep cloned by somatic cell
nuclear transfer. BMC Dev Biol 7:141.
https://doi.org/10.1186/1471-213X-
7-141

12. Ma H, Van Dyken C, Darby H et al (2020)
Germline transmission of donor, maternal and
paternal mtDNA in primates. Hum Reprod
36:493–505. https://doi.org/10.1093/
humrep/deaa308

13. John S (2019) Genomic balance: two genomes establishing synchrony to modulate cellular fate and function. Cell 8:1306. https://
doi.org/10.3390/cells8111306

14. Eyre-walker A (2017) Mitochondrial replacement therapy: are mito-nuclear interactions
likely to be a problem? Genetics 205:1365–
1372. https://doi.org/10.1534/genetics.
116.196436/-/DC1.1

15. Martin WF, Garg S, Zimorski V (2015)
Endosymbiotic theories for eukaryote origin.
Philos Trans R Soc Lond Ser B Biol Sci 370:
20140330. https://doi.org/10.1098/rstb.
2014.0330

16. Spinelli JB, Haigis MC (2018) The multifaceted contributions of mitochondria to cellular
metabolism. Nat Cell Biol 20:745–754.
https://doi.org/10.1038/s41556-018-
0124-1

17. Gordaliza-Alaguero I, Cantó C, Zorzano A
(2019) Metabolic implications of organelle–
mitochondria communication. EMBO Rep

20:e47928. https://doi.org/10.15252/
embr.201947928

18. Chan DC (2020) Mitochondrial dynamics
and its involvement in disease. Annu Rev
Pathol: Mech Dis 15:235–259. https://doi.
org/10.1146/annurev-pathmechdis-
012419-032711

19. Desai R, East DA, Hardy L et al (2020) Mitochondria form contact sites with the nucleus
to couple prosurvival retrograde response. Sci
Adv 6:eabc9955. https://doi.org/10.1126/
sciadv.abc9955

20. Giacomello M, Pyakurel A, Glytsou C, Scorrano L (2020) The cell biology of mitochondrial membrane dynamics. Nat Rev Mol Cell
Biol 21:204–224. https://doi.org/10.1038/
s41580-020-0210-7

21. Sebastián D, Palacín M, Zorzano A (2017)
Mitochondrial dynamics: coupling mitochondrial fitness with healthy aging. Trends Mol
Med 23:201–215. https://doi.org/10.
1016/j.molmed.2017.01.003

22. Jansen RP, de Boer K (1998) The bottleneck:
mitochondrial imperatives in oogenesis and
ovarian follicular fate. Mol Cell Endocrinol
145:81–88. https://doi.org/10.1016/
s0303-7207(98)00173-7

23. Cao L, Shitara H, Horii T et al (2007) The
mitochondrial bottleneck occurs without
reduction of mtDNA content in female
mouse germ cells. Nat Genet 39:386–390.
https://doi.org/10.1038/ng1970

24. Motta PM, Nottola SA, Makabe S, Heyn R
(2000) Mitochondrial morphology in human
fetal and adult female germ cells. Hum
Reprod 15(Suppl 2):129–147. https://doi.
org/10.1093/humrep/15.suppl_2.129

25. Wassarman PM, Josefowicz WJ (1978)
Oocyte development in the mouse: an ultra-
structural comparison of oocytes isolated at
various stages of growth and meiotic competence. J Morphol 156:209–235. https://doi.
org/10.1002/jmor.1051560206

26. Udagawa O, Ishihara T, Maeda M et al
(2014) Mitochondrial fission factor Drp1
maintains oocyte quality via dynamic rear-
rangement of multiple organelles. Curr Biol
24:2451–2458. https://doi.org/10.1016/j.
cub.2014.08.060

27. Wakai T, Harada Y, Miyado K, Kono T
(2014) Mitochondrial dynamics controlled
by mitofusins define organelle positioning
and movement during mouse oocyte maturation. Mol Hum Reprod 20:1090–1100.
https://doi.org/10.1093/molehr/gau064

28. Carvalho KF, Machado TS, Garcia BM et al
(2020) Mitofusin 1 is required for oocyte
growth and communication with follicular

somatic cells. FASEB J 34:7644–7660. https://doi.org/10.1096/fj.201901761R

29. Chen Z, Wang ZH, Zhang G et al (2020) Mitochondrial DNA segregation and replication restrict the transmission of detrimental mutation. J Cell Biol 219:e201905160. https://doi.org/10.1083/JCB.201905160

30. Johnson MT, Freeman EA, Gardner DK, Hunt PA (2007) Oxidative metabolism of pyruvate is required for meiotic maturation of murine oocytes in vivo. Biol Reprod 77:2–8. https://doi.org/10.1095/biolreprod.106.059899

31. Su Y-Q, Sugiura K, Wigglesworth K et al (2007) Oocyte regulation of metabolic cooperativity between mouse cumulus cells and oocytes: BMP15 and GDF9 control cholesterol biosynthesis in cumulus cells. Development 135:111–121. https://doi.org/10.1242/dev.009068

32. Su Y-Q, Sugiura K, Eppig JJ (2009) Mouse oocyte control of granulosa cell development and function: paracrine regulation of cumulus cell metabolism. Semin Reprod Med 27:32–42. https://doi.org/10.1055/s-0028-1108008

33. Son MJ, Kwon Y, Son M-Y et al (2015) Mitofusins deficiency elicits mitochondrial metabolic reprogramming to pluripotency. Cell Death Differ 22:1957–1969. https://doi.org/10.1038/cdd.2015.43

34. Prigione A, Rohwer N, Hoffmann S et al (2014) HIF1 alpha modulates cell fate reprogramming through early glycolytic shift and upregulation of PDK1-3 and PKM2. Stem Cells 32:364–376. https://doi.org/10.1002/stem.1552

35. Chen H, Chan DC (2017) Mitochondrial dynamics in regulating the unique phenotypes of cancer and stem cells. Cell Metab 26:39–48. https://doi.org/10.1016/j.cmet.2017.05.016

36. Chiaratti MR, Garcia BM, Carvalho KF et al (2018) The role of mitochondria in the female germline: implications to fertility and inheritance of mitochondrial diseases. Cell Biol Int 42:711. https://doi.org/10.1002/cbin.10947

37. Cho YM, Kwon S, Pak YK et al (2006) Dynamic changes in mitochondrial biogenesis and antioxidant enzymes during the spontaneous differentiation of human embryonic stem cells. Biochem Biophys Res Commun 348:1472–1478. https://doi.org/10.1016/j.bbrc.2006.08.020

38. Ma H, Folmes CDL, Wu J et al (2015) Metabolic rescue in pluripotent cells from patients with mtDNA disease. Nature 524:234–238. https://doi.org/10.1038/nature14546

39. Chiaratti MR, Meirelles FV (2010) Mitochondrial DNA copy number, a marker of viability for oocytes. Biol Reprod 83:1–2. https://doi.org/10.1095/biolreprod.110.084269

40. Wai T, Ao A, Zhang X et al (2010) The role of mitochondrial DNA copy number in mammalian fertility. Biol Reprod 83:52–62. https://doi.org/10.1095/biolreprod.109.080887

41. Stewart JB, Chinnery PF (2020) Extreme heterogeneity of human mitochondrial DNA from organelles to populations. Nat Rev Genet 22:106–118. https://doi.org/10.1038/s41576-020-00284-x

42. Lieber T, Jeedigunta SP, Palozzi JM et al (2019) Mitochondrial fragmentation drives selective removal of deleterious mtDNA in the germline. Nature 570:380–384. https://doi.org/10.1038/s41586-019-1213-4

43. Kowald A, Kirkwood TBL (2011) Evolution of the mitochondrial fusion-fission cycle and its role in aging. Proc Natl Acad Sci U S A 108:10237–10242. https://doi.org/10.1073/pnas.1101604108

44. Russell OM, Gorman GS, Lightowlers RN, Turnbull DM (2020) Mitochondrial diseases: hope for the future. Cell 181:168–188. https://doi.org/10.1016/j.cell.2020.02.051

45. Miller B, Kim SJ, Kumagai H et al (2020) Peptides derived from small mitochondrial open reading frames: genomic, biological, and therapeutic implications. Exp Cell Res 393:112056. https://doi.org/10.1016/j.yexcr.2020.112056

46. Pozzi A, Dowling DK, Sloan D (2019) The genomic origins of small mitochondrial RNAs: are they transcribed by the mitochondrial DNA or by mitochondrial pseudogenes within the nucleus (NUMTs)? Genome Biol Evol 11:1883–1896. https://doi.org/10.1093/gbe/evz132

47. Rath S, Sharma R, Gupta R et al (2020) Mito-Carta3.0: an updated mitochondrial proteome now with sub-organelle localization and pathway annotations. Nucleic Acids Res 49:D1541–D1547. https://doi.org/10.1093/nar/gkaa1011

48. Quirós PM, Mottis A, Auwerx J (2016) Mito-nuclear communication in homeostasis and stress. Nat Rev Mol Cell Biol 17:213–226. https://doi.org/10.1038/nrm.2016.23

49. Milenkovic D, Matic S, Kühl I et al (2013) Twinkle is an essential mitochondrial helicase required for synthesis of nascent D-loop strands and complete mtDNA replication. Hum Mol Genet 22:1983–1993. https://doi.org/10.1093/hmg/ddt051

50. Gustafsson CM, Falkenberg M, Larsson NG (2016) Maintenance and expression of mammalian mitochondrial DNA. Annu Rev Biochem 85:133–160. https://doi.org/10.1146/annurev-biochem-060815-014402

51. Larsson NG, Wang J, Wilhelmsson H et al (1998) Mitochondrial transcription factor A is necessary for mtDNA maintenance and embryogenesis in mice. Nat Genet 18:231–236. https://doi.org/10.1038/ng0398-231

52. Agaronyan K, Morozov YI, Anikin M, Temiakov D (2015) Replication-transcription switch in human mitochondria. Science 347:548–551. https://doi.org/10.1126/science.aaa0986

53. Ojala D, Montoya J, Attardi G (1981) TRNA punctuation model of RNA processing in human mitochondria. Nature 290:470–474. https://doi.org/10.1038/290470a0

54. Morozov YI, Agaronyan K, Cheung ACM et al (2014) A novel intermediate in transcription initiation by human mitochondrial RNA polymerase. Nucleic Acids Res 42:1365–1372. https://doi.org/10.1093/nar/gkt1356

55. Clayton DA (1982) Replication of animal mitochondrial DNA. Cell 28:693–705

56. Kang E, Wu J, Gutierrez NM et al (2016) Mitochondrial replacement in human oocytes carrying pathogenic mitochondrial DNA mutations. Nature 540:270–275. https://doi.org/10.1038/nature20592

57. Wolf DP, Hayama T, Mitalipov S (2017) Mitochondrial genome inheritance and replacement in the human germline. EMBO J 36:2177–2181. https://doi.org/10.15252/embj.201797606

58. Jemt E, Persson Ö, Shi Y et al (2015) Regulation of DNA replication at the end of the mitochondrial D-loop involves the helicase TWINKLE and a conserved sequence element. Nucleic Acids Res 43:9262–9275. https://doi.org/10.1093/nar/gkv804

59. Doda JN, Wright CT, Clayton DA (1981) Elongation of displacement-loop strands in human and mouse mitochondrial DNA is arrested near specific template sequences. Proc Natl Acad Sci U S A 78:6116–6120. https://doi.org/10.1073/pnas.78.10.6116

60. Brown TA, Clayton DA (2002) Release of replication termination controls mitochondrial DNA copy number after depletion with 2′,3′-dideoxycytidine. Nucleic Acids Res 30:2004–2010. https://doi.org/10.1093/nar/30.9.2004

61. Holt AG, Davies AM (2020) The significance of mitochondrial DNA half-life to the lifespan of post-mitotic cells. bioRxiv:1–27. https://doi.org/10.1101/2020.02.15.950410

62. Burgstaller JP, Johnston IG, Jones NS et al (2014) MtDNA segregation in heteroplasmic tissues is common in vivo and modulated by haplotype differences and developmental stage. Cell Rep 7:2031–2041. https://doi.org/10.1016/j.celrep.2014.05.020

63. Krasich R, Copeland WC (2017) DNA polymerases in the mitochondria: a critical review of the evidence. Front Biosci Landmark 22:692–709. https://doi.org/10.2741/4510

64. Whitehall JC, Greaves LC (2020) Aberrant mitochondrial function in ageing and cancer. Biogerontology 21:445–459. https://doi.org/10.1007/s10522-019-09853-y

65. Rishishwar L, Jordan IK (2017) Implications of human evolution and admixture for mitochondrial replacement therapy. BMC Genomics 18:140. https://doi.org/10.1186/s12864-017-3539-3

66. Wei W, Tuna S, Keogh MJ et al (2019) Germline selection shapes human mitochondrial DNA diversity. Science 364:eaau6520. https://doi.org/10.1126/science.aau6520

67. Miao YW, Peng MS, Wu GS et al (2013) Chicken domestication: an updated perspective based on mitochondrial genomes. Heredity 110:277–282. https://doi.org/10.1038/hdy.2012.83

68. Wallace DC, Chalkia D (2013) Mitochondrial DNA genetics and the heteroplasmy conundrum in evolution and disease. Cold Spring Harb Perspect Biol 5:a021220. https://doi.org/10.1101/cshperspect.a021220

69. Meadows JRS, Hiendleder S, Kijas JW (2011) Haplogroup relationships between domestic and wild sheep resolved using a mitogenome panel. Heredity 106:700–706. https://doi.org/10.1038/hdy.2010.122

70. Achilli A, Olivieri A, Soares P et al (2012) Mitochondrial genomes from modern horses reveal the major haplogroups that underwent domestication. Proc Natl Acad Sci U S A 109:2449–2454. https://doi.org/10.1073/pnas.1111637109

71. Achilli A, Bonfiglio S, Olivieri A et al (2009) The multifaceted origin of taurine cattle reflected by the mitochondrial genome. PLoS One 4. https://doi.org/10.1371/journal.pone.0005753

72. Wu GS, Yao YG, Qu KX et al (2007) Population phylogenomic analysis of mitochondrial DNA in wild boars and domestic pigs revealed multiple domestication events in East Asia. Genome Biol 8. https://doi.org/10.1186/gb-2007-8-11-r245

73. Hiendleder S, Schmutz SM, Erhardt G et al (1999) Transmitochondrial differences and varying levels of heteroplasmy in nuclear transfer cloned cattle. Mol Reprod Dev 54: 24–31. https://doi.org/10.1002/(SICI)1098-2795(199909)54:1<24::AID-MRD4>3.0.CO;2-S

74. Hiendleder S, Mainz K, Plante Y, Lewalski H (1998) Analysis of mitochondrial DNA indicates that domestic sheep are derived from two different ancestral maternal sources: no evidence for contributions from urial and argali sheep. J Hered 89:113–120. https://doi.org/10.1093/jhered/89.2.113

75. Naderi S, Rezaei HR, Taberlet P et al (2007) Large-scale mitochondrial DNA analysis of the domestic goat reveals six haplogroups with high diversity. PLoS One 2:e1012. https://doi.org/10.1371/journal.pone.0001012

76. Yonekawa H, Moriwaki K, Gotoh O et al (1982) Origins of laboratory mice deduced from restriction patterns of mitochondrial DNA. Differentiation 22:222–226. https://doi.org/10.1111/j.1432-0436.1982.tb01255.x

77. Lei L, Spradling AC (2016) Mouse oocytes differentiate through organelle enrichment from sister cyst germ cells. Science 352:95–99. https://doi.org/10.1126/science.aad2156

78. Chinnery PF, Samuels DC (1999) Relaxed replication of mtDNA: a model with implications for the expression of disease. Am J Hum Genet 64:1158–1165. https://doi.org/10.1086/302311

79. Diaz F, Bayona-Bafaluy MP, Rana M et al (2002) Human mitochondrial DNA with large deletions repopulates organelles faster than full-length genomes under relaxed copy number control. Nucleic Acids Res 30:4626–4633. https://doi.org/10.1093/nar/gkf602

80. Davis AF, Clayton DA (1996) In situ localization of mitochondrial DNA replication in intact mammalian cells. J Cell Biol 135:883–893. https://doi.org/10.1083/jcb.135.4.883

81. Johnston IG, Burgstaller JP, Havlicek V et al (2015) Stochastic modelling, Bayesian inference, and new in vivo measurements elucidate the debated mtDNA bottleneck mechanism. eLife 4:e07464. https://doi.org/10.7554/eLife.07464

82. Cree LM, Samuels DC, de Sousa Lopes SC et al (2008) A reduction of mitochondrial DNA molecules during embryogenesis explains the rapid segregation of genotypes.

Nat Genet 40:249–254. https://doi.org/10.1038/ng.2007.63

83. Wai T, Teoli D, Shoubridge EA (2008) The mitochondrial DNA genetic bottleneck results from replication of a subpopulation of genomes. Nat Genet 40:1484–1488. https://doi.org/10.1038/ng.258

84. Stewart JB, Freyer C, Elson JL et al (2008) Strong purifying selection in transmission of mammalian mitochondrial DNA. PLoS Biol 6:e10. https://doi.org/10.1371/journal.pbio.0060010

85. Hill JH, Chen Z, Xu H (2014) Selective propagation of functional mitochondrial DNA during oogenesis restricts the transmission of a deleterious mitochondrial variant. Nat Genet 46:389–392. https://doi.org/10.1038/ng.2920

86. Zhang Y, Wang ZH, Liu Y et al (2019) PINK1 inhibits local protein synthesis to limit transmission of deleterious mitochondrial DNA mutations. Mol Cell 73:1127–1137.e5. https://doi.org/10.1016/j.molcel.2019.01.013

87. Floros VI, Pyle A, Dietmann S et al (2018) Segregation of mitochondrial DNA heteroplasmy through a developmental genetic bottleneck in human embryos. Nat Cell Biol 20:144–151. https://doi.org/10.1038/s41556-017-0017-8

88. Fan W, Waymire KG, Narula N et al (2008) A mouse model of mitochondrial disease reveals germline selection against severe mtDNA mutations. Science 319:958–962. https://doi.org/10.1126/science.1147786

89. Zaidi AA, Wilton PR, Su MSW et al (2019) Bottleneck and selection in the germline and maternal age influence transmission of mitochondrial DNA in human pedigrees. Proc Natl Acad Sci U S A 116:25172–25178. https://doi.org/10.1073/pnas.1906331116

90. Olivo PD, Van de Walle MJ, Laipis PJ, Hauswirth WW (1983) Nucleotide sequence evidence for rapid genotypic shifts in the bovine mitochondrial DNA D-loop. Nature 306:400–402. https://doi.org/10.1038/306400a0

91. Cox RT, Spradling AC (2006) Milton controls the early acquisition of mitochondria by drosophila oocytes. Development 133:3371–3377. https://doi.org/10.1242/dev.02514

92. Cox RT (2003) A Balbiani body and the fusome mediate mitochondrial inheritance during Drosophila oogenesis. Development 130:1579–1590. https://doi.org/10.1242/dev.00365

93. Ashley MV, Laipis PJ, Hauswirth WW (1989) Rapid segregation of heteroplasmic bovine mitochondria. Nucleic Acids Res 17:7325–7331. https://doi.org/10.1093/nar/17.18.7325

94. Hauswirth WW, Laipis PJ (1982) Mitochondrial DNA polymorphism in a maternal lineage of Holstein cows. Proc Natl Acad Sci U S A 79:4686–4690. https://doi.org/10.1073/pnas.79.15.4686

95. Koehler CM, Lindberg GL, Brown DR et al (1991) Replacement of bovine mitochondrial DNA by a sequence variant within one generation. Genetics 129:247–255

96. Jenuth J, Peterson A, Fu K, Shoubridge E (1996) Random genetic drift in the female germline explains the rapid segregations of mammalian mitochondrial DNA. Nat Genet 14:146–151. https://doi.org/10.1038/ng1096-146

97. Rebolledo-Jaramillo B, Su MS-W, Stoler N et al (2014) Maternal age effect and severe germ-line bottleneck in the inheritance of human mitochondrial DNA. Proc Natl Acad Sci U S A 111:15474–15479. https://doi.org/10.1073/pnas.1409328111

98. Burgstaller JP, Kolbe T, Havlicek V et al (2018) Large-scale genetic analysis reveals mammalian mtDNA heteroplasmy dynamics and variance increase through lifetimes and generations. Nat Commun 9:2488. https://doi.org/10.1038/s41467-018-04797-2

99. Meirelles FV, Smith LC (1997) Mitochondrial genotype segregation in a mouse heteroplasmic lineage produced by embryonic karyoplast transplantation. Genetics 145:445–451

100. Lee H-S, Ma H, Juanes RC et al (2012) Rapid mitochondrial DNA segregation in primate preimplantation embryos precedes somatic and germline bottleneck. Cell Rep 1:506–515. https://doi.org/10.1016/j.celrep.2012.03.011

101. Hämäläinen RRH, Manninen T, Koivumäki H et al (2013) Tissue-and cell-type–specific manifestations of heteroplasmic mtDNA 3243A> G mutation in human induced pluripotent stem cell-derived disease model. Proc Natl Acad Sci U S A 110:E3622–E3630. https://doi.org/10.1073/pnas.1311660110

102. Fujikura J, Nakao K, Sone M et al (2012) Induced pluripotent stem cells generated from diabetic patients with mitochondrial DNA A3243G mutation. Diabetologia 55:1689–1698. https://doi.org/10.1007/s00125-012-2508-2

103. Rand DM (2008) Mitigating mutational meltdown in mammalian mitochondria. PLoS Biol 6:e35. https://doi.org/10.1371/journal.pbio.0060035

104. Freyer C, Cree LM, Mourier A et al (2012) Variation in germline mtDNA heteroplasmy is determined prenatally but modified during subsequent transmission. Nat Genet 44:1282–1285. https://doi.org/10.1038/ng.2427

105. Folmes CDL, Martinez-Fernandez A, Perales-Clemente E et al (2013) Disease-causing mitochondrial heteroplasmy segregated within induced pluripotent stem cell clones derived from a MELAS patient. Stem Cells 31:1298–1308. https://doi.org/10.1002/stem.1389

106. Latorre-Pellicer A, Lechuga-Vieco AV, Johnston IG et al (2019) Regulation of mother-to-offspring transmission of mtDNA heteroplasmy. Cell Metab 30:1120–1130.e5. https://doi.org/10.1016/j.cmet.2019.09.007

107. Ma H, Xu H, O'Farrell PH (2014) Transmission of mitochondrial mutations and action of purifying selection in Drosophila melanogaster. Nat Genet 46:393–397. https://doi.org/10.1038/ng.2919

108. Cherry ABC, Gagne KE, McLoughlin EM et al (2013) Induced pluripotent stem cells with a pathological mitochondrial DNA deletion. Stem Cells 31:1287–1297. https://doi.org/10.1002/stem.1354

109. Kauppila JHK, Baines HL, Bratic A et al (2016) A phenotype-driven approach to generate mouse models with pathogenic mtDNA mutations causing mitochondrial disease. Cell Rep 16:2980–2990. https://doi.org/10.1016/j.celrep.2016.08.037

110. Ma H, O'Farrell PH (2016) Selfish drive can trump function when animal mitochondrial genomes compete. Nat Genet 48:798–802. https://doi.org/10.1038/ng.3587

111. Gitschlag BL, Kirby CS, Samuels DC et al (2016) Homeostatic responses regulate selfish mitochondrial genome dynamics in C. elegans. Cell Metab 24:91–103. https://doi.org/10.1016/j.cmet.2016.06.008

112. Phillips WS, Coleman-Hulbert AL, Weiss ES et al (2015) Selfish mitochondrial DNA proliferates and diversifies in small, but not large, experimental populations of Caenorhabditis briggsae. Genome Biol Evol 7:2023–2037. https://doi.org/10.1093/gbe/evv116

113. Kang E, Wu G, Ma H et al (2014) Nuclear reprogramming by interphase cytoplasm of

two-cell mouse embryos. Nature 509:101–104. https://doi.org/10.1038/nature13134

114. Klucnika A, Ma H (2019) A battle for transmission: the cooperative and selfish animal mitochondrial genomes. Open Biol 9:180267. https://doi.org/10.1098/rsob.180267

115. Samuels DC, Li C, Li B et al (2013) Recurrent tissue-specific mtDNA mutations are common in humans. PLoS Genet 9:e1003929. https://doi.org/10.1371/journal.pgen.1003929

116. Zhang Y, Chen Y, Gucek M, Xu H (2016) The mitochondrial outer membrane protein MDI promotes local protein synthesis and mtDNA replication. EMBO J 35:1045–1057. https://doi.org/10.15252/embj.201592994

117. Pickles S, Vigié P, Youle RJ (2018) Mitophagy and quality control mechanisms in mitochondrial maintenance. Curr Biol 28:R170–R185. https://doi.org/10.1016/j.cub.2018.01.004

118. Lima A, Lubatti G, Burgstaller J et al (2020) Differences in mitochondrial activity trigger cell competition during early mouse development. bioRxiv. https://doi.org/10.1101/2020.01.15.900613

119. Lewis SC, Uchiyama LF, Nunnari J (2016) ER-mitochondria contacts couple mtDNA synthesis with mitochondrial division in human cells. Science 353:aaf5549. https://doi.org/10.1126/science.aaf5549

120. Prigione A, Fauler B, Lurz R et al (2010) The senescence-related mitochondrial/oxidative stress pathway is repressed in human induced pluripotent stem cells. Stem Cells 28:721–733. https://doi.org/10.1002/stem.404

121. Chung YG, Eum JH, Lee JE et al (2014) Human somatic cell nuclear transfer using adult cells. Cell Stem Cell 14:777–780. https://doi.org/10.1016/j.stem.2014.03.015

122. Tachibana M, Amato P, Sparman M et al (2013) Human embryonic stem cells derived by somatic cell nuclear transfer. Cell 153:1228–1238. https://doi.org/10.1016/j.cell.2013.05.006

123. Burgstaller JP, Johnston IG, Poulton J (2015) Mitochondrial DNA disease and developmental implications for reproductive strategies. Mol Hum Reprod 21:11–22. https://doi.org/10.1093/molehr/gau090

124. Steinborn R, Schinogl P, Wells DN et al (2002) Coexistence of Bos taurus and B. indicus mitochondrial DNAs in nuclear transfer-derived somatic cattle clones. Genetics 162:823–829

125. Steinborn R, Schinogl P, Zakhartchenko V et al (2000) Mitochondrial DNA heteroplasmy in cloned cattle produced by fetal and adult cell cloning. Nat Genet 25:255–257. https://doi.org/10.1038/77000

126. Meirelles FV, Bordignon V, Watanabe Y et al (2001) Complete replacement of the mitochondrial genotype in a Bos indicus calf reconstructed by nuclear transfer to a Bos taurus oocyte. Genetics 158:351–356

127. Evans MJ, Gurer C, Loike JD et al (1999) Mitochondrial DNA genotypes in nuclear transfer-derived cloned sheep. Nat Genet 23:90–93. https://doi.org/10.1038/12696

128. Liu Z, Cai Y, Wang Y et al (2018) Cloning of macaque monkeys by somatic cell nuclear transfer. Cell 172:881–887.e7. https://doi.org/10.1016/j.cell.2018.01.020

129. Hiendleder S, Zakhartchenko V, Wenigerkind H et al (2003) Heteroplasmy in bovine fetuses produced by intra- and inter-subspecific somatic cell nuclear transfer: neutral segregation of nuclear donor mitochondrial DNA in various tissues and evidence for recipient cow mitochondria in fetal blood. Biol Reprod 68:159–166. https://doi.org/10.1095/biolreprod.102.008201

130. Takeda K, Tasai M, Iwamoto M et al (2006) Transmission of mitochondrial DNA in pigs and progeny derived from nuclear transfer of Meishan pig fibroblast cells. Mol Reprod Dev 73:306–312. https://doi.org/10.1002/mrd.20403

131. Takeda K, Kaneyama K, Tasai M et al (2008) Characterization of a donor mitochondrial DNA transmission bottleneck in nuclear transfer derived cow lineages. Mol Reprod Dev 75:759–765. https://doi.org/10.1002/mrd.20837

132. Takeda K, Akagi S, Kaneyama K et al (2003) Proliferation of donor mitochondrial DNA in nuclear transfer calves (Bos taurus) derived from cumulus cells. Mol Reprod Dev 64:429–437. https://doi.org/10.1002/mrd.10279

133. Lechuga-Vieco AV, Latorre-Pellicer A, Johnston IG et al (2020) Cell identity and nucleo-mitochondrial genetic context modulate OXPHOS performance and determine somatic heteroplasmy dynamics. Sci Adv 6:eaba5345. https://doi.org/10.1126/sciadv.aba5345

134. Acton BM, Lai I, Shang X et al (2007) Neutral mitochondrial heteroplasmy alters physiological function in mice. Biol Reprod 77:

569–576. https://doi.org/10.1095/biolreprod.107.060806

135. Roubertoux PL, Sluyter F, Carlier M et al (2003) Mitochondrial DNA modifies cognition in interaction with the nuclear genome and age in mice. Nat Genet 35:65–69. https://doi.org/10.1038/ng1230

136. Moreno-Loshuertos R, Acín-Pérez R, Fernández-Silva R et al (2006) Differences in reactive oxygen species production explain the phenotypes associated with common mouse mitochondrial DNA variants. Nat Genet 38:1261–1268. https://doi.org/10.1038/ng1897

137. Ma H, Marti Gutierrez N, Morey R et al (2016) Incompatibility between nuclear and mitochondrial genomes contributes to an interspecies reproductive barrier. Cell Metab 24:283–294. https://doi.org/10.1016/j.cmet.2016.06.012

138. Steinborn R, Zakhartchenko V, Jelyazkov J et al (1998) Composition of parental mitochondrial DNA in cloned bovine embryos. FEBS Lett 426:352–356. https://doi.org/10.1016/s0014-5793(98)00350-0

139. Steinborn R, Zakhartchenko V, Wolf E et al (1998) Non-balanced mix of mitochondrial DNA in cloned cattle produced by cytoplast-blastomere fusion. FEBS Lett 426:357–361. https://doi.org/10.1016/s0014-5793(98)00351-2

140. Lloyd RE, Lee J-H, Alberio R et al (2006) Aberrant nucleo-cytoplasmic cross-talk results in donor cell mtDNA persistence in cloned embryos. Genetics 172:2515–2527. https://doi.org/10.1534/genetics.105.055145

141. Arbeithuber B, Hester J, Cremona MA et al (2020) Age-related accumulation of de novo mitochondrial mutations in mammalian oocytes and somatic tissues. PLoS Biol 18:e3000745. https://doi.org/10.1371/journal.pbio.3000745

142. del Mar González M, Ramos A, Aluja MP, Santos C (2020) Sensitivity of mitochondrial DNA heteroplasmy detection using Next Generation Sequencing. Mitochondrion 50:88–93. https://doi.org/10.1016/j.mito.2019.10.006

143. Srirattana K, St. John JC (2017) Manipulating the mitochondrial genome to enhance cattle embryo development. G3: Genes, Genomes, Genet 7:2065–2080. https://doi.org/10.1534/g3.117.042655

144. Lee JH, Peters A, Fisher P et al (2010) Generation of mtDNA homoplasmic cloned lambs. Cell Reprogram 12:347–355. https://doi.org/10.1089/cell.2009.0096

145. Min KK, Jang G, Hyun JO et al (2007) Endangered wolves cloned from adult somatic cells. Cloning Stem Cells 9:130–137. https://doi.org/10.1089/clo.2006.0034

146. Srirattana K, Imsoonthornruksa S, Laowtammathron C et al (2012) Full-term development of gaur-bovine interspecies somatic cell nuclear transfer embryos: effect of trichostatin A treatment. Cell Reprogram 14:248–257. https://doi.org/10.1089/cell.2011.0099

147. Vogel G (2001) Endangered species. Cloned gaur a short-lived success. Science 291:409. https://doi.org/10.1126/science.291.5503.409a

148. Gómez MC, Pope CE, Giraldo A et al (2004) Birth of African Wildcat cloned kittens born from domestic cats. Cloning Stem Cells 6:247–258. https://doi.org/10.1089/clo.2004.6.247

149. Loi P, Ptak G, Barboni B et al (2001) Genetic rescue of an endangered mammal by cross-species nuclear transfer using post-mortem somatic cells. Nat Biotechnol 19:962–964. https://doi.org/10.1038/nbt1001-962

150. Oh HJ, Kim MK, Jang G et al (2008) Cloning endangered gray wolves (Canis lupus) from somatic cells collected postmortem. Theriogenology 70:638–647. https://doi.org/10.1016/j.theriogenology.2008.04.032

151. Lagutina I, Fulka H, Lazzari G, Galli C (2013) Interspecies somatic cell nuclear transfer: advancements and problems. Cell Reprogram 15:374–384. https://doi.org/10.1089/cell.2013.0036

152. Bayona-Bafaluy MP, Müller S, Moraes CT (2005) Fast adaptive coevolution of nuclear and mitochondrial subunits of ATP synthetase in orangutan. Mol Biol Evol 22:716–724. https://doi.org/10.1093/molbev/msi059

153. Dey R, Barrientos A, Moraes CT (2000) Functional constraints of nuclear mitochondrial DNA interactions in xenomitochondrial rodent cell lines. J Biol Chem 275:31520–31527. https://doi.org/10.1074/jbc.M004053200

154. Yu G, Tian J, Yin J et al (2014) Incompatibility of nucleus and mitochondria causes xenomitochondrial cybrid unviable across human, mouse, and pig cells. Anim Biotechnol 25:139–149. https://doi.org/10.1080/10495398.2013.841709

155. Thongphakdee A, Kobayashi S, Imai K et al (2008) Interspecies nuclear transfer embryos reconstructed from cat somatic cells and bovine ooplasm. J Reprod Dev 54:142–147. https://doi.org/10.1262/jrd.19159

156. Kitiyanant Y, Saikhun J, Chaisalee B et al (2001) Somatic cell cloning in Buffalo (Bubalus bubalis): effects of interspecies cytoplasmic recipients and activation procedures. Cloning Stem Cells 3:97–104. https://doi.org/10.1089/153623001753205052

157. Chen D-Y, Wen D-C, Zhang Y-P et al (2002) Interspecies implantation and mitochondria fate of panda-rabbit cloned embryos. Biol Reprod 67:637–642. https://doi.org/10.1095/biolreprod67.2.637

158. Yang CX, Han ZM, Wen DC et al (2003) In vitro development and mitochondrial fate of macaca – rabbit cloned embryos. Mol Reprod Dev 65:396–401. https://doi.org/10.1002/mrd.10320

159. Yang CX, Kou ZH, Wang K et al (2004) Quantitative analysis of mitochondrial DNAs in macaque embryos reprogrammed by rabbit oocytes. Reproduction 127:201–205. https://doi.org/10.1530/rep.1.00088

160. Chen Y, He ZX, Liu A et al (2003) Embryonic stem cells generated by nuclear transfer of human somatic nuclei into rabbit oocytes. Cell Res 13:251–263. https://doi.org/10.1038/sj.cr.7290170

161. Hua S, Zhang Y, Li X-C et al (2007) Effects of granulosa cell mitochondria transfer on the early development of bovine embryos in vitro. Cloning Stem Cells 9:237–246. https://doi.org/10.1089/clo.2006.0020

162. Chang KH, Lim JM, Kang SK et al (2003) Blastocyst formation, karyotype, and mitochondrial DNA of interspecies embryos derived from nuclear transfer of human cord fibroblasts into enucleated bovine oocytes. Fertil Steril 80:1380–1387. https://doi.org/10.1016/j.fertnstert.2003.07.006

163. Mastromonaco GF, Favetta LA, Smith LC et al (2007) The influence of nuclear content on developmental competence of gaur × cattle hybrid in vitro fertilized and somatic cell nuclear transfer embryos. Biol Reprod 76:514–523. https://doi.org/10.1095/biolreprod.106.058040

164. Lanza RP, Cibelli JB, Diaz F et al (2000) Cloning of an endangered species (Bos gaurus) using interspecies nuclear transfer. Cloning 2:79–90. https://doi.org/10.1089/15204550436104

165. Sansinena MJ, Lynn J, Bondioli KR et al (2011) Ooplasm transfer and interspecies somatic cell nuclear transfer: heteroplasmy, pattern of mitochondrial migration and effect on embryo development. Zygote 19:147–156. https://doi.org/10.1017/S0967199410000419

166. Imsoonthornruksa S, Srirattana K, Phewsoi W et al (2012) Segregation of donor cell mitochondrial DNA in gaur-bovine interspecies somatic cell nuclear transfer embryos, fetuses and an offspring. Mitochondrion 12:506–513. https://doi.org/10.1016/j.mito.2012.07.108

167. Jiang Y, Chen T, Wang K et al (2006) Different fates of donor mitochondrial DNA in bovine-rabbit and cloned bovine-rabbit reconstructed embryos during preimplantation development. Front Biosci J Virtual Libr 11:1425–1432. https://doi.org/10.2741/1893

168. Poulton J, Steffann J, Burgstaller J, McFarland R (2019) 243rd ENMC international workshop: developing guidelines for management of reproductive options for families with maternally inherited mtDNA disease, Amsterdam, The Netherlands, 22–24 March 2019. Neuromuscul Disord 29:725–733. https://doi.org/10.1016/j.nmd.2019.08.004

169. Greenfield A, Braude P, Flinter F et al (2017) Assisted reproductive technologies to prevent human mitochondrial disease transmission. Nat Biotechnol 35:1059–1068. https://doi.org/10.1038/nbt.3997

170. Herbert M, Turnbull D (2018) Progress in mitochondrial replacement therapies. Nat Rev Mol Cell Biol 19:71–72. https://doi.org/10.1038/nrm.2018.3

171. Yamada M, Emmanuele V, Sanchez-Quintero MJ et al (2016) Genetic drift can compromise mitochondrial replacement by nuclear transfer in human oocytes. Cell Stem Cell 18:749–754. https://doi.org/10.1016/j.stem.2016.04.001

172. Hyslop LA, Blakeley P, Craven L et al (2016) Towards clinical application of pronuclear transfer to prevent mitochondrial DNA disease. Nature 534:383–386. https://doi.org/10.1038/nature18303

173. Tachibana M, Amato P, Sparman M et al (2013) Towards germline gene therapy of inherited mitochondrial diseases. Nature 493:627–631. https://doi.org/10.1038/nature11647

174. Ma H, O'Neil RC, Marti Gutierrez N et al (2017) Functional human oocytes generated by transfer of polar body genomes. Cell Stem Cell 20:112–119. https://doi.org/10.1016/j.stem.2016.10.001

175. Craven L, Tuppen HA, Greggains GD et al (2010) Pronuclear transfer in human embryos to prevent transmission of mitochondrial DNA disease. Nature 465:82–85. https://doi.org/10.1038/nature08958

176. Zhang J, Liu H, Luo S et al (2017) Live birth derived from oocyte spindle transfer to prevent mitochondrial disease. Reprod Biomed Online 34:361–368. https://doi.org/10.1016/j.rbmo.2017.01.013

177. Mitalipov S, Amato P, Parry S, Falk MJ (2014) Limitations of preimplantation genetic diagnosis for mitochondrial DNA diseases. Cell Rep 7:935–937. https://doi.org/10.1016/j.celrep.2014.05.004

178. Chinnery PF (2020) Mitochondrial replacement in the clinic. N Engl J Med 382:1855–1857. https://doi.org/10.1056/NEJMcibr2002015

179. Shock LS, Thakkar PV, Peterson EJ et al (2011) DNA methyltransferase 1, cytosine methylation, and cytosine hydroxymethylation in mammalian mitochondria. Proc Natl Acad Sci U S A 108:3630–3635. https://doi.org/10.1073/pnas.1012311108

180. Bellizzi D, D'aquila P, Scafone T et al (2013) The control region of mitochondrial DNA shows an unusual CpG and non-CpG methylation pattern. DNA Res 20:537–547. https://doi.org/10.1093/dnares/dst029

181. Dzitoyeva S, Chen H, Manev H (2012) Effect of aging on 5-hydroxymethylcytosine in brain mitochondria. Neurobiol Aging 33:2881–2891. https://doi.org/10.1016/j.neurobiolaging.2012.02.006

182. Wong M, Gertz B, Chestnut BA, Martin LJ (2013) Mitochondrial DNMT3A and DNA methylation in skeletal muscle and CNS of transgenic mouse models of ALS. Front Cell Neurosci 7:279. https://doi.org/10.3389/fncel.2013.00279

183. Feng S, Xiong L, Ji Z et al (2012) Correlation between increased ND2 expression and demethylated displacement loop of mtDNA in colorectal cancer. Mol Med Rep 6:125–130. https://doi.org/10.3892/mmr.2012.870

184. Gao J, Wen S, Zhou H, Feng S (2015) De-methylation of displacement loop of mitochondrial DNA is associated with increased mitochondrial copy number and nicotinamide adenine dinucleotide subunit 2 expression in colorectal cancer. Mol Med Rep 12:7033–7038. https://doi.org/10.3892/mmr.2015.4256

185. Devall M, Smith RG, Jeffries A et al (2017) Regional differences in mitochondrial DNA methylation in human post-mortem brain tissue. Clin Epigenetics 9:1–15. https://doi.org/10.1186/s13148-017-0337-3

186. Saini SK, Mangalhara KC, Prakasam G, Bamezai RNK (2017) DNA Methyltransferase1 (DNMT1) Isoform3 methylates mitochondrial genome and modulates its biology. Sci Rep 7:1525. https://doi.org/10.1038/s41598-017-01743-y

187. Van Der Wijst MGP, Van Tilburg AY, Ruiters MHJ, Rots MG (2017) Experimental mitochondria-targeted DNA methylation identifies GpC methylation, not CpG methylation, as potential regulator of mitochondrial gene expression. Sci Rep 7:177. https://doi.org/10.1038/s41598-017-00263-z

188. Meirelles F, Smith L (1998) Mitochondrial genotype segregation during preimplantation development in mouse heteroplasmic embryos. Genetics 148:877–883

189. Ferreira CR, Meirelles FV, Yamazaki W et al (2007) The kinetics of donor cell mtDNA in embryonic and somatic donor cell-derived bovine embryos. Cloning Stem Cells 9:618–629. https://doi.org/10.1089/clo.2006.0082

190. Takeda K (2019) Functional consequences of mitochondrial mismatch in reconstituted embryos and offspring. J Reprod Dev 65:485–489. https://doi.org/10.1262/jrd.2019-089

191. Pinkert CA, Irwin MH, Johnson LW, Moffatt RJ (1997) Mitochondria transfer into mouse ova by microinjection. Transgenic Res 6:379–383. https://doi.org/10.1023/A:1018431316831

192. Ingraham CA, Pinkert CA (2003) Developmental fate of mitochondria microinjected into murine zygotes. Mitochondrion 3:39–46. https://doi.org/10.1016/S1567-7249(03)00075-8

193. Gomes LC, Di Benedetto G, Scorrano L (2011) During autophagy mitochondria elongate, are spared from degradation and sustain cell viability. Nat Cell Biol 13:589–598. https://doi.org/10.1038/ncb2220

194. Takeda K, Tasai M, Akagi S et al (2010) Microinjection of serum-starved mitochondria derived from somatic cells affects parthenogenetic development of bovine and murine oocytes. Mitochondrion 10:137–142. https://doi.org/10.1016/j.mito.2009.12.144

195. Takeda K, Akagi S, Takahashi S et al (2002) Mitochondrial activity in response to serum starvation in bovine (Bos taurus) cell culture. Cloning Stem Cells 4:223–229. https://doi.org/10.1089/15362300260339502

Chapter 5

Stem Cells as Nuclear Donors for Mammalian Cloning

Carolina Gonzales da Silva and Carlos Frederico Martins

Abstract

Mammals are routinely cloned by introducing somatic nuclei into enucleated oocytes. Cloning contributes to propagating desired animals, to germplasm conservation efforts, among other applications. A challenge to more broader use of this technology is the relatively low cloning efficiency, which inversely correlates with donor cell differentiation status. Emerging evidence suggests that adult multipotent stem cells improve cloning efficiency, while the greater potential of embryonic stem cells for cloning remains restricted to the mouse. The derivation of pluripotent or totipotent stem cells from livestock and wild species and their association with modulators of epigenetic marks in donor cells should increase cloning efficiency.

Key words Multipotency, Pluripotency, Nuclear transplantation, Reprogramming

1 Introduction

Animal cloning by somatic cell nuclear transfer (SCNT) involves transferring a cell nucleus into an enucleated oocyte [1, 2]. The somatic nucleus will undergo nuclear reprogramming inside the oocyte cytoplasm, a process in which the somatic gene expression program shuts down followed by the reactivation of the embryonic genes. This transcriptional resetting establishes both embryonic and fetal developmental potential [3, 4]. The birth of the cloned sheep named Dolly was a scientific proof that cloning by SCNT reprograms terminally differentiated cells to a totipotent state [5].

Despite multiple applications of animal cloning, the overall efficiency of the technology (measured by the number of live born animals in relation to the transferred embryos) remains low in mammalian species [6, 7]. Cloning efficiency can vary from <1.0% to 20.0%, albeit commonly described below 5% in livestock [6, 8, 9]. Additionally, clones may display unusually high birthweights, developmental abnormalities, or low neonatal viability [6, 10].

Several lines of research attempted at improving cloning efficiency, such as choice of alternative donor cell types, improving

Marcelo Tigre Moura (ed.), *Somatic Cell Nuclear Transfer Technology*, Methods in Molecular Biology, vol. 2647,
https://doi.org/10.1007/978-1-0716-3064-8_5,
© The Author(s), under exclusive license to Springer Science+Business Media, LLC, part of Springer Nature 2023

oocyte quality, reducing the number of donor cell passages, and trying several cell cycle synchronization methods for preparing donor cells. Nonetheless, most of these attempts led to limited improvements in cloning efficiency. Moreover, the efficiency of SCNT cloning is based on the oocyte ability to reprogram the nucleus of the donor cell into an embryonic state within a few hours [11–13]. Therefore, the epigenetic status of donor cells might be one of the most important factors for the overall cloning efficiency [14, 15]. This means that epigenetic marks specific to the somatic cell type must be removed during nuclear reprogramming [3, 16]. If reprograming does not occur properly [17, 18], it leads to aberrant gene expression patterns in cloned embryos [16, 19], abnormalities during post-implantation development [20], and low cloning efficiency [21, 22].

Cells derived from embryos at early developmental stages lead to higher cloning efficiencies than using terminally differentiated somatic cells [10, 15, 23–26]. This fact suggests that donor cells in a less differentiated state (or perhaps in an undifferentiated state altogether) may be more suitable for cloning animals [27, 28]. Since most applications of animal cloning rely on using adult animals as cell donors, it motivated the search for less differentiated cells in adult organisms for SCNT.

Stem cells match the aforementioned role of less specialized cell types that may display greater reprogramming potential during animal cloning. Stem cells share two fundamental traits, namely, the ability to give rise to daughter stem cells (ability to self-renewal) and multi-lineage differentiation potential [29, 30]. Stem cells can be classified as embryonic stem cells (ESCs) (or pluripotent stem cells) and adult stem cells (or somatic stem cells—SSCs) [29]. ESCs are immortal (scape replicative crisis) and sustain pluripotency under ex vivo and in vivo conditions [31]. These ESCs derive from the inner cell mass of blastocysts and propagate under species-specific in vitro culture conditions [31]. Upon exposure to differentiation stimuli, ESCs can give rise to all cell types that form the developing fetus and adult organism both in vitro or in vivo [31]. The superior competence of mouse ESCs to produce high blastocysts rates in animal cloning has been widely proven [23, 24, 32]. Research on ESCs has been largely restricted due to the requirement of species-specific culture conditions, which remain known in the mouse, rats, some non-man primates, and humans [31, 33–35]. Their use in animal cloning is limited in other species because the definitive ESCs has not yet been established [36]. Livestock ESC-like cells fail to retain pluripotency traits during long-term culture and do not contribute to the germline upon their injection into blastocysts and embryo transfer. More recently, totipotent ESCs were found as rare cells in mouse pluripotent cultures (harboring a transcriptome similar to 2-cell stage embryos) but also after genetic perturbations in the genome of pluripotent cells [37, 38]. Putative cell lines with totipotent traits (e.g., extended pluripotent cells with both embryonic and extra-embryonic differentiation potential) have been described in live-stock [39], albeit their in vivo developmental potential remains poorly described.

This restricted availability of ESCs in most mammalian species motivated the investigation of alternative cell types for animal cloning. The development of direct reprogramming technology is an alternative approach for generation of pluripotent cells [40, 41]. Induced pluripotent cells (iPSCs) offer unlimited, ethi-cally acceptable, and tailored source of pluripotent cells from somatic cells [40, 42]. The application of iPSCs for animal cloning holds great potential, although these cells need adequate self-renewal conditions such as ESCs, thus limiting their potential in livestock. SSCs are multipotent cells found in several organs and contribute to their homeostasis. Similar to other stem cells, SSCs have the ability to self-renewal and differentiate into multiple cell types [43]. There are SSCs in the blood, skin, intestine, among other tissues and organs. SSCs are also found during fetal develop-ment, such as in the umbilical cord. Fetal SSCs may substitute adult SSCs for animal cloning, due to greater genomic stability and proliferative ability [44].

A cell type that has stood out in recent years for animal cloning is mesenchymal stem cells (MSCs), most notably those derived from adipose tissue or the bone marrow. MSCs are also defined as mesenchymal progenitor cells [45]. Although these cell types have been tested in many studies for therapeutic ability to repair or regenerate organs [46], their use in animal cloning is under more recent development [8, 10, 14, 15, 28, 30, 47–50]. Despite their applications, the origin of MSCs in the body remains elusive. However, these cells must meet a few criteria to be considered MSC [51, 52]. Firstly, MSCs must adapt to adherent in vitro culture, hold the expression of surface markers that are mesen-chyme-specific (e.g., CD29, CD44, CD73, CD90, CD105) and lack the expression of both hematopoietic (CD34) and lymphocyte (CD45) surface markers [52]. Another essential step for the char-acterization of MSCs is their differentiation into osteocytes, chon-drocytes, and adipocytes [51]. Alongside the progress in isolating and characterizing stem cells in several mammalian species, much progress was made toward using them as nuclear donors for animal cloning. Therefore, the aim of the chapter was to describe the application of stem cells for mammalian cloning and its potential impact on cloning efficiency.

2 Use of Stem Cells for Mammalian Cloning

2.1 Cattle (Bos taurus)

Several research groups described ESC-like cells that met several pluripotency criteria under in vitro assays [53–56]. For instance, Saito et al. [55] found ESC-like cells with a diploid karyotype under extensive culture and displayed several pluripotency markers, such as alkaline phosphatase activity, stage-specific embryonic antigen-1, and OCT4 expression. These ESC-like cells were differentiated in neural precursors upon directed differentiation and gave rise to cell types of the three germ layers under spontaneous differentiation. The use of ESC-like cells as nuclear donors for SCNT culminated in cloned calves more efficiently than using somatic cells [55].

Much ongoing efforts strive to generate iPSCs from cattle [57–59]. Moreover, germline transmission was not confirmed for such putative iPS cell lines, thus should be termed iPSC-like cells. Their use as nuclear donors for SCNT allows generating cloned blastocysts [57, 59]. It remains unknown if these iPSC-like cells may support full-term development after SCNT. The challenges posed by partially reprogrammed or iPSC-like cells is the inability to silence exogenous reprogramming factors and potential genomic instability [40, 60, 61].

Bone marrow MSCs are promising cells for bovine SCNT [62]. When these cells were used in SCNT with oocytes from fresh and cooled ovaries at 10 °C for 24 h, the proportion of cloned blastocysts decreased with oocytes from cooled ovaries (39% vs. 7%). After transfer of cloned blastocysts to recipient females, two recipients with SCNT embryos from cooled oocytes and one from fresh oocytes were pregnant at day 40 after embryo transfer. Curiously, one healthy cloned calf was obtained from cooled oocytes. These efficiencies were not impressive in the context of cattle cloning but the work demonstrated that cattle MSCs are amenable to nuclear reprogramming by SCNT. In another study, bone marrow MSCs and their osteocyte progenies were used for SCNT [63]. Cloned blastocyst rates were similar among bone marrow MSCs (63.7%), osteocytes (53.9%), and fibroblasts (52.4%).

MSCs from amniotic fluid, adipose tissue, and Wharton's jelly (umbilical cord) were characterized before use for SCNT [49, 50]. These cells were positive for surface markers enriched in MSCs (CD29+, CD73+, CD90+, and CD105+) and negative for hematopoietic (CD34) and lymphocyte antigen (CD45) markers analyzed by flow cytometry, immunocytochemistry, and RT-PCR. These MSCs differentiated into osteocytes, chondrocytes, and adipocytes [49, 50]. To test their potential for animal cloning, these cell lines were compared to skin fibroblasts upon SCNT [50]. Cleavage rate was higher using skin fibroblasts (70.64%) than MSCs (38.26–39.08%), albeit blastocyst rates were similar among groups

(17.27–19.00%). Finally, MSCs gave rise to two cloned calves [50]. The cloned calf from MSCs from amniotic fluid was born at 277 days of gestation by cesarean section, because the recipient cow had hydropsy. This cloned calf weighed 58.5 kg and died shortly after birth. The cloned calf from MSCs derived from adipose tissue was born after 291 days of gestation without veterinarian assistance. This later cloned calf was born healthy and weighed 35 kg. Another work tested MSCs from the Wharton jelly for SCNT [49]. Blastocyst rates after SCNT with MSCs from the Wharton's jelly (25.80%) were similar to SCNT using skin fibroblasts (19.00%). Furthermore, cloned blastocysts from such MSCs established early pregnancies after transfer to recipient females [49].

MSCs from amniotic fluid have less genome-wide DNA methylation and more hydroxymethylation than skin fibroblasts [64]. When reconstructed oocytes with MSCs from amniotic fluid were treated with a histone deacetylase inhibitor named trichostatin A (TSA), there was a decrease in the expression levels of histone deacetylase 3 and increase in lysine acetyltransferase 2A, which are epigenetic regulators that will likely facilitate nuclear reprogramming. These MSCs treated with TSA showed significant improvement in the production of cloned blastocysts [64].

2.2 Buffalo (*Bubalus bubalis*)

There are reports of attempting buffalo cloning using stem cells, thus including ESC-like cells [65] and MSCs from the amniotic membrane [28] and the amniotic fluid [14].

George et al. [65] derived buffalo ESC-like cells from in vitro *fertilization* (IVF) and SCNT blastocysts. These cell lines displayed pluripotency markers and gave rise to cell type derivative of the three germ layers using differentiation protocols after the formation of embryoid bodies. Upon oocyte reconstruction, ES-like cell lines (both IVF and SCNT derived) led to greater cleavage rates but similar blastocyst rates than fibroblast controls [65]. In addition, total cell number was greater in cloned blastocysts from ESC-like cells (mean number of 180.7 cells for IVF and 174.0 for SCNT) in contrast to fibroblast-derived blastocysts (average of 157.0 cells). A single cloned calf was born from ESC-like cells derived from IVF blastocysts [65], albeit the number of transferred blastocysts to recipients was small (12–16 per group).

The blastocyst rate after SCNT using amniotic membrane MSCs (28.9%) was similar to their IVF controls (30.6%) and higher than fibroblast-SCNT controls (19.5%) [28]. Gene expression analysis and total cell number in blastocysts reinforced that cloned embryos from MSC more closely resemble IVF embryos than fibroblasts derived counterparts. For instance, the relative abundance of epigenetic regulators (DNMT3A and HDAC2), BLC2 and GLUT1 mirrored IVF embryos, particularly at the blastocyst stage [28]. According to Yang and Rajamahendran [66], embryos of greater viability display higher levels of BCL2 in contrast to BAX.

The reduced expression of the glucose transporter triggers apoptosis in murine blastocysts [67]. Therefore, it is possible that the low level of GLUT1 expression observed in fibroblast-derived cloned embryos might lead to higher incidence of apoptosis. These facts reinforce the idea that both cellular and molecular analyses contribute to assessing the development potential of cloned embryos.

The disadvantage of using amniotic membrane derived MSCs is the requirement for an invasive collection that opens the uterus to expose the amniotic membrane. This approach may compromise the reproductive potential of the donor female. To circumvent this issue, cell retrieval may occur from the amniotic fluid and display similar reprogramming potential [14].

2.3 Swine (Sus scrofa)

Cloned piglets were obtained from iPSC-like cells [68]. Authors tested several iPSC-like cell lines generated using various reprogramming strategies. Two iPSC-like cell lines derived from ear skin fibroblasts and bone marrow MSCs using lentiviral vectors carrying doxycycline (DOX) inducible human pluripotency-associated transcription factors as reprogramming factors [69]. Four additional iPSC-like cell lines derived from fetal fibroblasts and MSCs using retroviral vectors carrying mouse, porcine, or human reprogramming factors [70]. Despite the transfer of >30,000 reconstructed oocytes with iPSC-like cells, it did not lead to live births [68]. Authors noticed that iPSC-like cells did not silence reprogramming factors, in contrast to fully reprogrammed mouse and human iPSCs [41, 71]. Therefore, authors decided to randomly differentiate iPSC-like cells to induce silencing of reprogramming factors, which formed epithelial-like cells that were used for SCNT [68]. Differentiated iPSC-like cells improved cloned blastocyst rates and led to one live birth and one stillbirth. Reconstructed oocytes with iPSC-like cells and treated with the histone deacetylase inhibitor Scriptaid also increased blastocyst rates and culminated in cloned piglets.

One report compared iPSC-like cells grown under different conditions and multipotent embryonic germ cells for SCNT cloning [72]. These cell lines were also compared to two fibroblast cell lines that gave rise to iPSC-like cells. Blastocyst rates for iPSC-like cells grown with GSK3β and MEK inhibitors (2i) and leukemia inhibitory factor (LIF) were 14.7%, iPSC-like cells grown in 2i with fibroblast grown factor were 10.1%, and embryonic germ cells were 34.5%, while fibroblast cell lines controls were intermediate (25.2–36.7%). Cloned embryos derived from fibroblasts and embryonic germ cells produced live offspring at similar efficiencies (3.2% and 4.0%, respectively) but some displayed malformations [72]. These results suggest that iPSC-like cells do not improve cloning efficiency as donor cells in comparison with multipotent or lineage committed cells.

Stem Cells and Animal Cloning   111

MSCs derived from adipose tissue and peripheral blood MSCs were used for pig cloning [15]. Reconstructed oocytes with adipose MSCs led to higher blastocyst rates on day 5 and 6 post-activation than those reconstructed with blood MSCs and fibroblasts (29.6% and 41.1% for adipose MSCs, 23.9% and 35.5% for blood MSCs, and 22.1% and 33.3% for fibroblasts). More importantly, live birth rate was higher using blood MSCs among the three groups.

Other groups demonstrated that bone marrow MSCs were better donor cells than fetal fibroblasts, due to their enhanced production of cloned embryos [73]. In this study, blastocyst rates did not differ between IVF (20.6%) and SCNT embryos from MSCs (18.4%) but were higher than SCNT using fetal fibroblasts (9.5 ± 2.1%). Another study showed that bone marrow MSCs used for SCNT enhanced both cleavage and blastocyst rates [74]. Further, Lee et al. [75] compared the reprogramming potential of fetal fibroblasts, bone marrow MSCs and MSCs differentiated progeny in miniature pigs. The blastocyst rate after SCNT with bone marrow MSCs was significantly higher than using osteocytes, adipocytes, chondrocytes, and fetal fibroblasts (47.7%, 34.5%, 31.1%, 36.8%, and 14.5%, respectively). From 523 two-cell SCNT embryos derived from bone marrow MSCs transferred surgically to five synchronized recipient sows, four viable cloned miniature pig offspring were born alive [75]. These evidences suggest that bone marrow MSCs have greater potential as donor cells for pig cloning than fibroblasts.

Li et al. [76] demonstrated that porcine neural stem cells remain multipotent during extended in vitro culture and useful for SCNT cloning, thus generating healthy cloned offspring. A total of 2020 two-cell SCNT embryos transferred to ten recipient females led to six pregnancies. Forty cloned piglets (18 males and 22 females) were born, and twenty-two clones reached sexual maturity and proven fertile.

2.4 Horses (Equus caballus)

The first report on using stem cells for equine cloning relied on MSCs [8]. Initially, it compared different donor cells (iPSC-like cells vs. adult fibroblasts) and fibroblasts fused to enucleated oocytes injected with vectors harboring reprogramming factors. In contrast to fibroblasts, iPSC-like cells did not lead to cloned blastocysts and oocyte injection with pluripotency-associated genes did not improve blastocyst production, pregnancy rates, nor lead to full-term development. Moreover, authors compared umbilical cord-derived MSCs, fetal fibroblasts from a cloned fetus, and adult fibroblasts for SCNT. Higher blastocyst rates were obtained using umbilical cord MSCs (15.6%) in comparison with both fetal and adult fibroblasts (8.9% and 9.3%, respectively). Despite similar pregnancy rates among groups, viable foals were from fetal fibroblasts only.

The follow-up work described the first cloned foal from bone marrow MSCs [77]. More importantly, it demonstrated that bone marrow MSCs generate more viable cloned foals than fibroblasts. This advantage was evident for cloned blastocyst rates (54.4% vs. 32.6%) and remained after embryo transfer to recipients. Foal viability after delivery was similar between clones from MSCs (95.2% of deliveries) and foals obtained by artificial insemination (AI) (98.4% of deliveries), while much higher than cloned foals generated from fibroblasts (52.9%). This AI-based contemporary group allowed assessing the neonatal viability of cloned foals in detail [77]. During hospitalization, cloned foals from fibroblasts needed more veterinary care than those from bone marrow MSCs (21.2 ± 5.4 versus 6.3 ± 4.0 days of care, respectively). Cloned foals from fibroblasts had high creatinine levels [77], which suggested dysfunctional excretory ability of their placenta.

This later report by Olivera et al. [77] represents of the most detailed research addressing clinical aspects of cloned pregnancies and neonatal survival of cloned horses, which included AI foals as more physiological controls. It also represents a milestone for future work with SCNT since post-implantation analyses are paramount for determining cloning efficiency using stem cells and identifying potential pitfalls for further improvements.

2.5 Goats (Capra hircus) and Sheep (Ovis aries)

Similar to other livestock mentioned in this chapter, the use of stem cells for cloning small ruminants begin after 2010 [10, 78, 79]. For instance, Kwong et al. [78] produced more cloned blastocysts using bone marrow MSCs (25.3%) than using fibroblasts (20.6%), which was the first report on using MSCs for goat cloning.

Another study reported the birth of cloned goats from transgenic satellite cells derived from skeletal muscle [10]. On the other hand, adipose tissue-derived MSCs for SCNT did not lead to cloned offspring. Despite similar blastocyst rates, SCNT embryos differed for histone H4K5 acetylation, thus suggesting differences in SCNT-mediated reprogramming between stem cell types [10]. Moreover, both stem cell types underwent in vitro culture for more than 60 passages without sings of replicative crisis, whereas fetal fibroblast cells displayed abnormal karyotypes and marked signs of aging after 15 passages [10].

The derivation of bone marrow MSCs from sheep allows establishing stable cell lines for genetic modification (e.g., transfection with DNA vector the carrying green fluorescent protein—GFP). These cell lines support the development of cloned GFP+ blastocysts [80]. The use of fetal MSCs from the bone marrow was less promising for sheep SCNT [79]. Although cleavage rates were similar to SCNT with fibroblast donor cells (75.7% for fibroblasts and 79.8% for fetal MSCs), blastocyst rates were low with both donor cell types (1.73% and 1.2% for fibroblasts and fetal MSCs, respectively). Authors suggest that technical factors were responsible for the low blastocyst rate after SCNT. Therefore, these initial results do not reflect the potential of fetal MSCs for sheep cloning.

2.6 Dogs (Canis lupus familiaris)

The use of MSCs from the adipose tissue of a transgenic beagle dog produced two healthy dogs [47]. This study presents data regarding the post-implantation development of cloned embryos from stem cells, which are extremely important because they reflect in vivo quality of cloned embryo obtained in the laboratory. Although the pregnancy rate was 20% (one of five recipient females), the overall efficiency rate was 2.4% (2 puppies from 82 embryos transferred to recipients). One puppy died soon after delivery, albeit it did not show visible abnormalities.

One putative advantage of MSCs for SCNT is their endogenous expression of pluripotency-associated genes (e.g., OCT4, SOX2) in contrast to other somatic cell types such as fibroblasts [81]. Further, cell culture conditions may favor the modulation of gene expression in MSCs, albeit it did not improve dog cloning using interspecies SCNT with pig oocytes [82].

2.7 Red Deer (Cervus elaphus)

The first report of red deer cloning used antler stem cells [83]. Antler stem cells form the male antlerogenic periosteum during horn growth, which represents a unique source of cells. These cells were chosen for SCNT due to their proliferative ability and less differentiated state (i.e., multi-lineage differentiation potential). In addition to testing multipotent stem cells for SCNT, authors differentiated antler stem cells into osteogenic and adipose cell fates to compare their reprogramming potential [83]. However, there was no difference in blastocyst and birth rates between stem cells and their differentiated progeny. The overall cloning efficiency considering all cell types was 13% (11/84) at birth and 10% (8/84) at weaning. All deer recipients developed functional mammary glands and initiated labor signaling without hormonal induction [83], which is standard practice in some livestock species.

2.8 Mouse (Mus musculus)

The first report on using stem cells was for mouse cloning [23]. In this study, reconstructed oocytes with late passage ESCs produced fewer blastocysts than those reconstructed with cumulus cells. Remarkably, cloning efficiency with ESCs (2.4%) was twice than using cumulus cells (1.2%), indicating the superiority of ESCs for full-term development of clones [23]. Other research groups applied ESCs for mouse cloning [84–87]. Cloned blastocysts may yield novel ESC lines for re-cloning (second round of nuclear transfer) as an alternative route [88]. Cloning using ESCs became attractive for propagating aged and infertile animals [88, 89] or deceased animals [90].

The first use of bone marrow MSCs in mouse cloning resulted in low preimplantation development [91], in contrast to the high blastocyst rates already described in this chapter for other species. Karyotype analysis by G-banding revealed frequent aneuploidy in the MSCs, which probably impaired development potential after SCNT [91]. In agreement with these findings, MSCs derived from adipose tissue also exhibited chromosomal instability at passage 13 [92]. Adipose MSCs showed morula/blastocyst production rate (37.8%) similar to cumulus cells (40.8%). MSCs from adipose tissue are an alternative as donor cell type for SCNT due to their easy collection but require karyotyping before use for SCNT in the mouse. Mouse neural stem cells did not show promising results, since cleavage rates were similar to reconstruct oocytes with cumulus cells or Sertoli cells [93], albeit cloned blastocysts rates were much lower (7.1%) than other cell types (50.2–54.0%). Full-term development was lower for cloning with neural stem cells (0.5%) in comparison with cumulus cells (2.7%) and Sertoli cells (2.2%) [93]. Most cloned embryos from neural stem cells (51%) develop beyond the 2-cell to 4-cell stage transition, which suggests better embryonic genome activation than fibroblast-derived SCNT embryos arrested mostly at the 2-cell stage [91].

Molecules that modulate epigenetic mechanisms have been used in association with stem cells to increase cloning efficiency and diminish neonatal losses. Reconstructed oocytes treated with TSA after chemical activation resulted in a 2- to 5-fold increase in blastocyst rates, despite using different somatic cell types (e.g., fibroblasts, neural stem cells) [21]. It also enhanced full-term development using cumulus cells by >5-fold [21]. Therefore, treating reconstructed oocytes with TSA can dramatically improve cloning efficiency by facilitating nuclear reprogramming in mice but also for cattle and swine, as previously described in this chapter.

3 Perspectives on Using Stem Cells for Mammalian Cloning

Stem cells improve cloning efficiency by increasing embryo yields and their quality that ultimately reflect in greater numbers of clones born alive and healthy. Despite these proof-of-principle studies, it is advisable to improve the understanding of nuclear reprogramming under this context, thus including better characterization of donor stem cells and resulting SCNT embryos. Analyses based on gene expression and epigenetic marks should reveal details of why stem cells are more amenable to nuclear reprogramming and what are the roadblocks to even greater cloning efficiencies. Another promising strategy is applying chromatin-modifying compounds to stem cells before oocyte reconstruction or during early development of SCNT embryos to facilitate nuclear reprogramming. This association will likely enhance nuclear reprogramming of stem cells to a totipotent state.

Implantation and post-implantation development are critical developmental time points for attaining cloned full-term development. Nonetheless, few studies monitored cloned pregnancies obtained from stem cells. More studies such as the one carried out by Olivera et al. [77], which monitored many cloned pregnancies in detail and also neonatal conditions, will provide clues for why stem cells improve post-implantation development. Under such circumstances, researchers will better assess and predict the viability of clones during pregnancy and after delivery.

References

1. Wolf E, Zakhartchenko V, Brem G (1998) Nuclear transfer in mammals: recent developments and future perspectives. J Biotechnol 65: 99–110

2. Campbell KHS (1999) Nuclear transfer in farm animal species. Semin Cell Dev Biol 10:245–252

3. Enright BP, Kubota C, Yang X, Tian XC (2003) Epigenetic characteristics and development of embryos cloned from donor cells treated by Trichostatin A or 5-aza-2′-deoxycytidine. Biol Reprod 69:896–901

4. Niemann H, Lucas-Hahn A (2012) Somatic cell nuclear transfer cloning: practical applications and current legislation. Reprod Domest Anim 47:2–10

5. Wilmut I, Schnieke AE, McWhir J, Kind AJ, Campbell KH (1997) Viable offspring derived from fetal and adult mammalian cells. Nature 385:810–813

6. Wilmut I, Beaujean N, de Sousa PA, Dinnyes A, King TJ, Paterson LA et al (2002) Somatic cell nuclear transfer. Nature 419:583–586

7. Thuan NV, Kishigami S, Wakayama T (2010) How to improve the success rate of mouse cloning technology. J Reprod Dev 56:20–30

8. Olivera R, Moro LN, Jordan R, Luzani C, Miriuka S, Radrizzani M et al (2016) In vitro and in vivo development of horse cloned embryos generated with iPSCs, mesenchymal stromal cells and fetal or adult fibroblasts as nuclear donors. PLoS One 11:1–14

9. Gugjoo MB, Amarpal Fazili MR, Shah RA, Sharma GT (2019) Mesenchymal stem cell: basic research and potential applications in cattle and buffalo. J Cell Physiol 234:8618–8635

10. Ren Y, Wu H, Ma Y, Yuan J, Liang H, Liu D (2014) Potential of adipose-derived mesenchymal stem cells and skeletal muscle-derived satellite cells for somatic cell nuclear transfer mediated transgenesis in Arbas Cashmere goats. PLoS One 9:1–12

11. Aston KI, Li GP, Hicks BA, Sessions BR, Pate BJ, Hammon D et al (2006) Effect of the time interval between fusion and activation on nuclear state and development in vitro and in vivo of bovine somatic cell nuclear transfer embryos. Reproduction 131:45–51

12. Beebe LFS, McIlfatrick SJ, Nottle MB (2009) Cytochalasin B and Trichostatin A treatment postactivation improves in vitro development of porcine somatic cell nuclear transfer embryos. Cloning Stem Cells 11:477–482

13. Niemann H (2016) Epigenetic reprogramming in mammalian species after SCNT-based cloning. Theriogenology 86:80–90

14. Sadeesh EM, Kataria M, Shah F, Yadav PS (2016) A comparative study on efficiency of adult fibroblasts and amniotic fluid-derived stem cells as donor cells for production of hand-made cloned buffalo (Bubalus bubalis) embryos. Cytotechnology 68:593–608

15. Yang Z, Vajta G, Xu Y, Luan J, Lin M, Liu C et al (2016) Production of pigs by hand-made cloning using mesenchymal stem cells and fibroblasts. Cell Reprogram 18:256–263

16. Cervera RP, Martin-Gutierrez N, Escorihuela E, Moreno R, Stojkovic M (2009) Trichostatin A affects histone acetylation and gene expression in porcine somatic cell nucleus transfer embryos. Theriogenology 72: 1097–1110

17. Wang YS, Xiong XR, An ZX, Wang LJ, Liu J, Quan FS et al (2011) Production of cloned calves by combination treatment of both donor cells and early cloned embryos with 5-aza-2′-deoxycytidine and trichostatin A. Theriogenology 75:819–825

18. Whitworth KM, Mao J, Lee K, Spollen WG, Samuel MS, Walters EM et al (2015) Transcriptome analysis of pig in vivo, in vitro–fertilized, and nuclear transfer blastocyst-stage embryos treated with histone deacetylase

inhibitors postfusion and activation reveals changes in the lysosomal pathway. Cell Reprogram 17:243–258

19. Blelloch R, Wang A, Meissner A, Pollard S, Smith A, Jaenisch R (2006) Reprogramming efficiency following somatic cell nuclear transfer is influenced by the differentiation and methylation state of the donor nucleus. Stem Cells 24:2007–2013

20. Li X, Li Z, Jouneau A, Zhou Q, Renard JP (2003) Nuclear transfer: progress and quandaries. Reprod Biol Endocrinol 1:84

21. Kishigami S, Mizutani E, Ohta H, Hikichi T, Thuan NV, Wakayama S et al (2006) Significant improvement of mouse cloning technique by treatment with trichostatin A after somatic nuclear transfer. Biochem Biophys Res Commun 340:183–189

22. Wee G, Shim J-J, Koo D-B, Chae J-I, Lee K-K, Han Y-M (2007) Epigenetic alteration of the donor cells does not recapitulate the reprogramming of DNA methylation in cloned embryos. Reproduction 134:781–787

23. Wakayama T, Rodriguez I, Perry ACF, Yanagimachi R, Mombaerts P (1999) Mice cloned from embryonic stem cells. Proc Natl Acad Sci U S A 96:14984–14989

24. Hochedlinger K, Jaenisch R (2002) Nuclear transplantation: lessons from frogs and mice. Curr Opin Cell Biol 14:741–748

25. Inoue K, Wakao H, Ogonuki N, Miki H, Seino K, Nambu-Wakao R et al (2005) Generation of cloned mice by direct nuclear transfer from natural killer T cells. Curr Biol 15:1114–1118

26. Makino H, Yamazaki Y, Hirabayashi T, Kaneko R, Hamada S, Kawamura Y et al (2005) Mouse embryos and chimera cloned from neural cells in the post natal cerebral cortex. Cloning Stem Cells 7:45–61

27. Wells DN, Oback B, Laible G (2003) Cloning livestock: a return to embryonic cells. Trends Biotechnol 21:428–432

28. Sadeesh EM, Shah F, Yadav PS (2016) Differential developmental competence and gene expression patterns in buffalo (*Bubalus bubalis*) nuclear transfer embryos reconstructed with fetal fibroblasts and amnion mesenchymal stem cells. Cytotechnology 68:1827–1848

29. Hemmat S, Lieberman DM, Most SP (2010) An introduction to stem cell biology. Facial Plast Surg 26:343–349

30. Gugjoo MB, Amarpal I, Chandra V, Wani MY, Dhama K, Sharma GT (2018) Mesenchymal stem cell research in veterinary medicine. Curr Stem Cell Res Ther 13:645–657

31. Smith AG (2001) Embryo-derived stem cells: of mice and men. Annu Rev Cell Dev Biol 17:435–462

32. Oback B (2009) Cloning from stem cells: different lineages, different species, same story. Reprod Fert Dev 21:83–94

33. Evans MJ, Kaufman MH (1981) Establishment in culture of pluripotential cells from mouse embryos. Nature 292:154–156

34. Thomson JA, Itskovitz-Eldor J, Shapiro SS, Waknitz MA, Swiergiel JJ, Marshall VS et al (1998) Embryonic stem cell lines derived from human blastocysts. Science 282:1145–1147

35. Thomson JA, Kalishman J, Golos TG, Durning M, Harris CP, Becker RA et al (1995) Isolation of a primate embryonic stem cell line. Proc Natl Acad Sci U S A 92:7844–7848

36. Ezashi T, Yuan Y, Roberts RM (2016) Pluripotent stem cells from domesticated mammals. Annu Rev Anim Biosci 4:223–253

37. Macfarlan TS, Gifford WD, Driscoll S, Lettieri K, Rowe HM, Bonanomi D et al (2012) Embryonic stem cell potency fluctuates with endogenous retrovirus activity. Nature 487:57–63

38. Wang Y, Na Q, Li X, Tee WW, Wu B, Bao S (2021) Retinoic acid induces NELFA-mediated 2C-like state of mouse embryonic stem cells associates with epigenetic modifications and metabolic processes in chemically defined media. Cell Prolif 54:e13049

39. Xiang J, Wang H, Zhang Y, Wang J, Liu F, Han X et al (2021) LCDM medium supports the derivation of bovine extended pluripotent stem cells with embryonic and extraembryonic potency in bovine-mouse chimeras from iPSCs and bovine fetal fibroblasts. FEBS J 288:4394–4411

40. Takahashi K, Yamanaka S (2006) Induction of pluripotent stem cells from mouse embryonic and adult fibroblast cultures by defined factors. Cell 126:663–676

41. Takahashi K, Tanabe K, Ohnuki M, Narita M, Ichisaka T, Tomoda K et al (2007) Induction of pluripotent stem cells from adult human fibroblasts by defined factors. Cell 131:861–872

42. Ogorevc J, Orehek S, Dovc P (2016) Cellular reprogramming in farm animals: an overview of iPSC generation in the mammalian farm animal species. J Anim Sci Biotechnol 7:1–10

43. He N, Dong Z, Zhu B, Nuo M, Bou S, Liu D (2016) Expression of pluripotency markers in Arbas cashmere goat hair follicle stem cells. In Vitro Cell Dev Biol Anim 52:782–788

44. Miyoshi K, Rzucidlo SJ, Pratt SL, Stice SL (2003) Improvements in cloning efficiencies may be possible by increasing uniformity in recipient oocytes and donor cells. Biol Reprod 68:1079–1086

45. Karahuseyinoglu S, Cinar O, Kilic E, Kara F, Akay GG, Demiralp DO et al (2007) Biology of stem cells in human umbilical cord stroma: in situ and in vitro surveys. Stem Cells 25:319–331

46. Ghorbani A, Feizpour A, Hashemzahi M, Gholami L, Hosseini M, Soukhtanloo M et al (2014) The effect of adipose derived stromal cells on oxidative stress level, lung emphysema and white blood cells of guinea pigs model of chronic obstructive pulmonary disease. DARU J Pharm Sci 22:1–12

47. Oh HJ, Park JE, Kim MJ, Hong SG, Ra JC, Jo JY et al (2011) Recloned dogs derived from adipose stem cells of a transgenic cloned beagle. Theriogenology 75:1221–1231

48. Nazari H, Shirazi A, Shams-Esfandabadi N, Afzali A, Ahmadi E (2016) The effect of amniotic membrane stem cells as donor nucleus on gene expression in reconstructed bovine oocytes. Int J Dev Biol 60:95–102

49. Silva CG, Martins CF, Cardoso TC, Cunha ER, Bessler HC, McManus CM et al (2016) Isolation and characterization of mesenchymal stem cells derived from bovine Wharton's jelly and their potential for use in cloning by nuclear transfer. Ciênc Rural 46:1830–1837

50. Silva CG, Martins CF, Cardoso TC, Cunha ER, Bessler HC, Martins GHL et al (2016) Production of bovine embryos and calves cloned by nuclear transfer using mesenchymal stem cells from amniotic fluid and adipose tissue. Cell Reprogram 18:127–136

51. Pittenger MF, Mackay AM, Beck SC, Jaiswal RK, Douglas R, Mosca JD et al (1999) Multilineage potential of adult human mesenchymal stem cells. Science 284:143–147

52. Abdulrazzak H, Moschidou D, Jones G, Guillot PV (2010) Biological characteristics of stem cells from foetal, cord blood and extraembryonic tissues. J R Soc Interface 7:689–706

53. Stice SL, Strelchenko NS, Keefer CL, Matthews L (1996) Pluripotent bovine embryonic cell lines direct embryonic development following nuclear transfer. Biol Reprod 54:100–110

54. Cibelli JB, Stice SL, Golueke PJ, Kane JJ, Jerry J, Blackwell C et al (1998) Transgenic bovine chimeric offspring produced from somatic cell-derived stem-like cells. Nat Biotechnol 16:642–646

55. Saito S, Sawai K, Ugai H, Moriyasu S, Minamihashi A, Yamamoto Y et al (2003) Generation of cloned calves and transgenic chimeric embryos from bovine embryonic stem-like cells. Biochem Biophys Res Commun 309: 104–113

56. Wang L, Duan E, Sung LY, Jeong BS, Yang X, Tian XC (2005) Generation and characterization of pluripotent stem cells from cloned bovine embryos. Biol Reprod 73:149–155

57. Pessôa LVF, Bressan FF, Freude KK (2019) Induced pluripotent stem cells throughout the animal kingdom: availability and applications. World J Stem Cells 11:491–505

58. Saidova AA (2019) Bovine stem cells: methodology and applications. SOJ Vet Sci 5:1–9

59. Su Y, Wang L, Fan Z, Liu Y, Zhu J, Kaback D et al (2021) Establishment of bovine-induced pluripotent stem cells. Int J Mol Sci 22:10489

60. Du X, Feng T, Yu D, Wu Y, Zou H, Ma S et al (2015) Barriers for deriving transgene-free pig iPS cells with episomal vectors. Stem Cells 33: 3228–3238

61. German SD, Campbell KH, Thornton E, McLachlan G, Sweetman D, Alberio R (2015) Ovine induced pluripotent stem cells are resistant to reprogramming after nuclear transfer. Cell Reprogram 17:19–27

62. Kato Y, Imabayashi H, Mori T, Tani T, Taniguchi M, Higashi M et al (2004) Nuclear transfer of adult bone marrow mesenchymal stem cells: developmental totipotency of tissue-specific stem cells from an adult mammal. Biol Reprod 70:415–418

63. Colleoni S, Donofrio G, Lagutina I, Duchi R, Galli C, Lazzari G (2005) Establishment, differentiation, electroporation, viral transduction, and nuclear transfer of bovine and porcine mesenchymal stem cells. Cloning Stem Cells 7:154–166

64. Silva CG, Martins CF, Bessler HC, Fonseca Neto AM, Cardoso TC, Franco MM et al (2019) Use of trichostatin A alters the expression of HDAC3 and KAT2 and improves in vitro development of bovine embryos cloned using less methylated mesenchymal stem cells. Reprod Domest Anim 54:289–299

65. George A, Sharma R, Singh KP, Panda SK, Singla SK, Palta P et al (2011) Production of cloned and transgenic embryos using buffalo (*Bubalus bubalis*) embryonic stem cell-like cells isolated from in vitro fertilized and cloned blastocysts. Cell Reprogram 13:263–272

66. Yang MY, Rajamahendran R (2002) Expression of BCL-2 and BAX proteins in relation to

quality of bovine oocytes and embryos produced *in vitro*. Anim Reprod Sci 70:159–169

67. Chi MM, Pingsterhaus J, Carayannopoulos M, Moley KH (2000) Decreased glucose transporter expression triggers BAX-dependent apoptosis in the murine blastocyst. J Biol Chem 275:40252–40257

68. Fan N, Chen J, Shang Z, Dou H, Ji G, Zou Q et al (2013) Piglets cloned from induced pluripotent stem cells. Cell Res 23:162–166

69. Wu Z, Chen J, Ren J, Bao L, Liao J, Cui C et al (2009) Generation of pig induced pluripotent stem cells with a drug-inducible system. J Mol Cell Biol 1:46–54

70. Esteban MA, Xu J, Yang J, Peng M, Qin D, Li E et al (2009) Generation of induced pluripotent stem cell lines from Tibetan miniature pig. J Biol Chem 284:17634–17640

71. Wernig M, Meissner A, Foreman R, Brambrink T, Ku M, Hochedlinger K et al (2007) In vitro reprogramming of fibroblasts into a pluripotent ES-cell-like state. Nature 448:318–324

72. Secher JO, Liu Y, Petkov S, Luo Y, Li D, Hall VJ et al (2017) Evaluation of porcine stem cells competence for somatic cell nuclear transfer and production of cloned animals. Anim Reprod Sci 178:40–49

73. Jin HF, Kumar BM, Kim JG, Song HJ, Jeong YJ, Cho SK et al (2007) Enhanced development of porcine embryos cloned from bone marrow mesenchymal stem cells. Int J Dev Biol 51:85–90

74. Song Z, Cong P, Ji Q, Chen L, Nie Y, Zhao H et al (2015) Establishment, differentiation, electroporation and nuclear transfer of porcine mesenchymal stem cells. Reprod Domest Anim 50:840–848

75. Lee SL, Kang EJ, Maeng GH, Kim MJ, Park JK, Kim TS et al (2010) Developmental ability of miniature pig embryos cloned with mesenchymal stem cells. J Reprod Dev 56:256–262

76. Li Q, Wang H, Zhou GB, Jinfei M, Yan L, Zhaihu Z et al (2014) Production of healthy cloned pigs with neural stem cells as nuclear donors. Anim Biotechnol 25:294–305

77. Olivera R, Moro LN, Jordan R, Pallarols N, Guglielminetti A, Luzzani C et al (2018) Bone marrow mesenchymal stem cells as nuclear donors improve viability and health of cloned horses. Stem Cells Cloning: Adv Appl 11:13–22

78. Kwong PJ, Nam HY, Khadijah WW, Kamarul T, Abdullah RB (2014) Comparison of *in vitro* developmental competence of cloned caprine embryos using donor karyoplasts from adult bone marrow mesenchymal

stem cells *vs* ear fibroblast cells. Reprod Domest Anim 49:249–253

79. Su X, Ling Y, Liu C, Meng F, Cao J, Zhang L et al (2015) Isolation, culture, differentiation, and nuclear reprogramming of Mongolian sheep fetal bone marrow–derived mesenchymal stem cells. Cell Reprogram 17:288–296

80. Czernik M, Fidanza A, Sardi M, Galli C, Brunetti D, Malatesta D et al (2013) Differentiation potential and GFP labeling of sheep bone marrow-derived mesenchymal stem cells. J Cell Biochem 114:134–143

81. Oh HJ, Park EJ, Lee SY, Soh JW, Kong IS, Choi SW et al (2012) Comparison of cell proliferation and epigenetic modification of gene expression patterns in canine foetal fibroblasts and adipose tissue-derived mesenchymal stem cells. Cell Prolif 45:438–444

82. Kim GA, Oh HJ, Lee TH, Lee JH, Oh SH, Lee JH et al (2014) Effect of culture medium type on canine adipose-derived mesenchymal stem cells and developmental competence of interspecies cloned embryos. Theriogenology 81:243–249

83. Berg DK, Li C, Asher G, Wells DN, Oback B (2007) Red deer cloned from antler stem cells and their differentiated progeny. Biol Reprod 77:384–394

84. Rideout WM 3rd, Wakayama T, Wutz A, Eggan K, Jackson-Grusby L, Dausman J et al (2000) Generation of mice from wild-type and targeted ES cells by nuclear cloning. Nat Genet 24:109–110

85. Amano T, Tani T, Kato Y, Tsunoda Y (2001) Mouse cloned from embryonic stem (ES) cells synchronized in metaphase with nocodazole. J Exp Zool 289:19–145

86. Amano T, Kato Y, Tsunoda Y (2002) The developmental potential of the inner cell mass of blastocysts that were derived from mouse ES cells using nuclear transfer technology. Cell Tissue Res 307:367–370

87. Bortvin A, Eggan K, Skaletsky H, Akutsu H, Berry DL, Yanagimachi R et al (2003) Incomplete reactivation of *Oct4*-related genes in mouse embryos cloned from somatic nuclei. Development 130:1673–1680

88. Wakayama S, Mizutani E, Kishigami S, Thuan NV, Ohta H, Hikichi T et al (2005) Mice cloned by nuclear transfer from somatic and ntEs cells derived from the same individuals. J Reprod Dev 51:765–772

89. Mizutani E, Ono T, Li C, Maki-Suetsugu R, Wakayama T (2008) Propagation of senescent mice using nuclear transfer embryonic stem cell lines. Genesis 46:478–483

90. Wakayama S, Wakayama T (2010) Improvement of mouse cloning using nuclear transfer-derived embryonic stem cells and/or histone deacetylase inhibitor. Int J Dev Biol 54:1641–1648

91. Inoue K, Noda S, Ogonuki N, Miki H, Inoue S, Katayama K et al (2007) Differential developmental ability of embryos cloned from tissue-specific stem cells. Stem Cells 25:1279–1285

92. Qin Y, Ji H, Wu Y, Liu H (2009) Chromosomal instability of murine adipose tissue-derived mesenchymal stem cells in long-term culture and development of cloned embryos. Cloning Stem Cells 11:445–452

93. Mizutani E, Ohta H, Kishigami S, Thuan NV, Hikichi T, Wakayama S et al (2006) Developmental ability of cloned embryos from neural stem cells. Reproduction 132:849–857

Chapter 6

Animal Transgenesis and Cloning: Combined Development and Future Perspectives

Melissa S. Yamashita and Eduardo O. Melo

Abstract

The revolution in animal transgenesis began in 1981 and continues to become more efficient, cheaper, and faster to perform. New genome editing technologies, especially CRISPR-Cas9, are leading to a new era of genetically modified or edited organisms. Some researchers advocate this new era as the time of synthetic biology or re-engineering. Nonetheless, we are witnessing advances in high-throughput sequencing, artificial DNA synthesis, and design of artificial genomes at a fast pace. These advances in symbiosis with animal cloning by somatic cell nuclear transfer (SCNT) allow the development of improved livestock, animal models of human disease, and heterologous production of bioproducts for medical applications. In the context of genetic engineering, SCNT remains a useful technology to generate animals from genetically modified cells. This chapter addresses these fast-developing technologies driving this biotechnological revolution and their association with animal cloning technology.

Key words Genetic engineering, Animal cloning, Animal transgenesis, Synthetic biology, Genome editing

1 Livestock Genomes as the Context for Transgenesis

The detailed knowledge about the structure of our genomes [1] indicates that transgenesis may be a natural, ancient, and more common process than we would expect. To begin our discussion, we must revisit the concept of a transgene: "A transgene is a gene or genetic material that has been transferred naturally or by genetic engineering technologies, from one organism to another." Therefore, transgenesis (TG) is a form of horizontal gene transfer between evolutionary distant species. In light of this concept, let us look at a few observations: dissociation and association analysis by DNA hybridization show a large fraction of our genome is composed of repetitive DNA [2]. Genome-wide sequencing of several species shows that more than two-thirds of their genomes contain noncoding elements [1, 3]. This noncoding DNA is mostly composed of repetitive sequences, such as transposable elements

Marcelo Tigre Moura (ed.), *Somatic Cell Nuclear Transfer Technology*, Methods in Molecular Biology, vol. 2647,
https://doi.org/10.1007/978-1-0716-3064-8_6,
© The Author(s), under exclusive license to Springer Science+Business Media, LLC, part of Springer Nature 2023

(TE) or transposons [3, 4]. Although the origin of TE is not completely known, due to their similarity to sequences found in virus genomes, they are believed to have originated from the natural insertion of retroviral DNA into the genome of our ancestors [5]. Other ways of incorporating DNA sequences from species distant in the evolutionary tree of life (named horizontal gene transfer), which is nothing more than a natural form of TG, come from the symbiosis between eukaryotic and prokaryotic cells. Further, organelles (e.g., mitochondria and chloroplasts) have probably evolved from symbiotic interactions between eubacteria and primitive eukaryotic cells, with the transfer of bacteria genes to the eukaryotic genome [6, 7]. We now acknowledge that more than half of the genes encoding mitochondrial and chloroplast proteins are located in the nuclear genome of animals and plants. These nuclear–cytoplasmic interactions are essential for organelle viability and organism survival [7].

Another known source of horizontal gene transfer is parasitism. Agrobacteria are parasitic prokaryotes that reside exclusively within plant roots. Studies have proven the transfer of genes from the genome of agrobacteria to plant chromosomes [8]. The most surprising fact is that these viral and bacterial genetic remnants introduced in plant genomes remained active and play roles in the metabolism of host cells, some of which appear to be essential for plant development and survival [8, 9]. Therefore, TG is a natural process and essential to organism survival throughout evolution. For instance, researchers identified *Agrobacterium rhizogenes* genes naturally transferred to the sweet potato (*Ipomea potatoes*) genome [9], which are transcriptionally active in the roots of this tuber.

According to the definition of TG, the sweet potato is a natural TG product and has been consumed by humans for over thousands of years. This observation corroborates with discoveries made by Barbara McClintock, who in the 1950s observed that more than 80% of domestic maize genome is composed of TE [10], with a probable viral origin [11]. This puts corn in the class of natural TG foods that have been consumed extensively over several millennia, well before the introduction of *Bacillus thuringiensis* corn (Bt-Corn) in our TG-crop menu in the 1990s [12]. Horizontal gene transfer is not documented only in bacteria, archaea, and plants, but also in placental mammals. The placenta evolved in the ancestors of mammals about 150 million years ago. It begins to form by the segregation of trophoblast and inner cell mass cells in the blastocyst-stage embryo. Trophoblast cells proliferate extensively and form the syncytiotrophoblast through an intense process of cell fusion. This cellular phenomenon is intermediated by cell surface proteins, named syncytins, which are protein-coding genes from endogenous retroviruses. These sequences were systematically incorporated into the genome during the evolution of placental mammals [13–15]. Therefore, the evolution of eutherian species (including humans) depended upon natural TG events.

Increasing genomic data support a scenario in which our genome behaves as an "ecosystem," in which several transgenic sequences combine over time to form novel genomes in evolving species [11]. These facts challenge the fiercest and most compelling arguments against the human consumption of TG-derived foods because these arguments lack support of scientific data. If most of what we eat is TG (if not all), what kind of threats could man-made TG food provoke? If society had more access to the data we discussed above, would it be so averse to the consumption of TG food? We should have a more critical and realistic view of our positions, especially those with scientific knowledge, before condemning any technology.

Since the evolution of DNA recombinant technology in the 1980s, we started to use our intelligence to execute horizontal gene transfer in a designed manner [16]. Therefore, we bring to this review some background on directed horizontal gene transfer applied to livestock genomes, with the intent to produce biomolecules in the so-called bioreactors (or biofactories), to improve animal production traits, to increase disease resistance, among other applications. As described below, we are witnessing an astonishing and quick development in the biotechnological toolbox for genome engineering, which is an inescapable process for our cultural and scientific evolution.

2   The Dawn of Recombinant DNA Technology and Animal Transgenesis

Restriction endonucleases isolated from several species of eubacteria were the first genetic scissors, which are enzymes that perform double-strand cleavage at specific DNA palindromic sequences [17, 18]. The phenomena of Lambda phage growth restriction presented by distinct strains of *Escherichia coli* were characterized in the 1950s [19, 20]. The identity of the "restriction factor" became clear with the characterization of enzymes responsible for bacterial defense against DNA invasions from Lambda bacteriophages [21, 22], collectively named restriction enzymes. Therefore, it took more than 10 years from the discovery of restriction enzymes to its first biotechnological application. However, the most interesting fact is the origin and evolution of restriction enzymes, which remains attributed to several events of horizontal gene transfer (or natural TG) among distinct prokaryotes species [23, 24]. The possibility of in vitro DNA manipulation made possible the construction of engineered plasmid vectors and transformation of *E. coli* with these synthetic auto-replicative DNA elements [25]. This event bookmarked the dawn of genetic engineering field. Soon after, Genentech produced the first

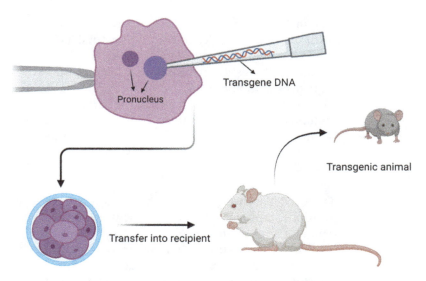

Fig. 1 Production of genetic-modified animals by pronuclear microinjection

biotechnological product in bacteria, which was the human somatostatin hormone [26]. This milestone opened the way for the company to clone and produce recombinant human insulin in *E. coli*, the first human-made TG product to reach the market [27]. The first biotechnological product targeted for the livestock market, which was developed by a joint venture between Genentech and Monsanto, was the recombinant bovine somatotrophin (BST or bovine growth hormone) also produced in *E. coli* [28, 29]. BST remains commercially available and holds intense demand to increase milk production in dairy cattle [30]. However, the first genetically modified (GM) animal was a mouse produced by transgene delivery into pronuclear stage embryos (i.e., pronuclear microinjection), which was pioneered by three independent research groups [31–33]. Since then, GM mice produced by pronuclear microinjection (Fig. 1) became a flagship of gene transfer in laboratory animals [34, 35]. Soon after, GM livestock animals (i.e., rabbits, pigs, and sheep) were described in a single work [36]. However, pronuclear microinjection is inefficient in large livestock animals such as cattle [37] and does not support gene targeting by DNA homologous recombination [34]. More than one decade later, the first cloned animal (sheep) by somatic cell nuclear transfer (SCNT) surged as an alternative to pronuclear microinjection [38, 39]. This promise was soon fulfilled by the report of transgenic sheep carrying the clotting factor IX protein gene driven by a mammary gland-specific promoter [40]. SCNT was appealing for transgenesis because primary somatic cells could be isolated, transfected with transgenes, and selected for transgene expression level and copy number before SCNT [41] (Fig. 2). Since transgene integration occurs randomly, TG cells generated by such means

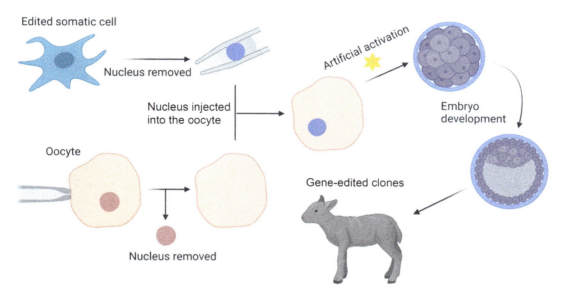

Fig. 2 Production of genetic-modified animals by nuclear transfer (SCNT)

may be screened for transgene integration sites [42], to discard cell clones carrying transgenes integrated into gene bodies or noncoding regulatory sequences, which may affect full-term developmental potential after SCNT. Gene targeting for livestock became viable a few years later, thus demonstrated by the targeted insertion into multiple loci in sheep [43, 44].

There are dozens of GM livestock documented in the scientific literature, although few reached the approval status (Table 1). The list of promising products that perished during the hard and long way from scientific labs to the final market is extensive. It is worth mentioning a few examples to illustrate some pitfalls that can hurdle the progress of GM animal products. The so-named "ecological pig" is an example of a brilliant idea and concept that was not adopted by commercial partners to put it on supermarket shelves. It was a GM pig expressing an *E. coli* phytase genes in the saliva, which would empower the animal with the capacity to digest dietary phosphorus, thus reducing phosphorus waste by 75% [45]. Therefore, the use of these transgenic pigs could result in a significant decrease in phosphorus pollution from the pig industry, with a great impact on the reduction of environmental pollution. Another case is the production of alpha 1-antitrypsin produced in the milk of TG sheep named Tracy [46]. Tracy was developed by a team of The Roslin Institute at the University of Edinburgh and the company PPL Therapeutics. However, the engineered protein caused unexpected breathing problems in human subjects during clinical trials, which ultimately lead to research and development discontinuation. Due to their high value in the market, the biomolecules for pharmaceutical applications are among the first targeted

Table 1
Genetically modified animals and their products available in the market

Product (brand name)	Company	Animal	Approval status
Fluorescent fish (GloFish)	Spectrum Brands	Fish	Taiwan and the United States in 2003
Antithrombin III (Atryn)	rEVO Biologics (LFB Group)	Goat	EU in 2006 and the United States in 2009
C1-Esterase Inhibitor (Ruconest)	Pharming	Rabbit	EU in 2010 and the United States in 2014
Sebelipase Alfa (Kanuma)	Alexion Pharmaceuticals	Chicken	EU and the United States in 2015
Salmon + GH (AquaAdvantage)	AquaBounty	Salmon	The United States in 2015 and Canada in 2016 but currently suspended
Factor VIIa (Sevenfact)	LFB Group	Rabbit	The United States in 2020

genes of interest to be produced in the mammary gland of GM livestock, [34]. Therefore, many attempts were recorded only in the patent databases, most of them linked to the Pharming group, like the human recombinant collagen produced in the milk of GM animals (USPTO Filing Date: 01/15/1994), and the recombinant human fibrinogen (rhFIB) (USPTO Filing Date: 01/15/1999). Both patents appear to have been abandoned.

Monoclonal antibodies (mAb) represent another type of biomolecules for biotechnological applications, such as for diagnosis, immunotherapy, compound purification, among others. The therapeutic use of mAb is by far its most valuable application, representing a market of hundreds of billion dollars for the pharmaceutical industry [47]. The mAb may treat several conditions, as diverse as migraine, Alzheimer, asthma, cancer, viral infection, and autoimmune diseases. However, cancer treatment is by far the hottest spot for mAb development, with more than 79 mAbs approved by the FDA in the recent years [48]. They were originally produced in hybridomas of mouse and human cells [49], and this process is still used frequently. However, limitations such as low scalability (low-throughput) and low automation of these hybridomas have propelled the development of recombinant monoclonal antibody technology by Fab/phage display [50]. Other routes to produce recombinant mAb are mammalian cell lines (e.g., CHO cell line) or in the milk of GM livestock [35].

An innovative and challenging approach to produce human polyclonal antibodies in cattle began in the late 1990s by Hematech

LCC [51], a company acquired by Kirin in 2005 and later by SAB Biotherapeutics in 2014. The first effort toward this goal was to transfer an artificial human mini-chromosome to cattle fetal fibroblasts and their subsequent use for SCNT to generate transchromosomic cattle (Tc Bovine™). These cloned Tc cattle were viable, carried the artificial mini-chromosome, and display human and cattle immunoglobulin in the blood plasma [51]. To abolish immunoglobulin production from cattle gene orthologs, researchers carried out sequential gene deletion of cattle immunoglobulin alleles in cattle fetal fibroblasts [52]. Since primary somatic cells may undergo replicative senescence after extensive culture [53], one allele was deleted at a time and used for SCNT to generate cloned fetuses and calves, thus allowing to rejuvenate cells for another round of gene deletion [52]. This re-cloning scenario allowed the deletion of several alleles in a single-cattle genome [52]. Further improvements in the construction of artificial minichromosomes allowed the production of Tc cattle with adequate production of human polyclonal antibodies for therapeutic applications [54–57]. This human polyclonal production platform was named DiversitAb™ and was used to produce neutralizing antibodies to the Ebola virus [58], and is currently submitted for trials to produce antibodies against the Coronavirus SARS-CoV-2. However, after 18 years of development, this promising technology has still not arrived in the marketplace.

3 GMO and Regulatory Aspects

To provide a perspective, the high rate of failure and bankruptcy observed in GMO companies (with emphasis on the ones focused on transgenic livestock or derivative products), requires understanding the intricate deregulatory process for approval of GMOs (or derivative products) for commercial purposes. The concerns about GMOs' biosafety began immediately after the report of the first GMO [25]. The initial document on the topic was entitled "Potential Biohazards of Recombinant DNA Molecules" [59]. One year later, a meeting at Asilomar (CA, USA) established the first proposal for a regulatory framework specific to GMOs and their derivatives. In 1976, the National Institutes of Health (NIH) formed the rDNA advisory committee to address the issues associated with GMOs. For decades, countries have adopted their own protocols and guidelines to access GMO risk and establish workflows to approve GMOs. Only in 1990 the joint venture of the World Health Organization (WHO) and the Food and Agriculture Organization (FAO) produced the report "Strategies for Assessing the Safety of Foods Produced by Biotechnology" with the objective to standardize protocols for assessing the safety of GM foods and establish an international guideline. However, the intention to unify the protocol of risk assessment and liberation of GMO was not easy, and many attempts were done until now. In 1993, the Organization for Economic Co-operation and Development (OECD) released the report "The safety evaluation of foods derived by modern technology—concepts and principles". This report recommended the assessment of the potential risks of GM food on a case-by-case basis using the principle of substantial equivalence. The principle of substantial equivalence establishes that the GMO food (or derivatives) should be assessed by comparing it with a similar traditional food that has proven to be safe in normal use, with a long history of safe use, the so-called conventional equivalent. The principle of substantial equivalence was a milestone in the GM food risk assessment and is used by many countries to some extent. Another landmark for the adoption of a unified protocol and guideline for GMOs food risk assessment and international commercialization, transfer, handling, and use was the Cartagena Protocol on Biosafety, inaugurated in 2000 and becoming mandatory for the 172 signatory countries in 2003 (https://bch.cbd.int/protocol/). In parallel with Cartagena Protocol, the FAO and WHO published the Codex Alimentarius International Food Standards (http://www.fao.org/fao-who-code xalimentarius/en/), a document to help countries coordinate and standardize the regulation of GM food and facilitate international trade. Since the Cartagena treaty and the Codex Alimentarius, the deregulation process involving GMOs became more standardized among the majority of countries. However, each country still has its autonomy to relax or toughen up its policies regarding GMOs and derivatives. For example, the EU and Brazil label products as "GM containing" if carrying >1.0% of GMO-derived content. Labeling is voluntary in other countries (e.g., The United States), although some US states are adopting the mandatory GM food labeling.

To release a GMO or its derivatives in the market, several years and substantial financial investments are spent in experiments, extensive product testing, technical reports, risk assessments, and product labeling, which are demanded by regulatory agencies. Several countries hold regulatory agencies for this purpose (Table 2). The extensive regulatory network and bureaucracy are prohibitive for small and medium biotech companies. Hence, these circumstances lead to high market concentration [60].

Twenty years after the production of human clotting factor IX protein in transgenic cloned sheep, the production of this protein remains restricted to transgenic mammalian cell lines [61]. This shows the uncertainty in producing biomolecules in GM animals and may explain the vast majority of attempts which do not lead to product approval and availability in the market [34, 35].

Unlike the expectations, the first marketable GM animal was a fluorescent fish intended for the pet market. Named GloFish, the first Green Fluorescent Protein (GFP) ubiquitously expressing

Table 2
Ten of the most relevant regulatory agencies in the world

Country or region	Deregulation agency
Africa	Common Market for Eastern and Southern Africa
Argentina	National Agricultural Biotechnology Advisory Committee (environmental impact), National Service of Health and Agrifood Quality (food safety), and National Agribusiness Direction (effect on trade)
Australia	Office of the Gene Technology Regulator (overseas all), Therapeutic Goods Administration (GM medicines), and Food Standards Australia New Zealand (GM food)
Brazil	National Biosafety Technical Commission (CTNBio) and Council of Ministers (only for socioeconomic issues if summoned by ministers or the president)
Canada	Health Canada and the Canadian Food Inspection Agency
China	Office of Agricultural Genetic Engineering Biosafety Administration (OAGEBA)
Europe Union	European Food Safety Authority (EFSA)
England	Department for Environment Food and Rural Affairs
India	Institutional Biosafety Committee (IBSC), Review Committee on Genetic Manipulation (RCGM), and Genetic Engineering Approval Committee (GEAC)
The United States	USDA (agriculture), FDA (food and drug safety), and EPA (environmental impact)

zebrafish, was developed and patented by a group from the National University of Singapore in 2000 (Patent WO/2000/049150). The rights to commercialize the GFP Zebrafish were acquired by Yorktown Technologies and more recently were transferred to Spectrum Brands Inc. in 2017. In parallel, the GFP Medaka fish was developed by the National Taiwan University and licensed by Taikong Corp under a different brand name. Since then, several fluorescent proteins were introduced in ornamental fish like Barbs, Tetras, and Rainbow sharks, besides the aforementioned Zebra and Medaka. The GloFish was approved to be sold in Taiwan and the United States by 2003, thus becoming the first GM animal to be sold directly to consumers. Several concerns and disputes have involved the commercialization of GloFish around the world due to the potential risks of contention scape and natural habitat colonization [62, 63]. However, studies have shown that GloFish are less efficient in mating and reproduce in natural environment than their non-GM counterparts [64], hold increased predator vulnerability [65], and display a low risk of habitat invasiveness [66]. Therefore, scientific evidence predicts a low risk of environmental damage for these fancy pets. Since GloFish were not intended for human or animal food chain or biomolecule production, it is considered an atypical case of GM organism (GMO) deregulation, which could explain its unexpected relatively fast approval. GMO deregulation usually takes much longer, with much greater associated costs before product approval for commercialization. For instance, recombinant Antitrobin III (Atryn^{TM}) produced in GM goats was the first bioproduct approved for pharmaceutical purposes. Atryn was originally developed by Genzyme Corp. and first documented in the early 1990s [67]. After three rounds of acquisitions (Genzyme Corp., GTC Biotherapeutics, and now rEVO Biologics) and 15 years of deregulation process, Atryn was approved to be commercialized in the European market, and 3 years later in the US market [35]. The long bureaucratic deregulation process, the high costs of development and clinical trials, make the business of selling recombinant bioproducts produced in GM livestock a very risky enterprise. The list of unsuccessful attempts to launch these biomolecules in the market, as well as the list of companies that disappeared or went bankrupt during the process of product development, is enormous. Another emblematic example is the AquaAdvantage^{TM} salmon, which is a fast-growing GM Atlantic salmon carrying a copy of the growth hormone gene ortholog from Chinook salmon and a gene promoter from the ocean pout, developed by the AquaBounty Technologies since 1989 [68, 69]. After 26 years, the AquaAdvantage was approved for commercialization (Table 1), albeit with several restrictions and safety precautions. However, due to the lobbying of salmon farming industry, the US Congress asked the U.S. Food and Drug Administration (FDA) agency to block AquaAdvantage salmon commercialization, while lawmakers were agreeing on labeling guidelines about the production of engineered foods. Only in 2019 the FDA lifted the restrictions for importing and commercializing the AquaAdvantage salmon, albeit this commercial dispute "disguised" of food safety or environmental risk concerns seems to be far from ending. This new "disguise" of international trade protectionism is becoming commonplace since it is easier to place commercial barriers justified on sanitary, food safety, or environmental risk concerns, instead of an allegation of internal market protection [70, 71]. In sharp contrast, genetically edited organisms (GEO) face softer deregulation pipeline, which may overcome some of these drawbacks, as described later in this chapter.

The beginning of the twenty-first-century eye witnessed novel technologies for precise editions of the genome, collectively called genome editing tools. These discoveries and further developments are thus revolutionizing the GMO-associated industries. Genome editing allows a drastic reduction in the regulatory pipeline in many countries and lowers the costs and timeline to generate GM organisms (microorganisms, plants, and animals), also known as GEO, and to potentially deliver them to the marketplace. This revolution

Animal Transgenesis and Cloning **131**

is opening the "doors" of the biotechnology industry to smaller companies or even startups around the world.

4 Genome-Edited Organisms (GEO)

One major goal of the biotechnology field was to develop molecular scissors capable of specific site-directed editions in eukaryotic genomes [72]. This approach is known as genome editing, and the enzymes able to reach this goal were named site-directed nucleases (SDNs). Rare-cutting endonucleases, homing endonucleases, or meganucleases were the initial candidates for editing complex genomes [72, 73]. These nucleases, which are mostly composed of introns of mobile genetic elements, thus have a recognition site of 18–30 nucleotides (Fig. 3a), which makes them a feasible candidate for a single edition in genomes with over 10^9 base pairs in size [74]. However, it was challenging to engineer them to edit a single targeted DNA sequence in the genome. The application of engineering approaches frequently diminished the endonuclease cleavage activity [74]. However, fusion of zinc-finger DNA-binding domains with the catalytic domain of *Fok I* nuclease, also known as zinc finger nucleases (ZFNs), made it possible to target, with relative high precision (Fig. 3b), a desirable site of the genome [75]. This improvement was recognized as the beginning of the genome editing era.

Among the most recent and promising techniques for targeted editing of eukaryotic genomes, we can mention ZFNs, transcription activator-like effector nucleases (TALENs), and, more recently, the nucleases directed by RNA guides of the Clustered Regularly Interspaced Short Palindromic Repeats (CRISPR)-associated system protein (CRISPR-Cas) technologies [76]. The ZFNs are considered the first generation of nucleases engineered to cut a specific location of plant and animal genomes [77], as described above. However, due to the lower specificity in DNA binding, ZFNs edit the intended site (on-target) and frequently in other nonintended sites (off-target), which is the main potential undesirable side effect of genomic editing technologies [77]. The second generation of targeted nucleases were TALENs (Fig. 3c), which are DNA-binding proteins produced by plant pathogenic bacteria to regulate gene expression in the host genome [78]. Each DNA recognition domain in TALENs binds to a single nucleotide in the targeted genome, a fact that made this genomic editing technology more accurate than ZFNs [79]. However, the construction of TALENs involves the concatenation of dozens (usually about 30) DNA-binding domains, where each domain of 34 amino acids recognizes a specific single nucleotide. Nonetheless, this process involves a great time and effort in vector construction and at a high cost [78]. More recently, a new genomic editing

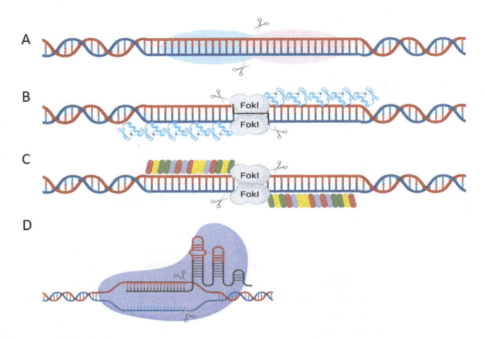

Fig. 3 Genome editing tools

technology called CRISPR-Cas (Fig. 3d) was developed based on the discovery of an adaptive immunity system present in bacteria and archaea against invading nucleic acids, commonly of viral origin [80]. The great advantage of this technology is that it relies on base pairing between a guide RNA (gRNA) and the genomic site, thus guiding the Cas9 nuclease to the target DNA sequence [81]. To reinforce the simplicity of CRISPR-Cas9 gene editing, during the construction of TALENs vectors, two domains of 30 repetitions of approximately 1800 nucleotides each (which will encode the protein domains of binding to the target DNA) are needed to give the desired specificity to the *Fok I* nuclease. In sharp contrast, CRISPR-Cas9 requires a sequence of 17–20 ribonucleotides in a context of 80 invariant ribonucleotides (scaffold sequence) to design the gRNA. These features allow researchers to rapidly create hundreds of customized (site-specific) gRNAs at a low cost [76]. These characteristics made CRISPR-Cas the prime technology for genome editing in the recent years, overtaking by far the other genome editing technologies [82].

The major concern about genome editing is the off-target effect (OTE), which is the impact within a cell of unintended editions on the recipient genome. These OTEs may vary from innocuous changes to disruptions of genes required for survival [83]. The OTE has been identified in all classes of SDN, thus including ZFNs [83, 84], TALENs [85], and CRISPR-Cas9 [86, 87]. Off-targets originate from base mismatch tolerance found in all SDNs regarding their DNA sequence recognition [84, 85, 88]. However, astonishing approaches to reduce off-targets during genome editing mediated by CRISPR-Cas9 have been developed recently. Creative strategies using Cas9 nickases [89] or Fok1-dead Cas9 fusion proteins [90] make the binding of two sgRNAs in tandem to produce the double-strand brake (DSB) in the targeted site necessary. This kind of approach increases the DSB specificity and can reduce detectable off-targets by 10,000-fold [91]. Another way to improve genome editing precision is by engineering the Cas9 nuclease to increase its specificity for target site recognition. This goal was reached with several mutations on Cas9 sequence that reduce off-target frequency [92]. High-fidelity Cas9 (SpCas9-HF1) engineered by four amino acid substitutions (N497A, R661A, Q695A, and Q926A) increases Cas9 specificity and shows no detectable OTE [93]. Furthermore, the truncation of the sgRNA from 20 bp in length to 17 or 18 bp could increase the target specificity up to 5000-fold [94]. Finally, all these strategies (e.g., Cas9 dimerization, Cas9 engineering, and sgRNA truncation) can be associated to make the DSB so specific that OTE could be abolished or become negligible. Even with an extremely high fidelity and a very low incidence of off-targets, there is still a need to screen engineered genomes for off-target editions. Thanks to the advancement of whole-genome sequencing with next-generation sequencing technologies, it is now possible to screen gene-edited cells for OTE with high efficiency and relatively low costs [95, 96]. Collectively, these features made CRISPR-Cas technology feasible, widely adopted, technically simple, "and of low-cost".

5 GEO and Regulatory Aspects

As mentioned earlier, the GMO deregulation process is expensive and long. This has a collateral effect of provoking a high concentration of GMO market in the hands of a few biotech conglomerate groups and limiting the free competition [60]. The popularization of CRISPR-Cas technology may revert this scenario and put small biotech companies or startups in the GEO market, with beneficial effects for consumers [97]. However, the breadth and depth of such impact of GEO on livestock and pharmaceutical markets will depend on the new biosafety regulatory framework adopted around the world. Therefore, if a GEO is considered equivalent to a GMO, it may have to undergo those costly and lengthy deregulation processes [98]. Since the revision of biosafety regulatory framework toward the inclusion of GEO in a distinct category of GMO is ongoing, there are different positions among governments on this topic [99]. For example, the EU adopted the most restrictive approach to GEO regulation, focusing the biosafety analysis on the process and equating it to GMO analysis [100]. Argentina was the first country to put GEO deregulation in a different status of GMOs regarding the biosafety in 2015, focusing on the absence of external (transgenic) DNA in the final product, not the technology itself [101]. In the following years, many other countries (mainly livestock and crop exporters) adopted different rules for GEOs' biosafety analysis and deregulation, also focusing on the final product not on the process [102, 103]. Despite differences between countries, there is already a consensual workflow and terminology for GEO deregulation [99, 102]. There are three different types of SDN approaches to produce a GEO: SDN-1, where the nuclease induces a DSB, which is repaired by the targeted organism mainly through non-homologous end joining (NHEJ) accompanied by a few random base substitutions, known as small insertions or small deletions (indels); SDN-2 is similar to SDN-1, albeit with the addition of a short single-stranded oligo DNA nucleotides (ssODNs) of 50–100 bp guiding the repair; and SDN-3 which is similar to SDN-2 but with the addition of DNA template with a sequence of interest flanked by DNA sequences homolog to the targeted site, generally obtained by DNA repair using homologous recombination (HR) [102]. Several countries are adopting a case-by-case process focused on the final GEO product, where SDN-1 and 2 are considered nontransgenic (non-GMO), and SDN-3 is considered a GMO and has to undergo the conventional GMO deregulation process [104, 105]. SDN-1 and SDN-3 are usually a consensus of non-GMO and GMO regulatory policies, respectively. The discrepancy between interpretations is whether SDN-2 is a GMO or not (and the size of the ssODN template accepted). Further, most regulatory documents make reference to ssODN of a "few bases," without a precise size for ssODN length [98, 105, 106].

The majority of new biosafety regulatory documents developed for GEOs reference SDN-2 as a non-OGM, with the exception of Australia and Canada [98, 107, 108]. Most species used to generate GMOs are commercialized internationally as commodities (livestock and crops), so a consensus on deregulatory process is highly desirable to facilitate the commercialization of OGM products or derivatives around the world. For this reason, several efforts were made to establish a standardization of the deregulation process. These efforts, described in detail in the Subheading 3, resulted in the Cartagena Protocol on Biosafety (https://bch.cbd.int/proto col/) and the Codex Alimentarius International Food Standards (http://www.fao.org/fao-who-codexalimentarius/en/). These two protocols signed by more than 170 countries standardize the biosafety workflow of OGMs in the world. The same effort is demanded for GEOs by the scientific community and the private sector [98–102, 105]. This process is already in progress since there was an Organization for Economic Co-operation and

Development (OECD) conference on "Genome Editing: Applications in Agriculture-Implications for Health, Environment and Regulation" held on June 2018 in Paris [106]. In this conference, policymakers, scientists, innovators, and other stakeholders debated to build a consensus policy and protocol to be adopted by OECD members for GEO commercialization in the foreseeable future.

6   Genome Editing Applied to Livestock Animals

Humans have shaped the genome of livestock species for centuries by applying selective breeding, with limited or no understanding of the genetic basis of heredity [109]. In the first half of the twentieth century, independent research groups showed the increase in phenotypic (and genotypic) variation by radiation or chemical-mediated mutagenesis. Despite its utility for plant breeding, such technologies cause random mutations and could not be directed to a specific *locus* [110].

In the second half of the twentieth century, the first targeted genomic changes in yeast and mice genomes were produced using homologous recombination (HR) [111–114]. The HR technology was precise but very inefficient. To tackle this hurdle, researchers relied on mouse "embryonic stem cells (ESCs)" to perform targeted gene modifications by HR. Despite the extremely low incidence of correct gene modifications, ES cells offered two unique advantages that characterize them: infinite proliferative capacity and pluripotency, which is the potential to give rise to all cell types of the body [115]. Therefore, ESCs carrying disrupted genes (knockout) could be expanded by clonal selection (without replicative crisis) and could contribute to mouse development by their introduction into preimplantation embryos [115]. The ES-based transgenesis technology placed the mouse at the center stage for investigating gene function and modeling human diseases [115]. However, no bona fide ESC lines have been described so far for livestock species, despite many efforts over decades of research. Nowadays, genome editing technologies surpassed these problems by their association with SCNT cloning or in some cases when delivered directly into one-cell embryos (zygotes). These approaches enabled direct genetic manipulation in essentially all cell types or complex animal genomes [109].

The ability to generate a DSB in the sequence of interest is the key step of a genome editing technique using SDN [109]. The different types of SDN (e.g., ZFNs, TALENs, CRISPR-Cas) share the same potential to induce gene disruption by NEHJ or homology-driven repair by HR, which consists of DNA cleavage at specific sites followed by cellular repair of the DSB [116]. This repair can be in the absence or presence of DNA molecules presenting homologous sequences surrounding the DSB. The NHEJ repair is activated in the absence of a DNA template, while HDR relies on homologous DNA templates [116]. These two pathways can be used for different genome editing applications, such as knockouts (NHEJ) or knockins by gene targeting (HDR). Knockout rats were initially produced using ZFNs due to the disruption of immunoglobulin M (IgM) and Rab38 genes [117]. Similar to ZFNs, the first animal generated using TALENs technology was a knockout rat for the IgM locus [118]. Soon after, the tyrosinase (Tyr) gene was disrupted by an exogenous CRISPR-Cas9 in the mouse, thus producing biallelic editions, leading to albino animals [119]. In a time span of less than 5 years, genomic editing evolved from ZFNs to CRISPR-Cas and became a widely used technology. In the following sections, we show examples of relevant livestock species that have been modified by genome editing for pharmacological research or to introduce traits that improve the economic relevance of such organisms.

6.1 Rabbits

Rabbits hold commercial interest in the context of transgenesis. Comparing rabbits with cows, rabbits have a gestation time of 1 month versus 9–10 months in the cows, and rabbit maturation (time to reach reproductive age) takes 5–6 months while cows take 15–20 months, and have a much smaller adult size [120]. Due to these features, rabbits became appealing for transgenesis and deliver results in a shorter period of time [120]. Furthermore, rabbit milk contains 2.5-fold more protein yield than sheep and 4.8-fold than goat milk, a fact that makes rabbits an attractive platform for producing heterologous proteins. Milk production in a lactating rabbit may reach 170–220 g/day and up to 10 kg/year, which may lead to up to 20 g/L of heterologous protein [121]. These factors make transgenic rabbits of lower cost compared with other livestock animals [120, 121]. Many heterologous proteins have been produced in transgenic rabbits, such as interleukin-2 [122], insulin-like growth factor-1 [123], growth hormone [124], lactoferrin [125], α1-antitrypsin [126], antithrombin [127], C1 inhibitor [128], erythropoietin [129], and α-glucosidase [130].

Recent studies used CRISPR/Cas9 for producing transgenic rabbits to model human disease, test pharmaceutical drugs, enhance economical traits, and produce heterologous proteins in their milk [131]. Transgenic rabbits offer the potential to mimic human diseases by replacing rabbit alleles with disease variants found in the human gene orthologs, which subsequently leads to pathophysiological conditions mirroring the human disease condition [132]. For example, a rabbit model was developed for safety drug testing with more reliable results in predicting proarrhythmia potential [131, 132]. Rabbit physiology also offers advantages to studying certain human diseases and test new drugs (e.g., atherosclerosis) due to their unique lipid metabolism much similar to humans [133]. Under this scenario, CRISPR-Cas9 was used to generate double knockout rabbits by inducing deletions in the apolipoprotein E and the low-density lipoprotein (LDL) receptor genes [134]. These animals showed hyperlipidemia, aortic and coronary atherosclerosis, and physiological alterations that supported their potential as models of atherosclerosis development and for testing new therapies [134]. Duchene muscular dystrophy (DMD) is another disease in which rabbits have potential as pre-clinical animal models. Using CRISPR-Cas9, dystrophin knockout rabbits displayed muscular dystrophy features with close resemblance to the human condition, thus suggesting that it may be a better model for DMD research than other species [135]. The interleukin 2 receptor subunit gamma (IL2RG) gene is essential for signal transduction between interleukins, and mutations in this gene can cause X-linked severe combined immunodeficiency in humans. Therefore, disrupting IL2RG may lead to an animal model with immunodeficiency [136]. CRISPR-Cas9 was used to knock out the IL2RG gene and generated immunodeficient rabbits, with potential for transplantation studies as xenograft models [136]. Growth differentiation factor 8 (GDF8 also known as myostatin) is an interesting gene for engineering livestock for improved meat production since its deletion causes muscle hypertrophy and further increases carcass yield. Guo et al. [137] were able to develop GDF8 knockout rabbits using CRISPR-Cas9, which displayed higher birth weight and improved growth rates. Rabbit coat color is another trait of economic value. The melanocortin 1 receptor (MC1R) locus defines the rabbit coat color, and the disruption of this gene using CRISPR-Cas9 led to a novel pale-yellow coat color [138].

6.2 Sheep and Goats

The first report of a transgenic livestock through pronuclear micro-injection was published nearly four decades ago [36]. The successful cloning of sheep from differentiated embryonic cells and the cloning of Dolly the sheep from adult somatic cells offered a new route to generating transgenic sheep from GM-cultured cells [37, 38]. Likewise, GM fetal fibroblast cells served as donor cells to generate the first transgenic cloned sheep [40]. A few years later, knockout sheep were produced disrupting Alpha1 (I) procollagen (COL1A1) and prion protein (PrP) genes in primary cells using HR and allocating them to SCNT cloning [43, 53]. This was a milestone because gene targeting by HR requires extensive cell culture expansion that may cause replicative crisis in these primary somatic cells. However, the use of fetal fibroblasts and extensive cell clone genotyping allowed us to expand cells sufficient for SCNT and ultimately the delivery of cloned knockout lambs [44]. After these seminal reports, several other knockout models were described in livestock species, most notably in pigs.

With the advancement of genetic engineering technologies and the development of CRISPR-Cas9, the generation of gene-edited sheep and goat focused on investigating gene function and enhancing traits of economic interest, such as muscle mass, coat color, milk composition, mature weight, wool fiber length, reproduction performance, among others [139]. Moreover, other applications that have been reported regarding gene editing in small ruminants involve their use as models for human disease, generation of disease-resistant animals, and hosting the growth of human organs for xenotransplantation [139].

ZFNs have been used to target the GDF8 gene to promote double musculature phenotype in sheep [140–142]. The β-lactoglobulin (BLG) is a major allergen in cow milk and 3% of human infants are susceptible [143, 144]. Therefore, the BLG gene is also a target for genome editing to produce BLG-free milk. Moreover, ZFNs were used to knock out BLG in dairy goats [145–147]. Even though there are studies showing the potential of ZFNs to produce transgenic animals, no reports indicate the actual generation of live edited sheep or goats. TALENs have also been used to produce genetically engineered small ruminants. Biallelic GDF8 knockout sheep showed increased muscle mass [148–150]. TALENs have also been applied to target the BLG gene in goats [151–154].

CRISPR-Cas9 technology has also been used to generate edited sheep and goat. In sheep, multiplex gene editing of GDF8, the agouti-signaling protein (ASIP) and the β-carotene oxygenase 2 (BCO2), was accomplished [155]. Another interesting study included an edition of the suppressor of cytokine signaling 2 (SOCS2) gene that resulted in the effects on milk production, body weight, and size [156]. Four important genes have been knocked out in goats: GDF8, BLG, PrP, and Nucleoporin 155 (NUP155) [139, 157]. A gene target interest to enhance wool fiber length is the Fibroblast growth factor 5 (FGF5) gene, a dominant inhibitor of fiber length and growth, and its disruption in sheep showed increase in wool length [158–160]. The disruption of the Agouti signaling protein (ASIP) gene by CRISPR-Cas9 caused variegation in coat color patterns in sheep, and this color variegation may become an important commercial trait for wool production [161]. Milk components are also an important target when discussing genetic modifications in small ruminants. The BLG gene has been disrupted by CRISPR-Cas9 in goats, generating animals with decreased expression of BLG and animals without BLG protein production in milk [157, 162]. Since melatonin has many nutritional and medicinal utilities, CRISPR-Cas9 was used to make gene-edited animals producing melatonin-enriched milk [163]. Another interesting target to genome editing is to enhance reproductive performance. Mutations in sheep BMPR-IB (FecB) gene and growth differentiation factor 9 (GDF9) gene have been

Animal Transgenesis and Cloning 139

performed to improve goat prolificacy since both genes increase ovulation rates in sheep [164, 165]. Improving animal health and welfare is also an important use of genome editing technologies. PrP-resistant animals can be produced by suppressing the expression of PrP, which is associated with the disease known as spongiform encephalopathy, that occur in humans and livestock species, and PrP-knockout goat fibroblasts cells were documented [157, 166, 167].

6.3 Pigs

The wide use of transgenic pigs in biomedicine and biopharming reflects their physiological similarities to humans [168]. Their applications range from research models for human disease and production of recombinant proteins to potentially providing tissue or organs for xenotransplantation [168]. Mutation in the cystic fibrosis transmembrane conductance regulator (CFTR) gene causes an autosomal-recessive disorder that causes bronchopulmonary failure and pancreatic enzyme insufficiency, hallmark symptoms of the cystic fibrosis disease. As a monogenic disease, the introduction of a functional CFTR allele could restore gene function and cure the patient [169, 170]. CFTR knockout pig models have been used to test gene therapy approaches for this disease. Using CRISPR-Cas9 to deliver a human CFTR allele at a safe harbor locus (i.e., not required for cell survival) in the porcine genome, it recovered CFTR gene expression in pig cells, thus showing proof-of-concept data on the potential of CRISPR-Cas9-driven gene therapy for cystic fibrosis [169].

The main drawback associated with xenotransplant is the hyperacute organ rejection. Pig organs express the GGTA-1 enzyme that synthesizes the Gal antigen, which is the main cause of xenograft rejection, since humans produce antibodies against the antigen. GGTA-1 knockout pigs have been generated using ZFNs, TALENs, and CRISPR-Cas9, aiming to prevent this problem [171–173]. Another antigen found in pig cells that causes rejection in humans are N-glycolylneuraminic acid (Neu5Gc), catalyzed by cytidine monophosphate-N-acetylneuraminic acid hydroxylase (CMAH) and the SDa antigen produced by beta-1,4-N-acetyl-galactosaminyltransferase 2 (β4GALNT2). Estrada et al. [174] generated a triple-knockout (3KO) pig using CRISPR-Cas9 (deletions for GGTA1, CMAH/β4, and GalNT2 genes). Cells from these 3KO pigs presented less human-anti-pig cytotoxicity response [175, 176]. However, xenografts in a pig-to-rhesus preclinical model, using pig 3KO kidneys, showed only a slight increase in organ survival [177]. This shows that, albeit the potential of these animals for xenotransplantation, much more advances need to be achieved before they are considered a reality for human medicine. Another interesting application of CRISPR-Cas9 in the context of xenotransplantation involves addressing biosafety concerns of potential cross-species infections. Porcine endogenous retroviruses (PERV) are integrated into multiple sites within the pig genome and may integrate into the human genome after xenotransplantation. In turn, pigs were generated lacking functional PERVs, which was accomplished by simultaneous deletions of all 25 functional PERVs using CRISPR-Cas9 [178]. These gene editions were performed in pig primary cells, which were destined for SCNT to generate cloned piglets lacking functional PERVs.

6.4 Cattle

Genetically engineered cattle have been produced for several applications, such as bioreactors for the production of heterologous proteins in milk, enhancement of traits with economic impact, and disease-resistant animals [179]. The use of transgenic cattle as bioreactors has the mammary gland as the main site for heterologous protein production. Exogenous protein expression in the milk facilitates its purification [180]. Transgenic dairy cows may produce from 1 to 14 g/L of heterologous proteins in the milk, while holding lactations of 305 days/year. In contrast, mammalian cell culture can yield 10 g/L during 10–12 days (production peak). This comparison demonstrates the higher efficiency of heterologous production in the milk, while dispensing additional costs with laboratory infrastructure, bioreactor apparatus, and culture requirements [181, 182].

Human lysozyme shows nonspecific immune response, anti-inflammatory properties, and both antifungal and antiviral activities [183]. This protein is studied for its potential as probiotic for food supplementation and cosmetics production, and its exogenous expression in the milk can also increase cow resistance to mastitis [183]. Since mastitis is an inflammatory process caused by *Staphylococcus aureus* that diminishes milk production in livestock species, many studies showed the exogenous expression of this protein in the mammary gland. A transgenic cow expressing human lysozyme was also generated using ZFNs by Liu et al. [184] and the animals presented resistance against *Staphylococcus aureus*. Furthermore, ZFNs were also used to insert the lysostaphin gene and express this protein in the milk of the transgenic cows [185]. Lysostaphinis are naturally produced by *Staphylococcus simulans* and are efficient against *S. aureus* infections. A mutation was introduced by CRISPR-Cas9 to disrupt the BLG gene in cattle. These cows produced milk lacking BLG [186, 187]. As mentioned above, BLG is the main allergen in the cow milk, but it is also the major protein, which accounts for approximately 12% of the total milk protein content [188]. Luo et al. [189] inserted the human serum albumin (HSA) gene into the BLG locus by TALEN nickase to generate animal expressing high levels of HSA without BLG (biallelic deletions) or diminished BLG content (monoallelic editions) in the milk. The HSA protein has great utility for several medical treatments because it is the most abundant protein in human plasma [190].

Wu et al. [191] produced cattle with high resistance to myco-bacterium by expressing the SP110 gene. The murine SP110 gene is a promising candidate to control *Mycobacterium tuberculosis* infections by limiting bacteria growth in macrophages and inducing apoptosis in infected bovine cells. Therefore, this gene was chosen to investigate the resistance against *Mycobacterium bovis*, which causes bovine tuberculosis. Using TALENs to knockin the murine SP110 gene into cattle genomes, it was possible to generate an animal able to abolish *M. bovis* growth and multiplication [191].

CRISPR with an Cas 9 nickase (Cas9n) has the potential to increase HR frequency by avoiding the NHEJ repair pathway [192, 193]. The Cas9n induced single-strand breaks (instead of DSB), further stimulating HR repair, and created cows with natural resistance-associated macrophage protein-1 (NRAMP1) gene allele knockins [193]. The NRAMP1 gene is associated with innate resistance to intracellular pathogens (e.g., *M. bovis*) since it is a natural mechanism driving the entry of this pathogen into macro-phages [194, 195]. CRISPR-Cas9 has also been applied to correct genetic variants associated with diseases in cattle. The Japanese Black cattle produce high-quality meat due to selective breeding for more than 60 years. Unfortunately, this also resulted in the accumulation of recessive mutations that leads to genetic disorders. The isoleucyl-tRNAsynthetase (IARS) syndrome is one such dis-ease caused by a mutation in the IARS gene, which leads to a reduction of 38% in its aminoacylation activity and diminished protein synthesis [196]. Using CRISPR-Cas9 to repair the mutant genetic variant, Ikeda et al. [196] generated cloned fetuses with corrected IARS alleles.

The horned phenotype represents the majority of the world's cattle population. This phenotype also represents a risk of injury to other animals and workers that manage them. Unfortunately, dehorning is labor-intensive and causes both pain and stress to the animal [197]. For the past few years, animal welfare is becom-ing a crucial aspect of livestock management practices. The polled phenotype is naturally found in some cattle breed and has been associated with genetic variants of the polled locus located on chromosome 1 [198]. There are four gene variants for the pooled locus: the Celtic mutation (Polled Celtic, Pc) positioned within an intergenic region of chromosome 1, the Polled Friesian (Pf) (restricted to dairy cattle populations), the Polled Mongolian, and the Polled Guarani [199]. The generation of polled dairy cattle using TALENs-mediated genome editing introduced the Pc variant into the cow genome [200, 201]. The Pc variant was also integrated into the genome of a horned Holstein–Friesian bull using CRISPR-Cas12a to generate a polled calf [202]. CRISPR-Cas12a is an RNA-guided endonuclease that has been recently harnessed as an alternative genome editing tool to the Cas9, differ-ing in the PAM (Protospacer Adjacent Motif) site context and gRNA requirements [203].

7 Concluding Remarks

The genomes of livestock species have become accessible to complex genetic editions. Initially, SCNT was pivotal to generate cloned transgenic animals from small batches of cells carrying specific genetic modifications. More recently, genome editing gained momentum, most notably by CRISPR-Cas system, and reached an efficiency threshold that motivates genetic editions directly into livestock preimplantation embryos. Despite the current ability to perform several genomic editions at once using CRISPR-Cas9, the genome must be screened for intended and nonintended (off-target) genetic editions. Perhaps this quality checking is the main reason why SCNT cloning remains a powerful technology assisting the generation of transgenic livestock carrying complex editions in their genomes.

Acknowledgments

We thank Marcelo Tigre Moura for the helpful assistance in reviewing this chapter.

Funding Melissa S. Yamashita is sponsored by CAPES scholarship.

References

1. Dunham I, Kundaje A, Aldred SF et al (2012) An integrated encyclopedia of DNA elements in the human genome. Nature 489:57–74

2. Biscotti MA, Olmo E, Pat Heslop-Harrison JS (2015) Repetitive DNA in eukaryotic genomes. Chromosom Res 23:415–420

3. de Koning APJ, Gu W, Castoe TA et al (2011) Repetitive elements may comprise over two-thirds of the human genome. PLoS Genet 7:e1002384

4. Kapusta A, Kronenberg Z, Lynch VJ et al (2013) Transposable elements are major contributors to the origin, diversification, and regulation of vertebrate long noncoding RNAs. PLoS Genet 9:e1003470

5. Fischer MG, Suttle CA (2011) A virophage at the origin of large DNA transposons. Science 332:231–234

6. McFadden GI (2001) Primary and secondary endosymbiosis and the origin of plastids. J Phycol 37:951–959

7. Keeling PJ, Archibald JM (2008) Organelle evolution: What's in a name? Curr Biol 18: R345–R347

8. Zupan J, Muth TR, Draper O, Zambryski P (2000) The transfer of DNA from agrobacterium tumefaciens into plants: a feast of fundamental insights. Plant J 23:11–28

9. Kyndt T, Quispe D, Zhai H et al (2015) The genome of cultivated sweet potato contains Agrobacterium T-DNAs with expressed genes: an example of a naturally transgenic food crop. Proc Natl Acad Sci U S A 112: 5844–5849

10. McClintock B (1950) The origin and behavior of mutable loci in maize. Proc Natl Acad Sci U S A 36:344–355

11. Biémont C (2010) A brief history of the status of transposable elements: from junk DNA to major players in evolution. Genetics 186: 1085–1093

12. Bawa AS, Anilakumar KR (2013) Genetically modified foods: safety, risks and public concerns-a review. J Food Sci Technol 50: 1035–1046

13. Imakawa K, Nakagawa S, Kusama K (2016) Placental development and endogenous retroviruses. Uirusu 66:1–10

14. Haig D (2012) Retroviruses and the placenta. Curr Biol 22:R609–R613

15. Denner J (2016) Expression and function of endogenous retroviruses in the placenta. APMIS 124:31–43
16. Melo EO (2017) Are we all transgenic? J Genet DNA Res 1:1–2
17. Arber W, Linn S (1969) DNA modification and restriction. Annu Rev Biochem 38:467–500
18. Roberts RJ (1976) Restriction endonucleases. CRC Crit Rev Biochem 4:123–164
19. Bertani G, Weigle JJ (1953) Host controlled variation in bacterial viruses. J Bacteriol 65:113–121
20. Luria SE, Human ML (1952) A nonhereditary, host-induced variation of bacterial viruses. J Bacteriol 64:557–569
21. Lederberg S, Meselson M (1964) Degradation of non-replicating bacteriophage DNA in non-accepting cells. J Mol Biol 8:623–628
22. Meselson M, Yuan R (1968) DNA restriction enzyme from E. coli. Nature 217:1110–1114
23. Jeltsch A, Pingoud A (1996) Horizontal gene transfer contributes to the wide distribution and evolution of type II restriction-modification systems. J Mol Evol 42:91–96
24. Naito T, Kusano K, Kobayashi I (1995) Selfish behavior of restriction-modification systems. Science 267:897–899
25. Cohen SN, Chang AC, Boyer HW, Helling RB (1973) Construction of biologically functional bacterial plasmids in vitro. Proc Natl Acad Sci U S A 70:3240–3244
26. Itakura K, Hirose T, Crea R et al (1977) Expression in Escherichia coli of a chemically synthesized gene for the hormone somatostatin. Science 198:1056–1063
27. Goeddel DV, Kleid DG, Bolivar F et al (1979) Expression in Escherichia coli of chemically synthesized genes for human insulin. Proc Natl Acad Sci U S A 76:106–110
28. Miller WL, Martial JA, Baxter JD (1980) Molecular cloning of DNA complementary to bovine growth hormone mRNA. J Biol Chem 255:7521–7524
29. Keshet E, Rosner A, Bernstein Y et al (1981) Cloning of bovine growth hormone gene and its expression in bacteria. Nucleic Acids Res 9:19–30
30. Bauman DE (1999) Bovine somatotropin and lactation: from basic science to commercial application. Domest Anim Endocrinol 17:101–116
31. Brinster RL, Chen HY, Trumbauer M et al (1981) Somatic expression of herpes thymidine kinase in mice following injection of a fusion gene into eggs. Cell 27:223–231

32. Costantini F, Lacy E (1981) Introduction of a rabbit beta-globin gene into the mouse germ line. Nature 294:92–94
33. Gordon JW, Ruddle FH (1981) Integration and stable germ line transmission of genes injected into mouse pronuclei. Science 214:1244–1246
34. Melo EO, Canavessi AMO, Franco MM, Rumpf R (2007) Animal transgenesis: state of the art and applications. J Appl Genet 48:47–61
35. Bertolini LR, Meade H, Lazzarotto CR et al (2016) The transgenic animal platform for biopharmaceutical production. Transgenic Res 25:329–343
36. Hammer RE, Pursel VG, Rexroad CEJ et al (1985) Production of transgenic rabbits, sheep and pigs by microinjection. Nature 315:680–683
37. Moura MT, Nascimento PS, Silva JCF, Deus PR, Oliveira MAL (2018) The evolving picture in obtaining genetically modified livestock. Anais Da Academia Pernambucana De Ciência Agronômica 13:145–169
38. Campbell KHS, McWhir J, Ritchie WA, Wilmut I (1996) Sheep cloned by nuclear transfer from a cultured cell line. Nature 380:64–66
39. Wilmut I, Schnieke AE, McWhir J et al (1997) Viable offspring derived from fetal and adult mammalian cells. Nature 385:810–813
40. Schnieke AE, Kind AJ, Ritchie WA et al (1997) Human factor IX transgenic sheep produced by transfer of nuclei from transfected fetal fibroblasts. Science 278:2130–2133
41. Vilceu BR, Anthoula K, Lazaris AS, Bilodeau Jose HF, Daniel P, Gilles A, Carol F, Keefer Lawrence C, Smith (2003) Transgene expression of green fluorescent protein and germ line transmission in cloned calves derived from in vitro-transfected somatic cells1. Biol Reprod 68(6):2013–2023. https://doi.org/10.1095/biolreprod.102.010066
42. Sharon FC, Lisauskas EL, Rech Francisco JL, Aragão (2007) Characterization of transgene integration loci in transformed Madin Darby bovine kidney cells. Cloning Stem Cells 9(4):456–460
43. McCreath KJ, Howcroft J, Campbell KHS et al (2000) Production of gene-targeted sheep by nuclear transfer from cultured somatic cells. Nature 405:1066–1069
44. Chris DP, Sarah D, Diana B, Judy W, John FA, Clark (2001) Gene targeting in primary fetal fibroblasts from sheep and pig. Cloning Stem Cells 3(4):221–231. https://doi.org/10.1089/15362300152725945

45. Golovan SP, Meidinger RG, Ajakaiye A et al (2001) Pigs expressing salivary phytase produce low-phosphorus manure. Nat Biotechnol 19:741–745

46. Wright G, Carver A, Cottom D et al (1991) High level expression of active human alpha-1-antitrypsin in the milk of transgenic sheep. Biotechnology (N Y) 9:830–834

47. Lu RM, Hwang YC, Liu IJ et al (2020) Development of therapeutic antibodies for the treatment of diseases. J Biomed Sci 27:1–30

48. Castelli MS, McGonigle P, Hornby PJ (2019) The pharmacology and therapeutic applications of monoclonal antibodies. Pharmacol Res Perspect 7:e00535

49. Schwaber J, Cohen EP (1973) Human × mouse somatic cell hybrid clone secreting immunoglobulins of both parental types. Nature 244:444–447

50. Siegel DL (2002) Recombinant monoclonal antibody technology. Transfus Clin Biol 9: 15–22

51. Kuroiwa Y, Kasinathan P, Choi YJ et al (2002) Cloned transchromosomic calves producing human immunoglobulin. Nat Biotechnol 20: 889–894

52. Kuroiwa Y, Kasinathan P, Matsushita H et al (2004) Sequential targeting of the genes encoding immunoglobulin-mu and prion protein in cattle. Nat Genet 36:775–780

53. Denning C, Priddle H New frontiers in gene targeting and cloning: success application and challenges in domestic animals and human embryonic stem cells. Reproduction:1–11. https://doi.org/10.1530/rep.0.1260001

54. Kuroiwa Y, Kasinathan P, Sathiyaseelan T et al (2009) Antigen-specific human polyclonal antibodies from hyperimmunized cattle. Nat Biotechnol 27:173–181

55. Akiko SH, Matsushita Hua W, Jin-An JP, Kasinathan EJ, Zhongde S, Yoshimi W, Glenn KJ, Knott (2013) Physiological level production of antigen-specific human immunoglobulin in cloned transchromosomic cattle. PLoS One 8(10):e78119. https://doi.org/10.1371/journal.pone.0078119

56. Hiroaki MA, Hua S, Jin-an W, Poothappillai J, Kasinathan Eddie J, Zhongde S, Yoshimi W, Glenn KJ, Knott (2014) Triple immunoglobulin gene knockout transchromosomic cattle: Bovine Lambda cluster deletion and its effect on fully human polyclonal antibody production. PLoS One 9(3):e90383. https://doi.org/10.1371/journal.pone.0090383

57. Hiroaki MA, Sano Hua W, Zhongde W J-a, Poothappillai J, Kasinathan EJ, Yoshimi S, Kuroiwa SD, Fugmann (2015) Species-specific chromosome engineering greatly improves fully human polyclonal antibody production profile in cattle. PLoS One 10(6):e0130699. https://doi.org/10.1371/journal.pone.0130699

58. Dye JM, Wu H, Hooper JW et al (2016) Production of potent fully human polyclonal antibodies against Ebola Zaire virus in transchromosomal cattle. Sci Rep 6:24897

59. Berg P, Baltimore D, Boyer HW et al (1974) Potential biohazards of recombinant DNA molecules published by: American Association for the Advancement of Science. Nature 185:303

60. Nepomuceno AL, Fuganti-Pagliarini R, Felipe MSS et al (2020) Brazilian biosafety law and the new breeding technologies. Front Agric Sci Eng 7:204–210

61. Kamilla SV, Tadeu P-CD, Covas (2017) Production of recombinant coagulation factors: are humans the best host cells? Bioengineered 8(5):462–470 5. https://doi.org/10.1080/21655979.2017.1279767

62. William M, Muir (2004) The threats and benefits of GM fish. EMBO Rep 5(7):654–659. https://doi.org/10.1038/sj.embor.7400197

63. Bratspies R (2005) Glowing in the dark: how America's first transgenic animal escaped regulation. Minnesota J Law Sci Technol 6:457–504

64. Richard D, Karl H, Yiyang R, Liu WM, Muir (2015) Evolution 69(5):1143–1157. https://doi.org/10.1111/evo.12662

65. Jeffrey E., Hill Anne R., Kapuscinski Tyler, Pavlowich (2011) Fluorescent transgenic Zebra Danio more vulnerable to predators than wild-type fish. Trans Am Fish Soc 140(4) 1001-1005 12. https://doi.org/10.1080/00028487.2011.603980

66. Jeffrey E, Hill LL, Scott L, Hardin (2014) Assessment of the risks of transgenic fluorescent ornamental fishes to the United States using the fish invasiveness screening kit (FISK). Trans Am Fish Soc 143(3):817–829 24. https://doi.org/10.1080/00028487.2014.880741

67. Julie DM, Christine H, Timothy O'D, Catherine E, Shirish B, Hirani KM, Katherine E, Gordon JM, McPherson (1991) Transgenic expression of a variant of human tissue-type Plasminogen activator in goat milk: purification and characterization of the recombinant enzyme. Nat Biotechnol 9(9):839–843. https://doi.org/10.1038/nbt0991-839

68. Henry C (2014) AquAdvantage® Salmon - a pioneering application of biotechnology in

aquaculture. BMC Proc 8(S4):O31. https://doi.org/10.1186/1753-6561-8-S4-O31

69. Jun S, Du Zhiyuan GGL, Fletcher Margaret A, Shears Madonna J, King David R, Idler Choy L, Hew (1992) Growth enhancement in transgenic Atlantic Salmon by the use of an "All Fish" Chimeric growth hormone gene construct. Nat Biotechnol 10(2):176–181. https://doi.org/10.1038/nbt0292-176

70. Erik MP, Zwanenberg (2003) Food and agricultural biotechnology policy: how much autonomy can developing countries exercise? Dev Policy Rev 21(5-6):655–667. https://doi.org/10.1111/j.1467-8659.2003.00230.x

71. Jacques, Peter T, Zuurbier (2008) Quality and safety standards in the food industry developments and challenges. Int J Prod Econ 113(1):107–122 S092552730700312X. https://doi.org/10.1016/j.ijpe.2007.02.050

72. Silva G, Poirot L, Galetto R et al (2011) Meganucleases and other tools for targeted genome engineering: perspectives and challenges for gene therapy. Curr Gene Ther 11:11–27

73. Belfort M, Roberts RJ (1997) Homing endonucleases: keeping the house in order. Nucleic Acids Res 25:3379–3388

74. Jasin M (1996) Genetic manipulation of genomes with rare-cutting endonucleases. Trends Genet 12:224–228

75. Kim YG, Cha J, Chandrasegaran S (1996) Hybrid restriction enzymes: zinc finger fusions to Fok I cleavage domain. Proc Natl Acad Sci U S A 93:1156–1160

76. Segal DJ, Meckler JF (2013) Genome engineering at the dawn of the golden age. Annu Rev Genomics Hum Genet 14:135–158

77. Zheng N, Li L, Wang X (2020) Molecular mechanisms, off-target activities, and clinical potentials of genome editing systems. Clin Transl Med 10:412–426

78. Perez-Pinera P, Ousterout DG, Gersbach CA (2012) Advances in targeted genome editing. Curr Opin Chem Biol 16:268–277

79. Sun N, Zhao H (2013) Transcription activator-like effector nucleases (TALENs): a highly efficient and versatile tool for genome editing. Biotechnol Bioeng 110:1811–1821

80. Jinek M, Chylinski K, Fonfara I et al (2012) A programmable dual-RNA-guided DNA endonuclease in adaptive bacterial immunity. Science (80-) 337:816–821

81. Pennisi E (2013) The CRISPR craze. Science 341:833–836

82. Fox R (2019) Too much compromise in today's CRISPR pipelines. CRISPR J 2:143–145

83. Cornu TI, Thibodeau-Beganny S, Guhl E et al (2008) DNA-binding specificity is a major determinant of the activity and toxicity of zinc-finger nucleases. Mol Ther 16:352–358

84. Pattanayak V, Ramirez CL, Joung JK, Liu DR (2011) Revealing off-target cleavage specificities of zinc-finger nucleases by in vitro selection. Nat Methods 8:765–770

85. Guilinger JP, Pattanayak V, Reyon D et al (2014) Broad specificity profiling of TALENs results in engineered nucleases with improved DNA-cleavage specificity. Nat Methods 11:429–435

86. Lin Y, Cradick TJ, Brown MT et al (2014) CRISPR/Cas9 systems have off-target activity with insertions or deletions between target DNA and guide RNA sequences. Nucleic Acids Res 42:7473–7485

87. Fu Y, Foden JA, Khayter C et al (2013) High-frequency off-target mutagenesis induced by CRISPR-Cas nucleases in human cells. Nat Biotechnol 31:822–826

88. Hsu PD, Scott DA, Weinstein JA et al (2013) DNA targeting specificity of RNA-guided Cas9 nucleases. Nat Biotechnol 31:827–832

89. Cho SW, Kim S, Kim Y et al (2014) Analysis of off-target effects of CRISPR/Cas-derived RNA-guided endonucleases and nickases. Genome Res 24:132–141

90. Tsai SQ, Wyvekens N, Khayter C et al (2014) Dimeric CRISPR RNA-guided FokI nucleases for highly specific genome editing. Nat Biotechnol 32:569–576

91. Wyvekens N, Topkar VV, Khayter C et al (2015) Dimeric CRISPR RNA-guided FokI-dCas9 nucleases directed by truncated gRNAs for highly specific genome editing. Hum Gene Ther 26:425–431

92. Kleinstiver BP, Prew MS, Tsai SQ et al (2015) Engineered CRISPR-Cas9 nucleases with altered PAM specificities. Nature 523:481–485

93. Kleinstiver BP, Pattanayak V, Prew MS et al (2016) High-fidelity CRISPR-Cas9 nucleases with no detectable genome-wide off-target effects. Nature 529:490–495

94. Fu Y, Sander JD, Reyon D et al (2014) Improving CRISPR-Cas nuclease specificity using truncated guide RNAs. Nat Biotechnol 32:279–284

95. Tsai SQ, Nguyen NT, Malagon-Lopez J et al (2017) CIRCLE-seq: a highly sensitive in vitro screen for genome-wide CRISPR-

Cas9 nuclease off-targets. Nat Methods 14: 607–614

96. Zischewski J, Fischer R, Bortesi L (2017) Detection of on-target and off-target mutations generated by CRISPR/Cas9 and other sequence-specific nucleases. Biotechnol Adv 35:95–104

97. Ledford H (2019) Gene-edited animal creators look beyond US market. Nature 566: 433–434

98. Lassoued R, Smyth SJ, Phillips PWB, Hesseln H (2018) Regulatory uncertainty around new breeding techniques. Front Plant Sci 9:1–10

99. Kleter GA, Kuiper HA, Kok EJ (2019) Gene-edited crops: towards a harmonized safety assessment. Trends Biotechnol 37:443–447

100. Davison J, Ammann K (2017) New GMO regulations for old: determining a new future for EU crop biotechnology. GM Crops Food 8:13–34

101. Whelan AI, Lema MA (2015) Regulatory framework for gene editing and other new breeding techniques (NBTs) in Argentina. GM Crops Food 6:253–265

102. Friedrichs S, Takasu Y, Kearns P et al (2019) An overview of regulatory approaches to genome editing in agriculture. Biotechnol Res Innov 3:208–220

103. Hamburger DJS (2018) Normative criteria and their inclusion in a regulatory framework for new plant varieties derived from genome editing. Front Bioeng Biotechnol 6:176

104. Custers R, Casacuberta JM, Eriksson D et al (2019) Genetic alterations that do or do not occur naturally; Consequences for genome edited organisms in the context of regulatory oversight. Front Bioeng Biotechnol 6:213

105. Friedrichs S, Takasu Y, Kearns P et al (2019) Policy considerations regarding genome editing. Trends Biotechnol 37:1029–1032

106. Friedrichs S, Takasu Y, Kearns P et al (2019) Meeting report of the OECD conference on "Genome Editing: Applications in Agriculture—Implications for Health, Environment and Regulation". Springer International Publishing

107. Smyth SJ (2017) Canadian regulatory perspectives on genome engineered crops. GM Crops Food 8:35–43

108. Mallapaty S (2019) Australian gene-editing rules adopt "middle ground". Nature:4–5. https://doi.org/10.1038/d41586-019-01282-8

109. Carroll D (2017) Genome editing: past, present, and future. Yale J Biol Med 90:653–659

110. Auerbach C, Robson JM (1946) Chemical production of mutations. Nature 157:302

111. Rothstein RJ (1983) One-step gene disruption in yeast. Methods Enzymol 101:202–211

112. Scherer S, Davis RW (1979) Replacement of chromosome segments with altered DNA sequences constructed in vitro. Proc Natl Acad Sci U S A 76:4951–4955

113. Smithies O, Gregg RG, Boggs SS et al (1985) Insertion of DNA sequences into the human chromosomal beta-globin locus by homologous recombination. Nature 317:230–234

114. Thomas KR, Folger KR, Capecchi MR (1986) High frequency targeting of genes to specific sites in the mammalian genome. Cell 44:419–428

115. Mario R, Capecchi (2005) Gene targeting in mice: functional analysis of the mammalian genome for the twenty-first century. Nat Rev Genet 6(6):507–512. https://doi.org/10.1038/nrg1619

116. Fernández A, Josa S, Montoliu L (2017) A history of genome editing in mammals. Mamm Genome 28:237–246

117. Geurts AM, Cost GJ, Freyvert Y et al (2009) Knockout rats via embryo microinjection of zinc-finger nucleases. Science 325:433

118. Tesson L, Usal C, Ménoret S et al (2011) Knockout rats generated by embryo microinjection of TALENs. Nat Biotechnol 29:695–696

119. Yen S-T, Zhang M, Deng JM et al (2014) Somatic mosaicism and allele complexity induced by CRISPR/Cas9 RNA injections in mouse zygotes. Dev Biol 393:3–9

120. Wang Y, Zhao S, Bai L et al (2013) Expression systems and species used for transgenic animal bioreactors. Biomed Res Int 2013: 580463

121. Bosze Z, Hiripi L, Carnwath JW, Niemann H (2003) The transgenic rabbit as model for human diseases and as a source of biologically active recombinant proteins. Transgenic Res 12:541

122. Bühler TA, Bruyére T, Went DF et al (1990) Rabbit β-casein promoter directs secretion of human interleukin-2 into the milk of transgenic rabbits. Biotechnology (N Y) 8(2): 140–143

123. Zinovieva N, Lassnig C, Schams D et al (1998) Stable production of human insulin-like growth factor 1 (IGF-1) in the milk of hemi- and homozygous transgenic rabbits over several generations. Transgenic Res 7(6):437–447

124. Lipiński D, Jura J, Kalak R et al (2003) Transgenic rabbit producing human growth hormone in milk. J Appl Genet 44(2):165–174

125. Han ZS, Li QW, Zhang ZY et al (2008) Adenoviral vector mediates high expression levels of human lactoferrin in the milk of rabbits. J Microbiol Biotechnol 18(1):153–159

126. Massoud M, Bischoff R, Dalemans W et al (1991) Expression of active recombinant human α1-antitrypsin in transgenic rabbits. J Biotechnol 18(3):193–203

127. Yang H, Li Q, Han Z, Hu J (2012) High level expression of recombinant human antithrombin in the mammary gland of rabbits by adenoviral vectors infection. Anim Biotechnol 23(2):89–100

128. Koles K, van Berkel PHC, Pieper FR et al (2004) N- and O-glycans of recombinant human C1 inhibitor expressed in the milk of transgenic rabbits. Glycobiology 14:51–64

129. Mikus T, Poplstein M, Sedláková J et al (2004) Generation and phenotypic analysis of a transgenic line of rabbits secreting active recombinant human erythropoietin in the milk. Transgenic Res 13(5):487–498

130. Jongen SP, Gerwig GJ, Leeflang BR et al (2007) N-glycans of recombinant human acid α-glucosidase expressed in the milk of transgenic rabbits. Glycobiology 17(6):600–619

131. Major P, Baczkó I, Hiripi L et al (2016) A novel transgenic rabbit model with reduced repolarization reserve: long QT syndrome caused by a dominant-negative mutation of the KCNE1 gene. Br J Pharmacol 173(12):2046–2061

132. Hornyik T, Castiglione A, Franke G et al (2020) Transgenic LQT2, LQT5, and LQT2-5 rabbit models with decreased repolarisation reserve for prediction of drug-induced ventricular arrhythmias. Br J Pharmacol 177(16):3744–3759

133. Fan J, Kitajima S, Watanabe T et al (2015) Rabbit models for the study of human atherosclerosis: from pathophysiological mechanisms to translational medicine. Pharmacol Ther 146:104–119

134. Yuan T, Zhong Y, Wang Y et al (2019) Generation of hyperlipidemic rabbit models using multiple sgRNAs targeted CRISPR/Cas9 gene editing system. Lipids Health Dis 18(1):69

135. Sui T, Lau YS, Liu D et al (2018) A novel rabbit model of Duchenne muscular dystrophy generated by CRISPR/Cas9. Dis Model Mech 11(6):dmm032201

136. Hashikawa Y, Hayashi R, Tajima M et al (2020) Generation of knockout rabbits with X-linked severe combined immunodeficiency (X-SCID) using CRISPR/Cas9. Sci Rep 10:9957

137. Guo R, Wan Y, Xu D et al (2016) Generation and evaluation of Myostatin knock-out rabbits and goats using CRISPR/Cas9 system. Sci Rep 6:29855

138. Xiao N, Li H, Shafique L et al (2019) A novel pale-yellow coat color of rabbits generated via MC1R mutation with CRISPR/Cas9 system. Front Genet 10:875

139. Kalds P, Zhou S, Cai B et al (2019) Sheep and goat genome engineering: from random transgenesis to the CRISPR era. Front Genet 10:1–27

140. Salabi F, Nazari M, Chen Q et al (2014) Myostatin knockout using zinc-finger nucleases promotes proliferation of ovine primary satellite cells in vitro. J Biotechnol 192 Pt A:268–280

141. Zhang C, Wang L, Ren G et al (2014) Targeted disruption of the sheep MSTN gene by engineered zinc-finger nucleases. Mol Biol Rep 41:209–215

142. Zhang X, Wang L, Wu Y et al (2016) Knockout of myostatin by zinc-finger nuclease in sheep fibroblasts and embryos. Asian-Australas J Anim Sci 29:1500–1507

143. Høst A (2002) Frequency of cow's milk allergy in childhood. Ann Allergy Asthma Immunol 89:33–37

144. Apps JR, Beattie RM (2009) Cow's milk allergy in children. BMJ 339:b2275

145. Xiong K, Li S, Zhang H et al (2013) Targeted editing of goat genome with modular-assembly zinc finger nucleases based on activity prediction by computational molecular modeling. Mol Biol Rep 40:4251–4256

146. Song Y, Cui C, Zhu H et al (2015) Expression, purification and characterization of zinc-finger nuclease to knockout the goat beta-lactoglobulin gene. Protein Expr Purif 112:1–7

147. Yuan Y, Cheng Y, Wang J, Peng Q (2016) Targeted mutagenesis of beta-lactoglobulin gene in caprine fetal fibroblasts by context-dependent assembly zinc-finger nucleases. Open Access Libr J 3:1–8

148. Proudfoot C, Carlson DF, Huddart R et al (2015) Genome edited sheep and cattle. Transgenic Res 24:147–153

149. Zhao X, Ni W, Chen C et al (2016) Targeted editing of myostatin gene in sheep by transcription activator-like effector nucleases. Asian-Australas J Anim Sci 29:413–418

150. Li H, Wang G, Hao Z et al (2016) Generation of biallelic knock-out sheep via gene-editing

and somatic cell nuclear transfer. Sci Rep 6: 33675

151. Ge H, Cui C, Liu J et al (2016) The growth and reproduction performance of TALEN-mediated β-lactoglobulin-knockout bucks. Transgenic Res 25:721–729

152. Yuan Y-G, Song S-Z, Zhu M-M et al (2017) Human lactoferrin efficiently targeted into caprine beta-lactoglobulin locus with transcription activator-like effector nucleases. Asian-Australas J Anim Sci 30:1175–1182

153. Zhu H, Liu J, Cui C et al (2016) Targeting human α-lactalbumin gene insertion into the goat β-lactoglobulin locus by TALEN-mediated homologous recombination. PLoS One 11:e0156636

154. Cui C, Song Y, Liu J et al (2015) Gene targeting by TALEN-induced homologous recombination in goats directs production of β-lactoglobulin-free, high-human lactoferrin milk. Sci Rep 5:10482

155. Wang X, Niu Y, Zhou J et al (2016) Multiplex gene editing via CRISPR/Cas9 exhibits desirable muscle hypertrophy without detectable off-target effects in sheep. Sci Rep 6:32271

156. Rachel RP, Julien S, Charlotte S, Christian A, Laeticia T, David L, Florent P, Olivier W, Guillaume B, Mathieu T, Cécile L, Gilles C, Gwenola F, James T-K, Kijas (2015) A point mutation in suppressor of Cytokine signalling 2 (Socs2) increases the susceptibility to inflammation of the mammary gland while associated with higher body weight and size and higher milk production in a sheep model. PLoS Genet 11(12):e1005629. https://doi.org/10.1371/journal.pgen.1005629

157. Ni W, Qiao J, Hu S et al (2014) Efficient gene knockout in goats using CRISPR/Cas9 system. PLoS One 9:e106718

158. Hu R, Fan ZY, Wang BY et al (2017) RAPID COMMUNICATION: generation of FGF5 knockout sheep via the CRISPR/Cas9 system. J Anim Sci 95:2019–2024

159. Li W-R, Liu C-X, Zhang X-M et al (2017) CRISPR/Cas9-mediated loss of FGF5 function increases wool staple length in sheep. FEBS J 284:2764–2773

160. Zhang R, Wu H, Lian Z (2019) Bioinformatics analysis of evolutionary characteristics and biochemical structure of FGF5 gene in sheep. Gene 702:123–132

161. Zhang X, Li W, Liu C et al (2017) Alteration of sheep coat color pattern by disruption of ASIP gene via CRISPR Cas9. Sci Rep 7:8149

162. Zhou W, Wan Y, Guo R et al (2017) Generation of beta-lactoglobulin knock-out goats using CRISPR/Cas9. PLoS One 12: e0186056

163. Ma T, Tao J, Yang M et al (2017) An AANAT/ASMT transgenic animal model constructed with CRISPR/Cas9 system serving as the mammary gland bioreactor to produce melatonin-enriched milk in sheep. J Pineal Res 63. https://doi.org/10.1111/jpi.12406

164. Fabre S, Pierre A, Mulsant P et al (2006) Regulation of ovulation rate in mammals: contribution of sheep genetic models. Reprod Biol Endocrinol 4:20

165. Zhang X, Li W, Wu Y et al (2017) Disruption of the sheep BMPR-IB gene by CRISPR/Cas9 in in vitro-produced embryos. Theriogenology 91:163–172.e2

166. Golding MC, Long CR, Carmell MA et al (2006) Suppression of prion protein in livestock by RNA interference. Proc Natl Acad Sci U S A 103:5285–5290

167. Yu G, Chen J, Yu H et al (2006) Functional disruption of the prion protein gene in cloned goats. J Gen Virol 87:1019–1027

168. Hryhorowicz M, Lipiński D, Hryhorowicz S et al (2020) Application of genetically engineered pigs in biomedical research. Genes (Basel) 11(6):670

169. Zhou ZP, Yang LL, Cao H et al (2019) In vitro validation of a CRISPR-mediated CFTR correction strategy for preclinical translation in pigs. Hum Gene Ther 30:1101–1116

170. Ruan J, Hirai H, Yang D et al (2019) Efficient gene editing at major CFTR mutation loci. Mol Ther Nucleic Acids 16:73–81

171. Kang J-T, Kwon D-K, Park A-R et al (2016) Production of α1,3-galactosyltransferase targeted pigs using transcription activator-like effector nuclease-mediated genome editing technology. J Vet Sci 17:89–96

172. Chuang C-K, Chen C-H, Huang C-L et al (2017) Generation of GGTA1 mutant pigs by direct pronuclear microinjection of CRISPR/Cas9 plasmid vectors. Anim Biotechnol 28: 174–181

173. Lipiński D, Nowak-Terpiłowska A, Hryhorowicz M et al (2019) Production of ZFN-mediated GGTA1 knock-out pigs by microinjection of gene constructs into pronuclei of zygotes. Pol J Vet Sci 22:91–100

174. Estrada JL, Martens G, Li P et al (2015) Evaluation of human and non-human primate antibody binding to pig cells lacking GGTA1/CMAH/β4GalNT2 genes. Xenotransplantation 22:194–202

175. Butler JR, Martens GR, Estrada JL et al (2016) Silencing porcine genes significantly reduces human-anti-pig cytotoxicity profiles: an alternative to direct complement regulation. Transgenic Res 25:751–759

176. Wang Z-Y, Martens GR, Blankenship RL et al (2017) Eliminating xenoantigen expression on swine RBC. Transplantation 101:517–523

177. Adams AB, Kim SC, Martens GR et al (2018) Xenoantigen deletion and chemical immunosuppression can prolong renal xenograft survival. Ann Surg 268:564–573

178. Niu D, Wei H-J, Lin L et al (2017) Inactivation of porcine endogenous retrovirus in pigs using CRISPR-Cas9. Science 357:1303–1307

179. Su F, Wang Y, Liu G et al (2016) Generation of transgenic cattle expressing human β-defensin 3 as an approach to reducing susceptibility to Mycobacterium bovis infection. FEBS J 283:776–790

180. Monzani PS, Adona PR, Ohashi OM et al (2016) Transgenic bovine as bioreactors: challenges and perspectives. Bioengineered 7:123–131

181. Demain AL, Vaishnav P (2009) Production of recombinant proteins by microbes and higher organisms. Biotechnol Adv 27:297–306

182. Kim JY, Kim Y-G, Lee GM (2012) CHO cells in biotechnology for production of recombinant proteins: current state and further potential. Appl Microbiol Biotechnol 93:917–930

183. Yang B, Wang J, Tang B et al (2011) Characterization of bioactive recombinant human lysozyme expressed in milk of cloned transgenic cattle. PLoS One 6:e17593

184. Liu X, Wang Y, Tian Y et al (2014) Generation of mastitis resistance in cows by targeting human lysozyme gene to β-casein locus using zinc-finger nucleases. Proc Biol Sci 281:20133368

185. Liu X, Wang Y, Guo W et al (2013) Zinc-finger nickase-mediated insertion of the lysostaphin gene into the beta-casein locus in cloned cows. Nat Commun 4:2565

186. Sun Z, Wang M, Han S et al (2018) Production of hypoallergenic milk from DNA-free beta-lactoglobulin (BLG) gene knockout cow using zinc-finger nucleases mRNA. Sci Rep 8:15430

187. Wei J, Wagner S, Maclean P et al (2018) Cattle with a precise, zygote-mediated deletion safely eliminate the major milk allergen beta-lactoglobulin. Sci Rep 8:7661

188. Kontopidis G, Holt C, Sawyer L (2004) Invited review: beta-lactoglobulin: binding properties, structure, and function. J Dairy Sci 87:785–796

189. Luo Y, Wang Y, Liu J et al (2016) Generation of TALE nickase-mediated gene-targeted cows expressing human serum albumin in mammary glands. Sci Rep 6:20657

190. Fanali G, di Masi A, Trezza V et al (2012) Human serum albumin: from bench to bedside. Mol Asp Med 33:209–290

191. Wu H, Zhang Y, Wang Y et al (2015) TALE nickase-mediated SP110 knockin endows cattle with increased resistance to tuberculosis. Proc Natl Acad Sci U S A 112:E1530–E1539

192. Cong L, Ran FA, Cox D et al (2013) Multiplex genome engineering using CRISPR/Cas systems. Science 339:819–823

193. Gao Y, Wu H, Wang Y et al (2017) Single Cas9 nickase induced generation of NRAMP1 knockin cattle with reduced off-target effects. Genome Biol 18:13

194. Vidal S, Tremblay ML, Govoni G et al (1995) The Ity/Lsh/Bcg locus: natural resistance to infection with intracellular parasites is abrogated by disruption of the Nramp1 gene. J Exp Med 182:655–666

195. Li HT, Zhang TT, Zhou YQ et al (2006) SLC11A1 (formerly NRAMP1) gene polymorphisms and tuberculosis susceptibility: a meta-analysis. Int J Tuberc Lung Dis 10:3–12

196. Ikeda M, Matsuyama S, Akagi S et al (2017) Correction of a disease mutation using CRISPR/Cas9-assisted genome editing in Japanese black cattle. Sci Rep 7:17827

197. Grondahl-Nielsen C, Simonsen HB, Lund JD, Hesselholt M (1999) Behavioural, endocrine and cardiac responses in young calves undergoing dehorning without and with use of sedation and analgesia. Vet J 158:14–20

198. Brenneman RA, Davis SK, Sanders JO et al (1996) The polled locus maps to BTA1 in a Bos indicus x Bos taurus cross. J Hered 87:156–161

199. Aldersey JE, Sonstegard TS, Williams JL, Bottema CDK (2020) Understanding the effects of the bovine POLLED variants. Anim Genet 51:166–176

200. Tan W, Carlson DF, Lancto CA et al (2013) Efficient nonmeiotic allele introgression in livestock using custom endonucleases. Proc Natl Acad Sci U S A 110:16526–16531

201. Carlson DF, Lancto CA, Zang B et al (2016) Production of hornless dairy cattle from genome-edited cell lines. Nat Biotechnol 34:479–481

202. Schuster F, Aldag P, Frenzel A et al (2020) CRISPR/Cas12a mediated knock-in of the Polled Celtic variant to produce a polled genotype in dairy cattle. Sci Rep 10:13570

203. Paul B, Montoya G (2020) CRISPR-Cas12a: functional overview and applications. Biom J 43:8–17

Chapter 7

Mouse Cloning Using Outbred Oocyte Donors and Nontoxic Reagents

Sayaka Wakayama, Yukari Terashita, Yoshiaki Tanabe, Naoki Hirose, and Teruhiko Wakayama

Abstract

Somatic cell nuclear transfer (SCNT) technology has become a useful tool for animal cloning, gene manipulation, and genomic reprogramming research. However, the standard mouse SCNT protocol remains expensive, labor-intensive, and requires hard work for many hours. Therefore, we have been trying to reduce the cost and simplify the mouse SCNT protocol. This chapter describes the methods to use low-cost mouse strains and steps from the mouse cloning procedure. Although this modified SCNT protocol will not improve the success rate of mouse cloning, it is a cheaper, simpler, and less tiring method that allows us to perform more experiments and obtain more offspring with the same working time as the standard SCNT protocol.

Key words Cloning, ICR, Latrunculin A, Nuclear transfer, Outbred strain, PLCζ

---

1 Introduction

Since the cloning of Dolly the sheep, which was first reported in 1997 [1], many mammalian species have been successfully cloned using somatic cell nuclear transfer (SCNT). However, due to the low success rate of cloning, this technique has not yet been applied to industrial or agricultural practices. For this reason, efforts have been made in many laboratories to understand the mechanisms of reprogramming and increasing cloning efficiency. One obstacle in studying reprogramming is its technical difficulty to perform SCNT. The process of SCNT is complicated and labor-intensive, which takes a long time. In general, the mouse SCNT method involves several steps in the nuclear transfer and in the activation of the reconstructed oocytes and culture of embryos with cytokinesis and histone deacetylase inhibitors (HDACi) [2, 3]. Nuclear transfer techniques were earlier significantly improved by developing a new method in which the nuclei were directly injected into

Marcelo Tigre Moura (ed.), *Somatic Cell Nuclear Transfer Technology*, Methods in Molecular Biology, vol. 2647,
https://doi.org/10.1007/978-1-0716-3064-8_7,
© The Author(s), under exclusive license to Springer Science+Business Media, LLC, part of Springer Nature 2023

oocytes using a Piezo impact drive unit rather than via cell fusion [3, 4]. However, the other processes have not been improved.

In the standard method, cytochalasin B (CB) is used as a cytokinesis inhibitor, essential for SCNT to prevent unwanted haploidization of the donor cell. However, due to its toxicity, the embryo must be washed completely from CB at 6 h after the oocyte activation. On the other hand, when the HDACi trichostatin A (TSA) was added in a culture medium, the efficiency of the mouse cloning could be enhanced by up to fivefold [2, 5–7]. However, TSA also has a toxicity to embryo development; therefore, the embryo must be washed completely by TSA at 10 h after oocyte activation [8]. Thus, to complete all SCNT processes, it takes 16 h, usually from 7 am to 11 pm (Fig. 1a), which can be tiring. To simplify this method, we have examined the effect of latrunculin A (Lat A) on SCNT because it has less toxicity to the cells. When CB was replaced with Lat A, the reconstructed embryos could be cultured with Lat A up to 10 h without changing the medium at 6 h after activation (Fig. 1b). Therefore, it would allow the experimentalist to rest for some time during the SCNT protocol. Interestingly, this method not only skipped one step but also slightly increased the mouse cloning success rate without epigenetic modifications [9, 10]. These results suggest that the cloning efficiency could be enhanced by either correcting epigenetic abnormalities or technical improvements.

Recently, we also succeeded in reducing one more step from the SCNT method, which is the oocyte activation treatment. Originally, reconstructed oocytes have to be activated with an artificial activation stimulus (usually $SrCl_2$) instead of fertilization stimulus by spermatozoa to initiate embryo development. By this step, some of the reconstructed oocytes die (approximately 10%). In addition, it is also conceivable that this artificial activation may contribute to the low success rate of mouse cloning. Therefore, we tried to activate reconstructed oocytes by sperm-specific phospholipase Cζ (PLCζ), which could mimic the natural process of fertilization [11, 12]. The cRNA of PLCζ was co-injected with donor nucleus into enucleated oocytes. As a result, it was found that most of all reconstructed oocytes were activated without any additional treatment and without oocytes lysis during activation (Fig. 1c). Unfortunately, although this method activates the reconstructed oocyte more naturally, the success rate of mouse cloning did not increase, suggesting that the artificial activation did not contribute to the low success rate of cloning [13].

On the other hand, cloned mice can now be generated from donor cells both from hybrid (F1) mice and inbred mouse strains [6]. However, recipient oocytes have to be collected only from F1 mice, such as BDF1 or BCF1 [14]. If F1 mice were produced in our mouse facility, there is a risk of mixing the parents and F1 offspring (BDF1 and B6 have the same coat color). Therefore, it is safer and

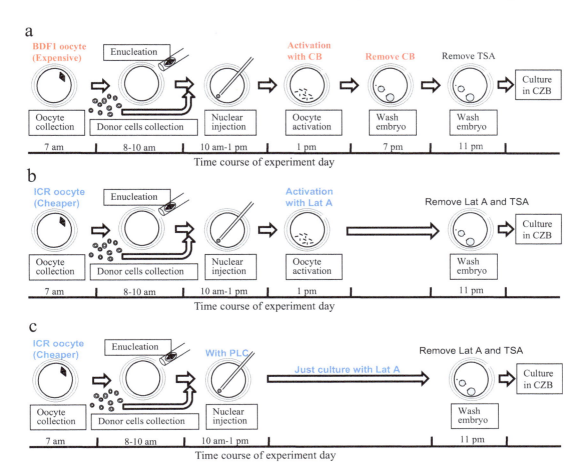

Fig. 1 Outline of the procedures on the day of the experiment. (**a**) Original method, (**b**) Lat A method, (**c**) Lat A and PLC method

more reliable to purchase F1 mice, although this is more expensive. For this reason, we examined the potential of ICR (outbred strain) oocytes and whether they can be used as recipient oocytes instead of F1 oocytes. The advantage of ICR is not only its ability to breed easily, but it also has a large litter size to produce many offspring. Moreover, ICR oocytes showed that the cytoplasm was transparent, the metaphase II spindle was clearly visible, and that it was easy to be removed from oocytes (Fig. 2). Although the success rate of mouse cloning was slightly reduced, the clone-specific epigenetic abnormality in cloned embryos was the same as when using F1 oocytes [15]. The use of ICR or outbred strains for research will allow laboratories to participate in the study of reprogramming without huge research funds. Here, we describe our mouse cloning protocol using outbred oocyte donors and nontoxic reagents.

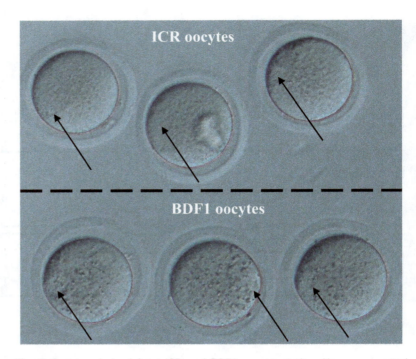

Fig. 2 Oocytes derived from ICR and BDF1 mouse strains. Upper suggested oocytes of ICR strain and lower suggested oocytes of BDF1 strain (black arrow: MII spindle)

2 Materials

2.1 Equipment

1. Inverted microscope with Hoffman or Nomarski optics.
2. Micromanipulator set.
3. Microforge.
4. Pipette puller.
5. Glass pipettes.
6. Warm plate.
7. Piezo impact drive system.
8. Humidified incubator set at 37 °C, 5% CO_2 in air.
9. Tissue culture hood.
10. Centrifuge.
11. Spectrophotometer.

2.2 Mouse Strains

1. Donor nucleus: The most popular hybrid mouse strains such as B6D2F1 (C57BL/6·DBA/2) or B6C3F1 (C57BL/6·C3H/He) can be used as donors. Inbred strains such as C57BL/6 or C3H/He can be used, but they have a lower embryonic development rate than hybrid strains [14, 16].

Simple Low-Cost Mouse Cloning   155

2. Recipient oocytes (B6D2F1): F1 mice (approximately 2–-3 months old) are the preferred sources of oocytes but are expensive and difficult to be produced in a laboratory by the care of students.

3. Recipient oocytes (ICR) (alternative method): ICR mice (approximately 2–3 months old, same as F1) can also be used as recipient oocytes [15]. The ICR oocytes were very translucent, and the metaphase spindles are easier to find than BDF1 oocytes. ICR mice can be produced from a mouse facility at a very low cost and even with the care of students.

4. Recipients, foster mother, and vasectomized male mice: The ICR strain is used to generate pseudopregnant surrogate mothers, lactating foster mothers, and vasectomized males.

2.3   Media and Solutions

1. Strontium chloride ($SrCl_2$) stock solution: 100 mM (20× stock solution). Dissolve $SrCl_2 \cdot 6H_2O$ in distilled water (DW) to make up a total volume of 100 mM and store in aliquots at room temperature. The final concentration will be 5 mM.

2. EGTA stock solution (option, if Ca-free CZB is not available): 40 mM (20× stock solution). Dissolve EGTA in DW to make up a total volume of 40 mM and store in aliquots at 4 °C. The final concentration will be 2 mM.

3. Cytochalasin B (CB) stock solution: 500 µg/mL (100× CB stock solution). Add 2 mL DMSO to a vial with 1 mg CB. Aliquot into small tubes (10–20 µL) and store at −30 °C. The final concentration will be 5 µg/mL.

4. Lat A stock solution: 500 µM (100× Lat A stock solution). Add 475 µL DMSO to a vial with 100 µg Lat A. Aliquot into small tubes (5–10 µL) and store at −30 °C. The final concentration will be 5 µM [9]. This Lat A stock solution is optional and is not used in this protocol.

5. TSA stock solution: 5 µM (100× TSA stock solution). Add 3.307 mL DMSO to a vial with 1 mg TSA to make a 1 mM stock solution (first stock solution). Aliquot into small tubes (5–10 µL and/or 200 µL) and store at −30 °C (1 mM TSA stock solution). Then, take 2 µL of the first stock solution and dilute with 398 µL of DMSO. Aliquot into small tubes (5–10 µL) and store at −30 °C. The final concentration will be 50 nM. This TSA stock solution is optional and is not used in this protocol.

6. L + T (Lat A and TSA) stock solution: 5 µM TSA and 500 µM Lat A (100× L + T stock solution). Take 3 µL of the first stock solution of TSA and dilute it with 597 µL of DMSO (5 µM). Add 475 µL of this solution to a vial with 100 µg Lat A (500 µM) to make a final stock solution containing both TSA and Lat A. Aliquot into small tubes (5–10 µL) and store at −

156   Sayaka Wakayama et al.

30 °C. The final concentration will be 50 nM of TSA and 5 μM of Lat A.

7. Mouse PLCζ-cRNA: cDNA sequences encoding PLCζ were cloned into pCS2 [17, 18] and were used as templates for cRNA synthesis [13]. cRNA was synthesized from the linearized template plasmid by in vitro transcription. The synthesized cRNA was polyadenylated with a poly(A) tailing kit. The cRNA with poly(A) tail was precipitated using lithium chloride and dissolved in nuclease-free water. After measuring the concentration, aliquots of 400 ng/μL were stored at −80 °C. The final concentration will be 20 ng/μL.

8. Hyaluronidase: 10% (100× stock solution). Dissolve 0.1 mg of hyaluronidase in 1 mL of HEPES-buffered Chatot, Ziomek, and Bavister (H-CZB) medium. Aliquot into small tubes (5–10 μL) and store at −30 °C. The final concentration will be 0.1% concentration.

9. Pregnant mare serum gonadotropin (PMSG) and human chorionic gonadotropin (hCG): 50 IU/mL. Dissolve PMSG and hCG individually in normal saline and store in aliquots at −30 °C.

10. Anesthetic stock solution: 0.75/4/5 mg/kg medetomidine/ midazolam/butorphanol mixed agents. Dissolve 1.875 mL medetomidine (1 mg/mL), 2 mL midazolam (5 mg/mL), and 2.5 mL of butorphanol (5 mg/mL) in 18.625 mL of normal saline. The final volume will be 25 mL. Use 0.1 mL/ 10 g of body weight. 0.75 mg/kg atipamezole (5 mg/mL) in normal saline can be used to wake up the mice [19].

2.4 Culture Media

1. Oocytes and embryos culture media: CZB medium [20] is used for oocyte and embryo culture. H-CZB medium is used for gamete handling and SCNT under air. H-CZB is not suitable for use in a CO_2 incubator. Ca^{2+}-free CZB medium is used for oocyte activation.

2. Donor cell diffusion and nuclear collection medium: 12% polyvinylpyrrolidone (PVP: 360 kDa).

3. PLCζ-cRNA containing donor cell diffusion and nuclear collection medium (PLC-PVP, alternative method): Add 2 μL of PLCζ-cRNA stock solution to 38 μL of 12% PVP. The final concentration will be 20 ng/μL (see **Note 1**).

4. Enucleation media: Add 2 μL of CB stock solution to 198 μL of H-CZB medium. The final concentration of CB will be 5 μg/ mL.

5. Reconstructed oocyte activation, prevention of DNA loss and reprogramming media (activation medium): Add 10 μL of $SrCl_2$ stock solution (final concentration, 5 mM) and 2 μL of

L + T stock solution (final concentration, 5 μM Lat A and 50 nM TSA) to 188 μL of Ca^{2+}-free CZB medium. If Ca^{2+}-free medium is not available, add 10 μL of EGTA stock solution (final concentration, 2 mM) to 178 μL of CZB medium [21] before adding the $SrCl_2$ stock solution and L + T stock solution.

6. Reconstructed oocyte prevention of DNA loss and reprogramming media (reprogramming medium): If donor nucleus was injected into oocyte with PLCζ-cRNA, reconstructed oocytes could be activated without $SrCl_2$ treatment. In this case, 2 μL of L + T stock solution (final concentration, 5 μM Lat A and 50 nM TSA) to 198 μL of CZB medium should be used for embryo culture.

2.5 Preparation of Holding, Enucleation, and Injection Micropipettes

1. Pull the glass pipette by pipette puller.

2. Cut the tip of pipette at the appropriate diameter (*see* **Note 2**).

3. Bend all pipettes close to the tip (approximately 300 μm away) at 10–25° (this will depend on the set up of the micromanipulator) using a microforge.

4. Store pipettes in a 10 cm dish at room temperature for many days.

5. Insert a small volume of PMM Operation Liquid into the back of the pipette using a 1 mL syringe and a 30 G needle (approximately 1–2 mm-long column), which will enhance the power of the piezo unit. PMM Operation Liquid should be injected into the pipette immediately before use because it will evaporate within a day.

2.6 Preparation of Media Dishes

1. Oocyte culture dish during manipulation: Place 20 (e.g., 4 × 5) or more droplets of CZB medium (approximately 10 μL each) on a 6 cm cell culture dish and cover this with sterile mineral oil [22]. This dish is used for oocyte collection to artificial activation; thus, it must be prepared at the beginning of the protocol and warmed in a 37 °C incubator with 5% CO_2.

2. Manipulation chamber: The manipulation dish includes three types of media: H-CZB, enucleation medium, and 12% PVP medium. Each droplet should be 10–15 μL in volume, should be placed on the top of a lid of a 6 cm dish (the bottom of the dish cannot be used since the high edges will prevent the appropriate pipette set up) as shown (Fig. 3a), and should be covered with mineral oil. Drawing lines on the outside of the dish will allow the user to easily distinguish the different types of media.

3. Manipulation chamber with PLC-PVP (alternative method): The manipulation dish includes four types of media: H-CZB, enucleation medium, 12% PVP medium, and PLC-PVP

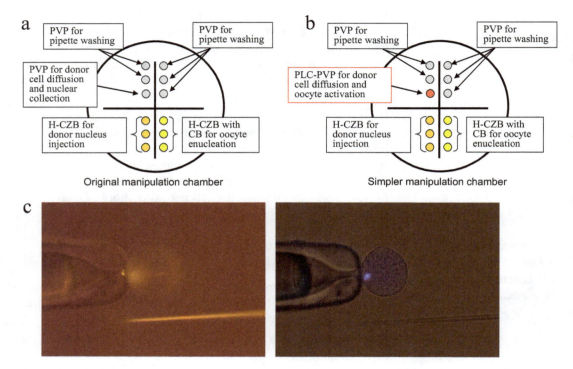

Fig. 3 Manipulation dish and volume of the injected PVP. The lid of a 6 cm dish is used as a micromanipulation chamber for SCNT. (**a**) Original manipulation chamber: this will be used for the Lat A method. (**b**) Simpler manipulation chamber: this will be used for PLCζ-mRNA co-injection with donor nucleus

medium. Each droplet should be 10–15 μL in volume, should be placed on the top of a lid of a 6 cm dish as shown (Fig. 3b), and should be covered with mineral oil as same as above.

4. Oocyte activation and reprogramming dish: Place 12 (e.g., 4 × 3) or more droplets of oocyte activation and reprogramming medium droplets and 8 (e.g., 4 × 2) droplets of CZB medium for washing embryos on a 6 cm dish [22]. Each droplet should be 10–15 μL in volume. Following this, cover this dish with mineral oil and draw lines to separate the droplet types on the outside of the dish.

5. Reprogramming dish (alternative method): Place 12 (e.g., 4 × 3) or more droplets of reprogramming medium droplets and 8 (e.g., 4 × 2) droplets of CZB medium for washing embryos on a 6 cm dish [22]. Each droplet should be 10–15 μL in volume. Following this, cover this dish with mineral oil and draw lines to separate the droplet types on the outside of the dish.

6. Culture dish: Place 12 or more droplets of CZB medium on a cell culture dish. The top 8 droplets are used for embryo washing and the last droplet is used for long-term culture of

Simple Low-Cost Mouse Cloning   159

the embryos until they are transferred to recipient mice. Each droplet should be 10–15 μL in volume. Following this, cover this dish with mineral oil.

2.7 Micro-manipulator Setup

1. Attach the holding pipette to a side of the micromanipulator.

2. On the opposite side, attach the enucleation pipette to the pipette holder of the piezo unit and fix the piezo unit onto the micromanipulator. The top of the pipette holder must be screwed on tightly. If this is not screwed tightly, the cutting of the zona pellucida of the oocyte by the piezo pulses will not be smooth.

3. Expel any air and oil along with a few drops of PMM Operation Liquid from the enucleation pipette for washing in PVP medium.

4. Wash both the inside and outside of the pipette with PVP medium until no oil remains (*see* **Note 3**).

3 Methods

3.1 Recipient Oocyte Preparation

1. PMSG injection: BDF1 or ICR female mice (8–10 weeks) are superovulated by an injection of 5 IU of PMSG into the abdominal cavity 3 days before the experiment. Typically, this is performed between 5 pm and 6 pm.

2. hCG injection: Those mice are then injected with hCG (5 IU) 48 h later (1 day before the experiment). Typically, this is performed between 5 pm and 6 pm, the same as PMSG injection.

3. Oocyte collection: Retrieve oocyte–cumulus cell complexes from the oviducal ampullae at 14–15 h after hCG injection.

4. Hyaluronidase treatment: To remove cumulus cells from oocyte, transfer oocyte-cumulus complexes into a 50 μL droplet of H-CZB containing 0.1% hyaluronidase (use 100× stock solution) for 5 min.

5. Oocyte selection: Wash oocytes from hyaluronidase medium using H-CZB twice and place them onto an oocyte culture dish prepared as mentioned above (*see* **Note 4**). The number of oocytes in a drop will depend on the level of skill and the type of experiment (*see* **Note 5**).

3.2 Donor Cell Preparation

3.2.1 Preparation of Cumulus Cells (Optional)

1. Cumulus cells are the easiest to prepare as nucleus donors because they can be collected and used immediately after oocyte selection without washing. In addition, there is no need to remove hyaluronidase from the medium [3].

160   Sayaka Wakayama et al.

2. Pick 1–3 µL of condensed cell suspension from hyaluronidase medium by mouth pipette.

3. Place the cells in a PVP medium droplet inside the micromanipulation chamber (Fig. 3a).

4. Immediately after introducing the cells to PVP medium, mix the cells with PVP medium gently but completely using sharp forceps (*see* **Note 6**). Take care not to scratch the bottom of the chamber.

3.2.2 Preparation of Tail Tip Fibroblasts (Optional)

1. Tail tip cells, probably fibroblast, have to be prepared 10 days before the day of the experiment, and is not so easy to perform nuclear transfer due to the quite hard cell membrane; these cells can easily collect either male or female, young or adult, wild or mutant mouse without special treatment [23].

2. Cut a part of the tail (at least 2 cm long) 2 weeks before the planned nuclear transfer experiment.

3. Wash the tails carefully in 70% ethanol.

4. Remove the skin in a sterile tissue culture hood and cut the tail into as many small pieces as possible (1–2 mm) in a 6 cm plastic dish.

5. Culture the fragments in 10 mL DMEM in a 5% CO_2, 37 °C incubator until they are used. There is no need to passage the cells if used within 2 weeks.

6. On the day of the experiment, remove the culture medium from the dish or flask and wash them in Dulbecco's phosphate-buffered saline (D-PBS(−)).

7. Remove PBS and incubate in trypsin-EDTA for 20–30 min in a 37 °C incubator with 5% CO_2 (*see* **Note 7**).

8. Add cell culture medium and triturate the cells to produce a single-cell suspension. Spin down the cells in a centrifuge at $300 \times g$ for 5 min.

9. Wash the cells at least three times by centrifugation with cell culture medium. Trypsin is very toxic at the time of nuclear injection; therefore, the donor cells must be washed thoroughly.

10. Prepare a very concentrated cell suspension in cell culture medium. The final volume should be less than 10 µL.

11. Mix the cells into PVP medium in the same manner as that described for cumulus cells.

3.2.3 Preparation of Embryonic Stem Cells (Optional)

Embryonic stem cells (ESCs) are the popular cell type for NT experiments because they have been demonstrated to yield the best success rate in the production of full-term offspring. However, ESCs are pluripotent cells and not differentiated somatic cells;

therefore, they are not appropriate for genomic reprogramming experiments. In addition, each ESC line, even if from the same genetic background, will react to this procedure to a different degree and will yield different results for embryonic development [24]. The number of passages of ESC lines will also affect the success rate. The ESC preparation is the same as the tail tip fibroblasts above (*see* Subheading 3.2.2, **steps 6–11**).

3.3 Donor Cell Preparation Using PLC-PVP Medium (Alternative Method)

1. This method is almost the same as above except for the preparation of the donor cell. When this method was used, the oocyte activation step could be skipped.

2. Add 2 μL of PLCζ stock solution to 38 μL of 12% PVP. The final concentration will be 20 ng/μL (*see* **Note 1**). Take about 10–15 μL of this mixed medium and place it on the manipulation chamber (*see* Subheading 3.2.3).

3. Pick 1–3 μL of condensed donor cell suspension the same as cumulus cells or fibroblast.

4. Place the cells in a PLC-PVP medium droplet inside the micromanipulation chamber (Fig. 3b, PLC-PVP for donor cell diffusion and oocyte activation).

5. Immediately after introducing the cells to PVP medium, mix the cells with PVP medium gently but completely using sharp forceps (*see* **Note 6**). Take care not to scratch the bottom of the chamber.

3.4 Enucleation of Oocytes

1. Place one group of oocytes in a droplet of enucleation medium into the micromanipulation chamber. Incubate for 5 min before starting the enucleation because CB in the enucleation medium can make the oolemma more flexible and reduce the risk of oocyte lysis.

2. Find oocytes spindle inside ooplasm and rotate the oocyte so that the spindle is placed between 8 and 10 o'clock positions and then attach the oocyte firmly to the holding pipette (*see* **Note 8**). Oocyte transparency is dependent on the mouse strain; oocytes from B6D2F1 and ICR are more visible than those from other strains (Fig. 2).

3. Using a few piezo pulses, cut through the zona pellucida; the addition of a slight negative pressure inside the pipette will increase the power of the pulses. To avoid damaging the oocyte, ensure there is a large space between the zona pellucida and the oolemma; this space should be approximately as thick as the zona pellucida itself (*see* **Note 9**).

4. Insert the enucleation pipette into the oocyte without using piezo pulses to avoid breaking the oolemma.

162 Sayaka Wakayama et al.

5. Remove the MII spindle by aspiration with a minimal volume of cytoplasm. Aspirating with a large volume of cytoplasm may hamper embryonic development.

6. Pinch off the oocyte membrane and the MII spindle slowly, drawing the needle out carefully until the oolemma seals (*see* **Note 10**).

7. When all oocytes in the group have been enucleated, wash them twice in CZB medium to remove CB completely.

8. Return them to the incubator in CZB medium for at least 30 min before starting donor cell injection. If CB is not completely washed out, many oocytes will lyse after injection.

9. Intense concentration will be required in the next step (injection of somatic or ESCs); therefore, once all the enucleation of oocytes is complete, short rest is advised before starting the injection.

3.5 Donor Nucleus Injection

1. Transfer a group (10–20) of enucleated oocytes into H-CZB medium. The number of oocytes per droplet should depend on the skill level of an individual, such that each group of injections is completed within 15 min.

2. Remove the donor nuclei from the cells by gently aspirating them in and out of the injection pipette until each nucleus is clearly separated from any visible cytoplasmic material. We usually take up 5–10 nuclei simultaneously into each injection pipette.

3. Move the injection pipette to the H-CZB droplet containing the enucleated oocytes.

4. Stabilize an enucleated oocyte using a holding pipette and cut the zona pellucida with a few piezo pulses (power level 2–3 and speed 5–6).

5. Reduce the power level of the piezo unit (power level 1–2 and speed 1) because the oolemma is weaker than the zona pellucida and the survival rate of oocytes after injection will be better with this reduced power.

6. Push one nucleus near the tip of the pipette while simultaneously advancing the pipette until it almost reaches the opposite side of the oocyte cortex.

7. Apply only one reduced piezo pulse to puncture the oolemma at the pipette tip. The puncture will be indicated by a rapid relaxation of the oocyte membrane (*see* **Note 11**).

8. Immediately release the donor nucleus into the ooplasm with a minimal amount of PVP medium. Gently withdraw the injection pipette from the oocyte (*see* **Note 12**).

Simple Low-Cost Mouse Cloning   163

9. (Option) When donor nucleus were picked up from the PLC-PVP medium, a small amount of PLCζ-cRNA (<1 pL) was co-injected with donor nucleus into enucleated oocytes (Fig. 3c) [13]. Therefore, reconstructed oocytes could be activated without any treatment (*see* **Note 13**).

10. When all the nuclei in the injection pipette taken up at **step 2** procedure have been injected into the oocytes, wash the injection pipette in a PVP dish by expelling some PMM Operation Liquid and applying power from the piezo unit. This washing step is important to prevent the pipette from getting sticky (*see* **Note 3**).

11. When a group of oocytes has been injected, keep the oocytes in this drop for 10–15 min before transferring them into CZB medium. During this period, the injection damage of oocyte will be fixed completely (*see* **Note 14**).

12. Culture the surviving reconstructed oocytes for at least 30 min in a 5% CO_2 incubator before activation.

13. When the donor cell or nucleus is too large to inject into oocytes, such as a nucleus from a cadaver, many injected oocytes lyse immediately after injection. In this case, nuclear injection should be performed in H-CZB with CB medium and the hole of the membrane should be removed when withdrawing the injection pipette, similar to that in case of MII spindle enucleation.

3.6 Oocyte Activation and Embryo Culture

1. The medium and culture dish should be prepared at least 30 min before use, and the dish must be placed in a CO_2 incubator for equilibration.

2. Transfer the oocytes into the activation medium (*see* Subheading 2.4, **item 5**), wash them twice, and then culture them for 10 h in a 5% CO_2 incubator at 37 °C.

3. After 10 h of activation, all embryos must be washed twice in CZB medium droplets (*see* **Note 15**). This should be followed by examination of the rate of oocyte activation. If SCNT and activation are done properly, each oocyte should possess two or three pseudo-pronuclei.

4. Move the cloned embryos to a new CZB medium dish for long-term culture until the two-cell or blastocyst stage is reached (*see* **Note 16**).

3.7 Embryo Culture Without Oocyte Chemical Activation

1. The medium and culture dish should be prepared at least 30 min before use, and the dish must be placed in a CO_2 incubator for equilibration.

2. 2. Transfer the reconstructed oocytes into the reprogramming, wash them twice, and then culture them for 10 h in a 5% CO_2 incubator at 37 °C.

3. Then, all embryos must be washed twice in CZB medium droplets. If SCNT are done properly, each oocyte should possess two or three pseudo-pronuclei.

4. Move the cloned embryos to a new CZB medium dish for long-term culture until the two-cell or blastocyst stage is reached.

3.8 Embryo Transfer

1. Foster mothers will be generated by mating the estrous ICR female mice with normal males on the same day or 1–2 days before the experiment. These mice would have delivered offspring before the full-term development of cloned pups. Then, lactating mothers will be used to care for the cloned pups delivered by cesarean section at E19.5. Foster mothers are necessary for cloned pups because the cloned mice litters will be too small to stimulate natural delivery in surrogate mothers [25, 26].

2. Pseudopregnant mothers will be prepared by mating estrous ICR female mice with vasectomized males on the same day as that of the experiment.

3. Cloned embryos can be transferred at different stages. Embryos at the two-cell (24 h after NT) or four- to eight-cell (48 h after NT) stage can be transferred into the oviducts of the recipients at 0.5 days post copulation (dpc). On the other hand, morulae/blastocysts (72 h after NT) or blastocysts (96 h after NT) can be transferred into the uterus of pseudopregnant mice at 2.5 dpc.

3.9 Cesarean Section

1. Euthanize the surrogate mother at 18.5 or 19.5 dpc (usually at 19.5 dpc for cloned fetuses but at 18.5 dpc for fertilized fetuses).

2. Quickly open the abdomen with a pair of sharp scissors.

3. Remove the uterus and dissect the cloned pups with their placentas.

4. Carefully wipe away the amniotic fluid from the skin, mouth, and nostrils.

5. Stimulate the pups to breathe by rubbing their backs or pinching them gently with blunt forceps (*see* **Note 17**).

6. Place the pups warm on a 37 °C hotplate until the pup become red and active.

7. Remove the foster mother from her cage.

Simple Low-Cost Mouse Cloning 165

8. Take some soiled bedding from the cage and nestle the cloned pups in the bedding material so that they take on the odor of the bedding.

9. Mix the cloned pups with some of the pups from the foster mother's litter and then retrun the foster mother.

4 Notes

1. PVP is too dense to be measured accurately. However, the concentration of PLCζ-cRNA is sufficient to activate oocytes; thus, some error is not a problem.

2. For the holding pipette, the outside diameter (OD) should be smaller than that of the oocyte (e.g., OD, 70–80 μm; inner diameter (ID), 10 μm). ID of the enucleation pipette should be 8–10 μm. ID of the injection pipette depends on the nucleus donor cell type: 5–6 μm for blood-derived cells, 6–7 μm for cumulus cells, 7–8 μm for fibroblasts or ESCs, and 8–9 μm for naked nucleus injection. Modified pipettes can also be ordered from several companies to save time.

3. During micromanipulation, the pipette could become sticky. The condition of the pipette is very important; it will affect not only oocyte survival rate but also embryo development after NT. To avoid this problem, wash the pipette in PVP medium completely; PVP will cover both the inside and the outside of the pipette to keep the surface slick. Without this step, the pipette soils rapidly and needs to be changed often.

4. Some practice is necessary to select good oocytes. Those oocytes possess large cytoplasm compared to bad ones.

5. We recommend that all the oocytes in a droplet should be manipulated within 15 min. If you are a beginner, place approximately 5 oocytes per drop, and if you have enough experience, place approximately 20 oocytes per drop.

6. When donor cells are transferred to the PVP droplet, mix them with PVP medium using sharp tweezers for at least 30 s. If the donor cells are not mixed thoroughly with PVP medium, they will aggregate and it will be difficult to isolate single cells. ESCs are particularly sensitive and fragile in PVP medium; therefore, it is better to make a new ES cell suspension every 30 min.

7. Fibroblasts have one of the toughest cell membranes, which are very difficult to break. If you fail to break the cell membrane before injection into oocytes, the reconstructed oocytes will not form pseudo-pronuclei after activation. Fibroblasts treated with trypsin for 20–30 min will have cell membranes that are more capable of breaking, thereby increasing the rate of

pseudo-pronuclear formation compared with that by usual trypsin treatment.

8. The key point in the enucleation step is finding the metaphase II (MII) spindle of the oocyte. It can be identified using Nomarski differential interference or Hoffman modulation contrast optics. Although you can use Hoechst DNA dyes to stain the nuclei, they are harmful to the oocyte and hamper embryonic development [27]. Sometimes, if you cannot detect the MII spindle at the enucleation step, then the oocytes should be cultured in a 37 °C incubator for approximately 30 min. It has been shown that when oocytes are exposed to room temperature, the spindle microtubules de-polymerize and become difficult to visualize. However, when the oocytes are warmed, the spindles become visible again.

9. When you use the piezo unit for the first time, it should be tested whether you can cut the zona pellucida smoothly. If you cannot do so, check the connection between the pipette and the pipette holder. The top of the pipette holder must be screwed on tightly. Expel all oil inside the pipette because this might have reduced the power transmission of the piezo unit. There should be a slight negative pressure inside the pipette to enhance the power of the piezo unit. In contrast, if the power is too strong even at the lowest setting, add a few drops of oil inside the pipette to reduce the power.

10. The MII spindle will appear like a small sphere under Nomarski or Hoffman optics. Moreover, the spindle is harder than the cytoplasm; therefore, the change in consistency will be felt through the micromanipulator.

11. No piezo pulses should be applied until the pipette is near the opposite side. If the piezo pulse is applied with the tip of the pipette in the middle of the oocyte, the oocyte will die after injection.

12. Sometimes, it becomes difficult to release the donor nucleus from the pipette because the pipette may be too dirty; hence, the pipette must be washed frequently using PVP medium. This can be done by expelling some PMM Operation Liquid and applying power from the piezo unit. Otherwise, the pipette should be changed.

13. When PLCζ-mRNA was co-injected with a donor nucleus using a large pipette, an extra PVP with PLCζ-mRNA was recovered for the survival of the injected oocytes, with only a small amount of PLCζ-mRNA remaining in the oocytes.

14. There are many factors that can induce oocyte lysis after injection, such as the pipette having a very large ID, the room temperature being excessively high, or the pipette insertion being too shallow. A large pipette causes the power of the

piezo unit to greatly increase, which can increase the rate of oocyte lysis. The process of NT should be performed at room temperature (25–26 °C). Do not use a warm plate for the microinjection process. The injection pipette must be inserted deeply into the oocyte before applying the piezo pulse. The rate of lysis will improve with training. If you are a novice, all the oocytes will lyse immediately after injection. Typically, after 1 month of training, approximately 50% survival may be observed, and after 1 year of practice, approximately 80% survival will be observed. Reconstructed oocytes should be placed in H-CZB medium for at least 10 min. If they are transferred into CZB medium immediately after injection, nearly 15% of them will undergo lysis from the damage of injection.

15. It is impossible to completely avoid cloned embryo death during activation. During strontium treatment, up to 10% of the oocytes will die and the medium will become dirty. However, this is normal and the surviving oocytes are usually undamaged. Therefore, check the activation medium with fresh intact oocytes as necessary.

16. When the activation is complete, check the rate of pseudo-pronuclear formation in the cloned embryos. If you cannot confirm this formation, it means that the procedure has failed. There are several reasons why oocytes do not form pseudo-pronuclei. Usually, this is because of failure to break the donor cell membrane or failure to activate oocytes. The injection pipette must be smaller than the donor cell. If the donor cell has a tough cell membrane (e.g., tail tip fibroblasts), apply a piezo pulse to break the donor cell membrane at the time of cell pickup or treat the cells with trypsin for a longer period.

17. To date, all cloned mice have been born with abnormal and hypertrophic placentas, and they often die just after birth from respiratory failure. At present, there is no way to avoid this lethal phenotype.

References

1. Wilmut I, Schnieke AE, McWhir J, Kind AJ, Campbell KH (1997) Viable offspring derived from fetal and adult mammalian cells. Nature 385:810–813

2. Kishigami S, Mizutani E, Ohta H, Hikichi T, Thuan NV, Wakayama S et al (2006) Significant improvement of mouse cloning technique by treatment with trichostatin A after somatic nuclear transfer. Biochem Biophys Res Commun 340:183–189

3. Wakayama T, Perry AC, Zuccotti M, Johnson KR, Yanagimachi R (1998) Full-term development of mice from enucleated oocytes injected with cumulus cell nuclei. Nature 394: 369–374

4. Kishigami S, Wakayama S, Thuan NV, Ohta H, Mizutani E, Hikichi T et al (2006) Production of cloned mice by somatic cell nuclear transfer. Nat Protoc 1:125–138

5. Kishigami S, Bui HT, Wakayama S, Tokunaga K, Van Thuan N, Hikichi T et al (2007) Successful mouse cloning of an outbred strain by trichostatin A treatment after somatic nuclear transfer. J Reprod Dev 53:165–170

6. Van Thuan N, Bui HT, Kim JH, Hikichi T, Wakayama S, Kishigami S et al (2009) The histone deacetylase inhibitor scriptaid enhances nascent mRNA production and rescues full-term development in cloned inbred mice. Reproduction 138:309–317

7. Ono T, Li C, Mizutani E, Terashita Y, Yamagata K, Wakayama T (2010) Inhibition of class IIb histone deacetylase significantly improves cloning efficiency in mice. Biol Reprod 83:929–937

8. Kishigami S, Ohta H, Mizutani E, Wakayama S, Bui HT, Thuan NV et al (2006) Harmful or not: trichostatin A treatment of embryos generated by ICSI or ROSI. Cent Eur J Biol 1:376–385

9. Terashita Y, Wakayama S, Yamagata K, Li C, Sato E, Wakayama T (2012) Latrunculin a can improve the birth rate of cloned mice and simplify the nuclear transfer protocol by gently inhibiting actin polymerization. Biol Reprod 86:180

10. Terashita Y, Yamagata K, Tokoro M, Itoi F, Wakayama S, Li C et al (2013) Latrunculin a treatment prevents abnormal chromosome segregation for successful development of cloned embryos. PLoS One 8:e78380

11. Yamamoto Y, Hirose N, Kamimura S, Wakayama S, Ito J, Ooga M et al (2020) Production of mouse offspring from inactivated spermatozoa using horse PLCzeta mRNA. J Reprod Dev 66:67–73

12. Saunders CM, Larman MG, Parrington J, Cox LJ, Royse J, Blayney LM et al (2002) PLC zeta: a sperm-specific trigger of Ca(2+) oscillations in eggs and embryo development. Development 129:3533–3544

13. Hirose N, Wakayama S, Inoue R, Ito J, Ooga M, Wakayama T (2020) Birth of offspring from spermatid or somatic cell by co-injection of PLCzeta-cRNA. Reproduction 160:319–330

14. Wakayama T, Yanagimachi R (2001) Mouse cloning with nucleus donor cells of different age and type. Mol Reprod Dev 58:376–383

15. Tanabe Y, Kuwayama H, Wakayama S, Nagatomo H, Ooga M, Kamimura S et al (2017) Production of cloned mice using oocytes derived from ICR outbred strain. Reproduction 154:859–866

16. Inoue K, Ogonuki N, Mochida K, Yamamoto Y, Takano K, Kohda T et al (2003) Effects of donor cell type and genotype on the efficiency of mouse somatic cell cloning. Biol Reprod 69:1394–1400

17. Ito J, Parrington J, Fissore RA (2011) PLCzeta and its role as a trigger of development in vertebrates. Mol Reprod Dev 78:846–853

18. Ito M, Shikano T, Oda S, Horiguchi T, Tanimoto S, Awaji T et al (2008) Difference in Ca2+ oscillation-inducing activity and nuclear translocation ability of PLCZ1, an egg-activating sperm factor candidate, between mouse, rat, human, and medaka fish. Biol Reprod 78:1081–1090

19. Kawai S, Takagi Y, Kaneko S, Kurosawa T (2011) Effect of three types of mixed anesthetic agents alternate to ketamine in mice. Exp Anim 60:481–487

20. Chatot CL, Ziomek CA, Bavister BD, Lewis JL, Torres I (1989) An improved culture medium supports development of random-bred 1-cell mouse embryos in vitro. J Reprod Fertil 86:679–688

21. Kishigami S, Wakayama T (2007) Efficient strontium-induced activation of mouse oocytes in standard culture media by chelating calcium. J Reprod Dev 53:1207–1215

22. Wakayama S, Kishigami S, Wakayama T (2019) Improvement of mouse cloning from any type of cell by nuclear injection. Methods Mol Biol 1874:211–228

23. Wakayama T, Yanagimachi R (1999) Cloning of male mice from adult tail-tip cells. Nat Genet 22:127–128

24. Wakayama T, Rodriguez I, Perry AC, Yanagimachi R, Mombaerts P (1999) Mice cloned from embryonic stem cells. Proc Natl Acad Sci U S A 96:14984–14989

25. Hayashi E, Wakayama S, Ito D, Hasegawa A, Mochida K, Ooga M, Ogura A, Wakayama T (2022) Mouse in vivo-derived late 2-cell embryos have higher developmental competence after high osmolality vitrification and − 80°C preservation than IVF or ICSI embryos. J Reprod Dev 68(2):118–124. https://doi.org/10.1262/jrd.2021-115

26. Ushigome N, Wakayama S, Yamaji K, Ito D, Ooga M, Wakayama T (2022) Production of offspring from vacuum-dried mouse spermatozoa and assessing the effect of drying conditions on sperm DNA and embryo development. J Reprod Dev 68(4):262–270. https://doi.org/10.1262/jrd.2022-048

27. Terashita Y, Li C, Yamagata K, Sato E, Wakayama T (2011) Effect of fluorescent mercury light irradiation on in vitro and in vivo development of mouse oocytes after parthenogenetic activation or sperm microinjection. J Reprod Dev 57:564–571

Chapter 8

Somatic Cell Nuclear Transfer in Rabbits

Pengxiang Qu, Wenbin Cao, and Enqi Liu

Abstract

Somatic cell nuclear transfer (SCNT) is a technology that enables differentiated somatic cells to acquire a totipotent state, thus making it of great value in developmental biology, biomedical research, and agricultural applications. Rabbit cloning associated with transgenesis has the potential to improve the applicability of this species for disease modeling, drug testing, and production of human recombinant proteins. In this chapter, we introduce our SCNT protocol for the production of live cloned rabbits.

Key words Cloning, Embryo, Nuclear transplantation, *Oryctolagus cuniculus*, Rabbit, Somatic cell nuclear transfer

1 Introduction

Rabbits have great value in biomedical research as models for human disease or drug discovery and testing [1, 2]. A number of genetically modified rabbits, such as those expressing human liver esterase, apolipoprotein, C-reactive protein, and other recombinant proteins, have far-reaching significance in promoting the discovery of new drugs [3]. Rabbits also have great potential for the production of humanized therapeutic antibodies because of their relatively small size, short reproductive cycle, and strong innate immune response. The antibodies obtained from rabbits are specific, with high affinity, and easily purified [4]. However, several applications remain restricted by the low efficiency of rabbit transgenesis by pronuclear injection or cloning [5].

Before the birth of the first cloned mammal by somatic cell nuclear transfer (SCNT), rabbits were an important animal model for cloning studies using embryonic donor cells [6–9]. Hence, the first reported cloning experiment in mammals was carried out with rabbits in 1975 [6]. The first cloned rabbit using SCNT was successfully obtained in 2002 [10]. Similar to other animals, the efficiency of rabbit cloning is very low [11–14]. In recent years,

Marcelo Tigre Moura (ed.), *Somatic Cell Nuclear Transfer Technology*, Methods in Molecular Biology, vol. 2647,
https://doi.org/10.1007/978-1-0716-3064-8_8,
© The Author(s), under exclusive license to Springer Science+Business Media, LLC, part of Springer Nature 2023

170   Pengxiang Qu et al.

many achievements have been made in animal cloning, and the abnormal reprogramming of cloned embryos is considered the main reason for the low cloning efficiency [15, 16]. Previous studies showed that histone acetylation in donor cells and cloned embryos correlated with their developmental potential, while the synergistic effects of combined treatment with histone deacetylase inhibitors (i.e., trichostatin A (TSA) and scriptaid (SCP)) on cloned rabbit embryos may improve cloning efficiency [11, 17]. Our research group found that selecting oocytes by the brilliant cresyl blue staining and delivery of sperm-borne small RNAs increases the in vitro developmental potential of cloned rabbit embryos. Further, we found that melatonin protects cloned rabbit embryos from electrofusion-mediated oxidative damage [18–20]. In this chapter, we describe in detail our rabbit SCNT protocol.

2 Materials

2.1 Equipment

1. Two incubators with 5% CO_2.

2. Stereomicroscope with thermal plate.

3. Temperature-controlled thermal plate.

4. Micropuller (Sutter Instrument, Novato, CA, USA; P-97), microforge (Narishige, Tokyo, Japan; MF-900), and micro-grinder (Narishige, EG-401).

5. Micromanipulators for holding the oocyte (left) and the cell injection system (right). The holding system is equipped with an Eppendorf air pump (CellTram/air), and the injection system is equipped with an Eppendorf oil pump (CellTram/oil).

6. Inverted microscope equipped with Oosight imaging system (CRI, UK) and thermal plate.

7. BTX cell fusion system (BTX; Holliston, MA, USA; ECM830).

8. A microelectrode was constructed consisting of a 25 cm long copper wire with a diameter of 100 microns passed through a 20 cm long plastic tube with an outer diameter of 3.5 mm and inner diameter of 2.5 mm. The copper wire was fixed to the plastic tube with glue so that electrostatic interference from the plastic tubing and the micromanipulation system and operators could be avoided. The copper wire was soldered to a 5 cm long platinum wire curved at about 150°. A second electrode was made in the same way, and each was attached to a micromanipulator control system (Eppendorf, Saxony, Germany).

2.2 Tools and Consumables

1. Sterile four-well dishes (MW4).

2. Sterile 96-well plates with flat-bottom.

Rabbit Cloning by SCNT   171

3. Glass capillaries for making holding pipettes (0.6 mm inner diameter and 1.0 mm outer diameter) and injection pipettes (0.8 mm inner diameter and 1.0 mm outer diameter).

4. Surgical instruments (e.g., scalpels, scissors, forceps, electric shaver).

5. Microcapillary pipettes.

6. Glass-bottomed dishes (Will Co-Dish).

2.3 Media and Solutions

1. Primary cell culture medium (DMEM20): Supplement Dulbecco's modified Eagle's medium (DMEM) with 20% fetal bovine serum (FBS), 100 U/mL penicillin, and 0.1 mg/mL streptomycin.

2. Culture medium for cell passaging (DMEM10): Supplement DMEM with 10% FBS, 100 U/mL penicillin, and 0.1 mg/mL streptomycin.

3. Culture medium for serum starvation (DMEM05): Supplement DMEM with 0.5% FBS, 100 U/mL penicillin, and 0.1 mg/mL streptomycin.

4. Phosphate-buffered saline (PBS): PBS+ containing 100 IU/mL penicillin, and 0.1 mg/mL streptomycin (PBS$^+$).

5. 0.25% trypsin solution.

6. Oocyte collection medium (OCM): Dulbecco's phosphate-buffered saline (D-PBS) with 5% FBS.

7. M2 medium: 95 mM NaCl, 4.8 mM KCl, 1.2 mM KH$_2$PO$_4$, 1.2 mM MgSO$_4$·7H$_2$O, 23 mM sodium lactate, 0.3 mM sodium pyruvate, 5.6 mM Glucose, 4.15 mM NaHCO$_3$, 1.7 mM CaCl$_2$·2H$_2$O, 20.9 mM HEPES, and 4.0 mg/mL BSA in ultrapure water. Adjust pH to 7.2–7.4, sterile filter (0.22 μm), and store at 4 °C.

8. Hyaluronidase solution: Dissolve 0.5% (w/v) hyaluronidase in M2 medium.

9. Micromanipulation medium (MM): Mix 7.0 mL M199 with HEPES, 7.0 mL DPBS with 5% FBS, 1.0 mL FBS, 7.5 μg/mL cytochalasin B, and store at −20 °C.

10. Cell fusion medium (CFM): Cytofusion medium C (BTX press, Catalog no. 47).

11. Synthetic oviductal fluid (SOFaa): 110 mM NaCl, 7.168 mM KCl, 1.191 mM KH$_2$PO$_4$, 0.4926 mM MgSO$_4$·6H$_2$O, 0.02824% sodium lactate, 0.3 mM sodium pyruvate, 1.498 mM glucose, 25.07 mM NaHCO$_3$, 1.707 mM CaCl$_2$·2H$_2$O, 2.0% 50× essential amino acids solution (Gibco, Grand Island, NY, USA; Lot no.1916660), 1.0% 100× nonessential amino acids solution (Gibco, Lot no. 1927136), and 8.0 mg/mL BSA in ultrapure water. Adjust pH to 7.2–7.4, filter through a 0.22 μm filter, and store at 4 °C for up to 1 month.

172   Pengxiang Qu et al.

12. Oocyte activation medium (OAM): SOFaa containing 5.0 μg/mL cycloheximide and 2.0 mM 6-dimethylaminopurine (6-DMAP).

13. Cell freezing medium: 10% DMSO, 20% FBS, and 70% DMEM.

2.4 Preparation of Micromanipulation Pipettes

2.4.1 Holding Pipette

1. Place a glass capillary (0.6 mm inner diameter and 1.0 mm outer diameter) in the micropuller. Set the instrument to a pressure (P) of 500, heat (H) of 655, pull (p) of 30, velocity (V) of 120, and time (T) of 200 (see **Note 1**).

2. Start the instrument to pull the pipette, which leads to melting of the glass capillary with the heating plate, and division of the capillary into two pipettes with tips with adequate diameters.

3. Place the tapered pipette tip close to the end of the left index finger and scratch this part (150–180 μm diameter) with a grinding wheel held in the right hand.

4. Break off the tip of the pipette above the scratch and examine it under the microscope to see whether the diameter is appropriate and with a blunt tip.

5. Place the pipette in the microforge with its tip near the glass sphere. Turn on the instrument and place the pipette tip near the heated sphere (the contact between the two should be at site with adequate inner diameter). The pipette tip will melt down, when it is necessary to turn off the heating. The pipette will break on the site in contact with the sphere, thus making a blunt tip with a 20–30 μm inner diameter.

6. Bend the 2–3 mm pipette tip to 30° using the microforge, then subject it to heat at 150° for sterilization and keep it in a sealed box.

2.4.2 Injection Pipette

1. Place a glass capillary (0.8 mm inner diameter and 1.0 mm outer diameter) in the micropuller. Set the instrument to the following parameters ($P = 500$, $H = 665$, $p = 0$, $V = 30$, and $T = 250$), which are afterward changed to the following parameters ($P = 500$, $H = 665$, $p = 80$, $V = 60$, and $T = 200$). Start the instrument to pull the capillary into two injection pipettes (see **Note 2**).

2. Break the pipette at a 15–18 μm diameter using the microforge. Sharpen the pipette tip at 45° with the microgrinder, thus creating a beveled tip. Wash the pipette with distilled water and expel the liquid from the needle with a no-load syringe to remove glass fragments and other debris, and then pull the tip of the pipette head to make it sharp.

3. Bend the tip 2–3 mm from the end to 30° with the microforge, sterilize at 150°, and store the holding pipette.

2.5 Micromanipulation Setup

1. Prepare an embryo culture dish with nine 50 μL SOFaa droplets in a 60 mm dish. Cover the droplets with mineral oil and put them in the incubator with 5% CO_2, saturated humidity at 38 °C no less than 2 h.

2. Add 400 μL SOFaa in each well of a four-well dish, cover with mineral oil, and put the dish in the incubator with 5% CO_2, saturated humidity at 38 °C no less than 2 h.

3. Prepare the micromanipulation dish by placing three 20 μL MM droplets (labeled as A, B, and C) and one 5% polyvinylpyrrolidone (PVP) droplet (D) on the glass bottom of a 35 mm dish and cover with mineral oil. Put the dish on the thermal plate of the inverted microscope set to 38 °C. Turn on the Oosight imaging system and the micromanipulation system. Prepare the dish and turn on the equipment 2 h before experiment onset.

4. Mount the holding pipette in the left pipette holder of the micromanipulator, and the injection pipette in the right pipette holder of the micromanipulator (Fig. 1).

5. Insert the end of the holding pipette into an MM droplet and adjust it such that it lies horizontally on the bottom of the micromanipulation dish.

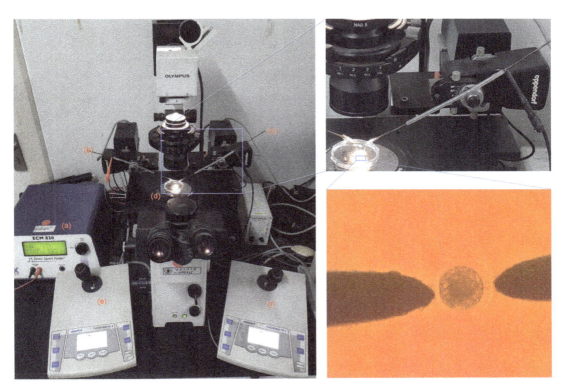

Fig. 1 Micromanipulation and cell fusion setup: (a) BTX cell fusion system; (b, c) microelectrodes; (d) thermal plate; (e, f) micromanipulators

174   Pengxiang Qu et al.

6. Place the injection pipette tip into the droplet and adjust it such that it lies horizontally on the bottom of the micromanipulation dish. Wash the injection pipette several times by aspirating and draining small amounts of PVP, thus lubricating the pipette's inner wall. Then, wash the pipette with mineral oil, and then with PVP once more (*see* **Note 3**).

7. Adjust the microscope stage so that both holding and injection pipettes are in the same MM droplet. Then, aspirate a small volume of MM into both pipettes.

3 Methods

Experiments with live animals must be in accordance with institutional and national regulation oversight. All oocyte or embryo washes described below must be in fresh medium droplets.

3.1 Preparation of Donor Cells

3.1.1 Isolation of Primary Fibroblast Cells

1. Several cell types (e.g., cumulus cells, fibroblasts) are suitable for rabbit cloning by SCNT. Fetuses or live animals may provide donor cells. For deriving primary fibroblast, skin samples are taken from the ear or the abdomen.

2. Disinfect the rabbit ear with 75% ethanol, and cut off a 10 mm^2 biopsy of skin tissue with sterilized scissors. Wash repeatedly in PBS$^+$.

3. Remove hair, fat, and connective tissue from the skin biopsy. Cut the skin fragment into 1.0 mm^3 pieces with ophthalmic scissors on a sterile 60 mm cell culture dish.

4. Digest the pieces with 0.25% trypsin in the incubator at 38 °C for 50 min, agitating gently in 10 min intervals. Apply gentle pipetting to dissociate the tissue.

5. Collect dissociated cells by centrifugation at 300 g for 5 min and resuspend in 3.0 mL DMEM20. Plate 1.0×10^6 cells/mL in a 60 mm cell culture dish and incubate in 5% CO_2, with saturated humidity at 38 °C.

6. After 2 days, observe the culture for cell adhesion (i.e., viability) and explant outgrowth.

7. Replace 1.5 mL of culture media with fresh DMEM20 every 2 days.

8. When the cell culture reaches 80–90% confluence, passage (subculture) using a 1:2 or 1:3 split as described above (*see* **step 5**). Transfer an aliquot into a freezing tube and place the freezing tube at 4 °C for 10 min, then at −20 °C for 30 min, and then at −80 °C for 16 h, and finally freeze aliquots in liquid nitrogen (−196 °C).

Rabbit Cloning by SCNT   175

3.1.2 Culture of Fibroblast Cells

1. Remove the DMEM20 with sterile glass pipette and wash twice with PBS⁻.

2. Add 1.0 mL 0.25% trypsin (trypsin) and culture in the incubator for 1–2 min. When the cells begin to round up and detach from the dish, add 1.0 mL DMEM10 to inhibit trypsin activity.

3. Apply gentle pipetting to dissociate cell clumps into single cells, centrifuge at 1000 g for 5 min, discard supernatant, and resuspend cells in 1.0 mL DMEM10.

4. Fibroblasts at passages 3–13 are cultured to 100% confluence on 96-well plates, and then serum-starved (cell synchronization at G0) by incubation in 100 μL DMEM05 for 2–5 days.

5. Prepare donor cells before the oocyte reconstruction step. Dissociate fibroblasts with trypsin for 1–2 min, then inhibit trypsin by adding 1.0 mL DMEM10, centrifuge at 1000 g for 5 min, and discard the supernatant.

6. Add 50 μL MM and resuspend the cells for oocyte enucleation and reconstruction.

3.2 Oocyte Collection and Denuding

Give preference to mature New Zealand rabbit does (>6 months old) weighing 3.5 kg as oocyte donors. Healthy multiparous does (>9 months old) are optimal as recipient females for cloned embryos.

1. Inject 80 IU pregnant mare serum gonadotropin (PMSG) subcutaneously into female rabbits, and inject 100 IU human chorionic gonadotropin (hCG) intravenously 96 h later (*see* **Note 4**).

2. Anesthetize the rabbits 12 h after the hCG injection with intravenous injection of 25 mg ketamine and 5.0 mg medetomidine.

3. Make sure animals are under anesthesia and perform an incision in the flank of the animal and pull out the oviduct from the body cavity (*see* **Note 5**).

4. Insert a syringe loaded with 5.0 mL OCM into the oviduct proximal to the uterotubal junction. Place the other end of the oviduct into a 15 mL tube and flush the contents of the oviduct into the tube.

5. Allow the flushed liquid containing cumulus-oocyte complexes stand for 5 min, then carefully remove the supernatant and incubate the sediment with hyaluronidase solution for 5 min.

6. Apply gentle pipetting with a 1.0 mL pipetter (~200 μm diameter of pipette tip) to remove cumulus cells (*see* **Note 6**).

7. Collect denuded oocytes under a stereomicroscope and transfer them into OCM. Place oocytes on a thermal plate at 38 °C for SCNT.

176 Pengxiang Qu et al.

3.3 Somatic Cell Nuclear Transfer

3.3.1 Oocyte Enucleation

1. Start the Oosight imaging system, adjust the microscopic field of vision at droplet A, and set the background.

2. Transfer 20–30 oocytes into droplet A and move the oocytes to the area above the holding and injection pipettes. Gently suction one oocyte to the holding pipette and rotate it to make the oocyte spindle at the 1 o'clock position in a clock's face (*see* **Note 7**).

3. Quickly insert the injection pipette from the 3 o'clock position and gently aspirate the oocyte spindle and the polar body with the injection pipette controlled by the oil-pressured system.

4. Release the oocyte gently from the holding pipette using the air-pressured system. Move it to the area below the holding pipette with the injection pipette and repeat the procedure with other oocytes (*see* **steps 3** and **4**).

5. Select all viable enucleated oocytes with intact membrane for the oocyte reconstruction step. Turn off the power to the Oosight imaging system.

3.3.2 Oocyte Reconstruction

1. Adjust the field of vision of the inverted microscope at droplet B on bright-field (200×).

2. Add 10–20 μL donor cell suspension into droplet B and allow it to stand for 5 min. Gently draw up 5–10 donor cells into the injection pipette (*see* **Note 8**).

3. Adjust the microscope stage to bring-field on droplet A, then inject one donor cell into the perivitelline space of each enucleated oocyte (*see* **Note 9**).

4. Release gently the cell couplet with the holding pipette. Move the cell couplet to the area above the holding pipette with the injection pipette and then proceed to the next enucleated oocyte (*see* **steps 3** and **4**).

5. Transfer all cell couplets to a 50 μL SOFaa droplet after completion of oocyte reconstruction. Wash couplets three times in 50 μL SOFaa droplets.

6. Transfer cell couplets to a 50 μL SOFaa droplet and incubate in 5% CO_2, saturated humidity at 38 °C for 30 min.

3.4 Cell Fusion

1. Place four CFM droplets (A–D) in a 60 mm dish and cover droplets with mineral oil. Put the dish on the microscope thermal plate set at 38 °C.

2. Mount the microelectrodes in the metal pipette holders of the micromanipulators and insert the microelectrode into the droplet and adjust it such that it lies horizontally on the bottom of the micromanipulation dish. Connect the microelectrodes to the BTX stimulator (Fig. 1) and program the instrument for three 20 μs DC pulses at 24 V (*see* **Note 10**).

Rabbit Cloning by SCNT   177

3. Transfer cell couplets from the SOFaa droplet to droplet A with microcapillary pipettes. Wash cell couplets with three CFM droplets (A–C) for 6 min (2 min in each droplet). Transfer cell couplets to droplet D for fusion (*see* **Note 11**).

4. Move cell couplets to the area above the microelectrodes, so that they line up vertically with the microelectrodes (*see* **Note 12**).

5. Gently clamp a cell couplet with the microelectrodes to maintain contact with the cell membrane and run the preset programs (Fig. 1). Then, move the cell couplet to the area below the microelectrodes (*see* **Note 13**).

6. Process the next cell couplet as described above (*see* **steps 4** and **5**).

7. Wash cell couplets three times in SOFaa droplets after cell fusion and transfer them to a SOFaa droplet at 38 °C for 30 min (*see* **Note 14**).

8. Check for fusion events (i.e., absence of donor cell attached to the enucleated oocyte) and subject fused couplets to chemical activation.

3.5 Oocyte Activation

1. Put an OAM droplet (A) and three SOFaa droplets (B–D) in a 35 mm dish, cover the droplets with mineral oil, and place them in the incubator with 5% CO_2, saturated at 38 °C.

2. Place fused couplets into the OAM droplet and incubate with 5% CO_2, saturated at 38 °C for 1 h.

3. Wash cloned embryos three times in SOFaa droplets B–D.

3.6 Embryo Culture

1. Transfer 30–40 cloned embryos per well in the four-well embryo culture dish and incubate with 5% CO_2, saturated at 38 °C for culture.

2. Perform embryo culture for 1 day (22–24 h) before embryo transfer or 5 days (120 h) to determine blastocyst rates.

3.7 Embryo Transfer

1. Prepare two recipient rabbits (one recipient for 20 cloned embryos) immediately after the onset of cloned embryo culture. Inject 80 IU hCG intravenously into each recipient doe (*see* **Note 15**).

2. Twenty-three hours after activation, place four SOFaa droplets (A–D) in a 35 mm dish and put them in the incubator with 5% CO_2, saturated humidity, and 38 °C at 23 h post-activation (*see* **Note 16**).

3. Select cloned embryos with normal cleavage and morphology for embryo transfer at 24 h post-activation. Wash embryos in droplets A–C to remove oil carryover. Maintain the dish with embryos on thermal plate at 38 °C (*see* **Note 17**).

4. Inject a mixture of 25 mg ketamine and 5 mg medetomidine intravenously into each recipient rabbit (*see* **Note 18**).

5. Place the rabbit under anesthesia on the operating table lying on its side. The operation field is usually located 3 cm from the posterior edge of the rib arch and 5 cm from the spine.

6. Shave the surface area to expose the skin, cover with a sterile drape, and disinfect with alcohol and iodophor. Make a small incision through the skin, muscle, and peritoneum in turn, and pull the adipose tissue surrounding the ampulla out of the body cavity.

7. Use a small amount of medium from droplet D followed by a small amount of air, and load the selected embryos from droplet C. Draw a small amount of air with the microcapillary pipette (*see* **Note 19**).

8. Insert the microcapillary pipette loaded with cloned embryos into the ampulla and inject them into the oviduct.

9. Put the tissue back into the body cavity and close the incision.

10. Place the recipient rabbit in a heat preservation box after the operation and monitor until normal activity has resumed.

3.8 Caesarean Section

1. Place pregnant recipient rabbits into breeding cages with padding 28 days after embryo transfer. Perform kit delivery by caesarean section on day 30 after embryo transfer.

2. Begin kid delivery by injecting a mixture of 25 mg ketamine and 5 mg medetomidine solution intravenously recipient rabbit. Place the recipient rabbit under anesthesia on the operating table in a supine position. Locate the operation field at the posterior midline of the umbilicus.

3. Shave the body surface area to expose the skin and cover with a sterile drape. Disinfect with 70% ethanol and iodophor. Make an incision through the skin, muscle layer, and peritoneum. Gently pull one uterine horn. Place a gauze pad soaked with warm saline solution between the uterine horn and the abdominal wall incision.

4. Make a longitudinal 2–3 cm incision at the curvature of the uterine horn near the uterine body (be careful to avoid damaging blood vessels).

5. Gently squeeze out the rabbit kits from the uterus and place them on a clean cotton towel. Remove mucus from mouths and noses.

6. Cut the umbilical cord 1.0 cm from the abdomen of the rabbit kit and disinfect the cut with 5% iodine tincture.

7. Squeeze uterine horns after recovering all cloned kits to discharge residual blood and placental fragments.

Rabbit Cloning by SCNT   179

8. Wash the area with warm saline solution at 37 °C. After washing, sprinkle about 100 mg penicillin and streptomycin on the uterus and uterine incision.

9. Insert the uterus into the body cavity and close the incision.

10. Place the recipient mother and cloned kits in heat preservation box after surgery and monitor them with good care.

11. Place the recipient mother next to cloned kits for suckling 2 h after the caesarean section. Feed the kits twice a day as described above until they are able to move about freely on their own. Wean kits 40 days after birth and house them in individual cages.

4   Notes

1. The parameters for making the holding pipettes are for reference only. The conditions (e.g., tension, heating, velocity, time) should be adjusted according to the type of glass and manufacturer of the micropuller. The holding pipettes can be purchased ready-made from companies, and they can also be made by a skilled hand without an instrument.

2. The parameters for making the injection pipettes are a starting point. The difficulty in making suction/injection pipettes is greater than for holding pipettes, and many researchers simply buy them from companies.

3. If enucleation or injection steps are not smooth, the injection pipette should be washed again with PVP.

4. Make sure rabbits are in a calm state before PMSG injections. If rabbits were purchased from suppliers, allowed them to acclimate for 1 week after arrival before injection. Superovulation may be adversely affected if the animals are under stress.

5. Oocyte collection should be carried out within 12–14 h after the hCG injection (not earlier or later). Females are used more than once.

6. Cumulus-oocyte complexes may adhere to the tube wall or the surface of the supernatant. It is best to check the supernatant under the microscope for oocytes before discarding it.

7. The spindle, polar body, and injection pipette should be in the same plane.

8. Before drawing up donor cells, injection pipette should be washed with PVP, and the number of cells should be limited to 5–10 each time. The best distance between cells in the injection pipette is ~100 μm.

9. Inject the donor cell into the perivitelline space and gently press it against the oocyte cytoplasm to ensure contact between them and to prevent cell adhesion to the *zona pellucida*.

10. Eliminate static electricity from operators and equipment before electrofusion to reduce its interference in fusion. Fusion dish may be purchased from several manufacturers.

11. When the cell couplets are placed in droplet A, they may float and stick to carryover oil droplets. Therefore, this step should be performed slowly and carefully to prevent embryo loss. Set the loading volume as small as possible when transferring embryos to another droplet.

12. Align the cell couplets and place them above the microelectrodes. This makes it more convenient for rapid operation and reduces damage from excessive electrical stimulation.

13. Gently contact the enucleated oocyte with the donor cell when performing the cell fusion. Afterward, gently withdraw the microelectrodes to release the cell couplets slowly. Never release them abruptly.

14. Set the loading volume as small as possible when transferring embryos to another droplet because residual fusion medium is detrimental to embryonic development.

15. Multiparous female rabbits are the most suitable embryo recipients. Rabbits can also be used during their natural estrous cycle without estrus synchronization using PMSG and hCG injections.

16. This droplet should not be covered with oil.

17. When transferring embryos to the droplet C, ensure that there is no residual oil on the surface of the droplet. Select cloned embryos with symmetrical blastomeres, in the absence of fragmentation.

18. The anesthetic dosage (25 mg ketamine and 5 mg medetomidine) is suitable for most rabbits within the 3.5–4.0 kg weight range. If the anesthetic dosage was not sufficient, it can be adjusted according to the actual situation.

19. The two-stage liquid loading method helps to avoid embryo loss in the transfer pipette. The number of transferred embryos should typically be 5–10 per ampulla, and the loading liquid should be 10–20 µL per embryo pool/uterus horn.

Acknowledgments

We thank our predecessors for their work. This work was partly carried out with the support of the project funded by the Natural Science Foundation of China under grant no, 32100649, and the China Postdoctoral Science Foundation under grant no. 2020TQ0240 and the Natural Science Foundation of Shaanxi Province under grant nos. 2020PT-001 and 2021PT-039.

References

1. Fan J, Watanabe T (2003) Transgenic rabbits as therapeutic protein bioreactors and human disease models. Pharmacol Ther 99:261–282
2. Esteves PJ, Abrantes J, Baldauf HM, BenMohamed L, Chen Y, Christensen N et al (2018) The wide utility of rabbits as models of human diseases. Exp Mol Med 50:1–10
3. Fan J, Wang Y, Chen YE (2021) Genetically modified rabbits for cardiovascular research. Front Genet 12:614379
4. Wang Y, Zhao S, Bai L, Fan J, Liu E (2013) Expression systems and species used for transgenic animal bioreactors. Biomed Res Int 2013:580463
5. Yang D, Xu J, Chen YE (2019) Generation of rabbit models by gene editing nucleases. Methods Mol Biol 1874:327–345
6. Bromhall JD (1975) Nuclear transplantation in the rabbit egg. Nature 258:719–722
7. Stice SL, Robl JM (1988) Nuclear reprogramming in nuclear transplant rabbit embryos. Biol Reprod 39:657–664
8. Collas P, Robl JM (1990) Factors affecting the efficiency of nuclear transplantation in the rabbit embryo. Biol Reprod 43:877–884
9. Collas P, Robl JM (1991) Relationship between nuclear remodeling and development in nuclear transplant rabbit embryos. Biol Reprod 45:455–465
10. Chesné P, Adenot PG, Viglietta C, Baratte M, Boulanger L, Renard JP (2002) Cloned rabbits produced by nuclear transfer from adult somatic cells. Nat Biotechnol 20(4):366–369
11. Yang F, Hao R, Kessler B, Brem G, Wolf E, Zakhartchenko V (2007) Rabbit somatic cell cloning: effects of donor cell type, histone acetylation status and chimeric embryo complementation. Reproduction 133:219–230
12. Meng Q, Polgar Z, Liu J, Dinnyes A (2009) Live birth of somatic cell-cloned rabbits following trichostatin A treatment and cotransfer of parthenogenetic embryos. Cloning Stem Cells 11:203–208
13. Li S, Chen X, Fang Z, Shi J, Sheng HZ (2006) Rabbits generated from fibroblasts through nuclear transfer. Reproduction 131:1085–1090
14. Yin M, Jiang W, Fang Z, Kong P, Xing F, Li Y et al (2015) Generation of hypoxanthine phosphoribosyltransferase gene knockout rabbits by homologous recombination and gene trapping through somatic cell nuclear transfer. Sci Rep 5:16023
15. Matoba S, Zhang Y (2018) Somatic cell nuclear transfer reprogramming: mechanisms and applications. Cell Stem Cell 23:471–485
16. Qu P, Wang Y, Zhang C, Liu E (2020) Insights into the roles of sperm in animal cloning. Stem Cell Res Ther 11:65
17. Chen CH, Du F, Xu J, Chang WF, Liu CC, Su HY et al (2013) Synergistic effect of trichostatin A and scriptaid on the development of cloned rabbit embryos. Theriogenology 79:1284–1293
18. Jia L, Ding B, Shen C, Luo S, Zhang Y, Zhou L et al (2019) Use of oocytes selected by brilliant cresyl blue staining enhances rabbit cloned embryo development in vitro. Zygote 27:166–172
19. Qin H, Qu P, Hu H, Cao W, Liu H, Zhang Y et al (2021) Sperm-borne small RNAs improve the developmental competence of pre-implantation cloned embryos in rabbit. Zygote 29:331–336
20. Qu P, Shen C, Du Y, Qin H, Luo S, Fu S et al (2020) Melatonin protects rabbit somatic cell nuclear transfer (SCNT) embryos from electrofusion damage. Sci Rep 10:2186

Chapter 9

Production of Cloned Pigs by Handmade Cloning

Gábor Vajta, Wen Bin Chen, and Zoltan Machaty

Abstract

Somatic cell nuclear transfer (SCNT) in pigs is a promising technology in biomedical research by association with transgenesis for xenotransplantation and disease modeling technologies. Handmade cloning (HMC) is a simplified SCNT method that does not require micromanipulators and facilitates the generation of cloned embryos in large quantities. As a result of HMC fine-tuning for porcine-specific requirements of both oocytes and embryos, HMC has become uniquely efficient (>40% blastocyst rate, 80–90% pregnancy rates, 6–7 healthy offspring per farrowing, and with negligible losses and malformations). Therefore, this chapter describes our HMC protocol to obtain cloned pigs.

Key words Cloning, Disease model, Handmade, Nuclear transfer, Nuclear transplantation, Pig, Vitrification, Xenotransplantation

---

1 Introduction

In contrast to larger domestic animals such as cattle and horses, cloning in pig production has limited value for breeding programs due to the short generation interval and large litters. However, pigs have unique physiological features that predispose them to be used as animal models for human medical purposes [1]. Pigs also have a relatively long lifespan, while their metabolism and organ features (e.g., structure and size) are similar to humans. Further, the pig genome is more similar to humans than commonly used laboratory animals. In addition, humans have thousands of years of experience in pig domestication and selective breeding, which is efficient and inexpensive.

Pigs bred using traditional methods (i.e., non-transgenic) can also be used for various purposes, including metabolic studies, drug testing, or developing new surgical procedures. However, genetic modification may considerably widen the fields of application. Pigs are ideal candidates to become universal donors for organ transplantation into humans (i.e., xenotransplantation) if the multiple defense system in both the host and the graft can be neutralized.

Marcelo Tigre Moura (ed.), *Somatic Cell Nuclear Transfer Technology*, Methods in Molecular Biology, vol. 2647,
https://doi.org/10.1007/978-1-0716-3064-8_9,
© The Author(s), under exclusive license to Springer Science+Business Media, LLC, part of Springer Nature 2023

Changes in the pig genome (sequence or function) to harbor human genetic disease variants may also help us to better understand a number of human diseases, including Alzheimer's disease [2], Parkinson's disease, diabetes [3, 4], and arteriosclerosis [5]. Until recently, the only efficient way to produce transgenic mammals was to perform the genetic modification in somatic cells and use them as donors for somatic cell nuclear transfer (SCNT). The introduction of new genome editing methods allows the production of transgenic pigs using a one-step manipulation (i.e., microinjection of gene editing components into the zygote). However, some reports show that the procedure may have unwanted effects due to off-target editing [6]. Hence, it seems safer to stick with a two-step approach: to establish and screen for gene-edited somatic cells without off-targets and to apply SCNT for generating transgenic offspring.

Handmade cloning (HMC) is a simple micromanipulator-free SCNT method, initially applied to cattle [7, 8], then to various domestic and wild species [9]. Its advantages are the outstanding cost efficiency, relative ease of the procedure, high productivity (i.e., reconstructed oocytes per cloning session), and low incidence of pregnancy losses or malformations. Healthy cloned offspring were also reported after oocyte delipation and cloned embryo vitrification [10]. After a 10-year systematic improvement in the pig HMC protocol, it has become uniquely suitable for the purposes outlined above [11, 12]. Therefore, this chapter provides detailed guidelines for the application of HMC to clone pigs from somatic cells.

2 Materials

2.1 Equipment

1. Tissue culture hood.

2. A stereomicroscope with strong, focused illumination and sharp contrast. A > 90 mm distance between the sample and the objective is indispensable for manual bisection.

3. Heated stage and bench adjusted to 38 °C.

4. CO_2 incubators: one with 5% CO_2 and high oxygen tension (~20% O_2) and another with 5% CO_2 and low oxygen tension (5% O_2).

5. Fusion machine (possible source: BLS, Budapest, Hungary; www.bls-ltd.com).

6. Vortex with hard rubber and strong motor.

7. Centrifuge.

Pig Handmade Cloning   185

2.2 Tools and Consumables

1. Embryo splitting blades (possible source: Minitüb, Germany).

2. Aggregation needles for microwell preparation, small diameter, DN-09/B, BLS (as above).

3. Pulled, fire-polished glass pipettes and appropriate pipette aids (*see* addendum).

4. Fusion chamber with 0.5 mm microslide (BTX model 450, 01-000209-01).

5. Small (35 mm) and medium (60 mm) plastic Petri dishes, Nunc four-well dishes (Thermo Fisher Scientific, Waltham, MA, USA).

2.3 Media and Solutions

Unless otherwise indicated, chemicals were obtained from Sigma-Aldrich, St. Louis, MO, USA.

1. T2 medium (T2M): Mix HEPES-buffered TCM-199 (H-TCM-199; cat. no. M7528) with 2.0% (v/v) adult cattle serum (CS; cat. no. B9433). T20 medium (T20M): Mix H-TCM-199 with 20% (v/v) CS.

2. PBS0: PBS without Ca^{++} and Mg^{++}.

3. Cell culture medium (CCM): Dulbecco's Modified Eagle's Medium (DMEM; Hyclone, SH30022.01B) supplemented with 15% (v/v) fetal bovine serum (FBS), 1.0% MEM NEAA (Gibco, 1228076), 1% glutamine (BBI, GB0224), and 1.0% penicillin and streptomycin.

4. Cell freezing medium (CFM): CCM with 10% dimethyl sulfoxide (DMSO; cat. no. D2650).

5. HEPES-buffered Tyrode's lactate medium (TL-HEPES): Add 114 mM NaCl, 3.2 mM KCl, 2.0 mM CaCl, 0.5 mM $MgCl_2 \cdot 6H_2O$, 2.0 mM $NaHCO_3$, 0.4 mM $NaH_2PO_4 \cdot H_2O$, 5.0 mM glucose, 10.0 mM sodium lactate, 0.1 mM sodium pyruvate, 240 mg/100 mL HEPES, 10,000 units/100 mL Sodium penicillin-G, 1 mg/100 mL phenol red, and BSA 3 mg/mL (add immediately before use).

6. In vitro maturation medium (IVMM): TCM-199 (Gibco 11150059) supplemented with 0.57 mM cysteine, 3.05 mM D-glucose, 0.91 mM sodium pyruvate, 10 ng/mL epidermal growth factor, 0.5 IU/mL ovine-luteinizing hormone, 0.5 IU/mL porcine follicle-stimulating hormone, 0.1% (w/v) polyvinyl alcohol (PVA), 75 mg/mL penicillin, and 50 mg/mL streptomycin.

7. Hyaluronidase solution: Dissolve 1.0 mg/mL hyaluronidase (cat. no. H4272) in H-TCM-199. Sterile filter (0.22 μM), prepare 500 μL aliquots, and store at −20 °C.

8. Pronase solution: Dissolve 10 mg/mL pronase (cat. no. P8811) in H-TCM-199. Make 200 μL aliquots and store

186 Gábor Vajta et al.

at -20 °C. Centrifuge at 16,000 g for 3 min after thawing and collect 150 µL of the supernatant for further use.

9. Cytochalasin B (CB) solution: Dissolve 5.0 mg (CB; cat. no. C6962) in 1.0 mL DMSO. Prepare 5.0 µL aliquots in Eppendorf tubes. The final concentration of the working solution is 5.0 µg/mL.

10. Phytohemagglutinin (PHA) solution: Dissolve 5.0 mg/mL (cat. no. L8754) in H-TCM-199. Make 20 µL aliquots and store at -20 °C (see **Note 1**). Centrifuge at 16,000 *g* for 3 min. Collect supernatant for further use (see **Note 2**).

11. Cell fusion medium (CFM): Dissolve 10.93 g D-mannitol and 0.2 g PVA in 196 mL ultrapure water. Sterile filter (0.22 µM) and store at -20 °C.

12. Embryo culture medium (PZM-3): Add 108 mM NaCl, 10 mM KCl, 0.35 mM KH_2PO_4, 0.4 mM $MgSO_4 \cdot 7H_2O$, 25.07 mM $NaHCO_3$, 0.2 mM Na-pyruvate, 2.0 mM Ca-(lactate)$_2 \cdot 5H_2O$, 1.0 mM L-glutamine, 5.0 mM hypotaurine, 20 µL/mL Basal Medium Eagle amino acids, 10 µL/mL Minimum Essential Medium nonessential amino acids, 0.05 mg/mL gentamicin, and 3.0 mg/mL fatty acid-free bovine serum albumin [13]. Adjust osmolarity (mOsm) to 288 ± 2 and pH to 7.3.

2.4 HMC Setup

1. Prepare IVMM dishes with 0.5 mL IVMM per well of a four-well dish and cover each well with 0.4 mL mineral oil. Incubate dishes overnight with 5% CO_2 and 5% O_2 under saturated humidity at 38.5 °C.

2. Prepare two embryo culture plates by adding 400 µL PZM-3 medium and 400 µL oil per well in NUNC four-well dishes on the day before cloning. Incubate dishes overnight with 5% CO_2 and 5% O_2 under saturated humidity at 38.5 °C.

3 Methods

3.1 Donor Cell and Embryo Culture Dish Preparation

These methods are used to establish porcine primary fibroblast cultures. Experiments with live animals must be performed in agreement with institutional and national guidelines.

1. Restrain the donor animal (usually adult pigs).

2. Shave and sterilize:
   Scrub an approximately 10×10 cm area of the ear lobe with alcohol spray, both the front and back sides. Avoid places with large blood vessels. Washed by hand, clean the dirt with a scraper and wash with alcohol. Use tweezers to take alcohol cotton balls to disinfect the scraped area three or four times, each time with a new cotton ball.

Pig Handmade Cloning   187

3. Biopsy:

Cut two pieces of tissue from the sterilized area with ear clippers and transfer them to a 1.5 mL Eppendorf tube (first tube) containing 1 mL sampling solution (PBS with 300 U/mL penicillin and 300 U/mL streptomycin).

4. Skin removal:

Place samples in a 30 mm diameter Petri dish and remove the skin from both sides by using two tweezers and a scalpel (for practical reasons, two operators are preferred for this task).

5. Washing:

Add 2 mL washing solution (PBS with 100 U/mL penicillin and 100 U/mL streptomycin) to a 30 mm Petri dish, transfer one sample to the dish, and cut off the unclean areas or those containing ear hair. Wash samples two or three times in the same solution, then transfer them to a 60 mm Petri dish containing a 500 μL drop of the same medium.

6. Cutting:

Cut the sample into 1 mm × 1 mm tissue pieces with scissors, then add 4 mL cell culture medium (CCM). Shake the dish to distribute the pieces evenly.

The whole sample processing has to be completed within 5 min.

7. Culture:

Culture in an incubator in 5% CO_2 in an air atmosphere at 37 °C and maximum humidity for 7 days.

8. Passage/freezing:

After 7 days, check cultures for cell growth under an inverted phase-contrast microscope. Perform passage, genetic modification, and/or cryopreservation according to cell growth and the purpose of your experiment.

3.2 Oocyte Collection and In Vitro Maturation (IVM)

It is indispensable to have an abattoir providing a safe supply of sow ovaries (50–100 per day, at least 3 days/week) within a 2–3 h driving distance. Collect enough ovaries considering the fact that approximately 200 oocytes can be safely handled by each operator per day.

1. Collect porcine ovaries at the abattoir and transport them to the laboratory in physiological saline at 35 °C.

2. Aspirate the follicular fluid from 3–6 mm follicles using a syringe attached to a 20-gauge needle and transfer the follicular content into a 50 mL tube.

3. Search for cumulus-oocyte complexes (COCs) in a 90 mm dish and select COCs with three intact and nonexpanded cumulus cell layers and oocytes with evenly dark cytoplasm.

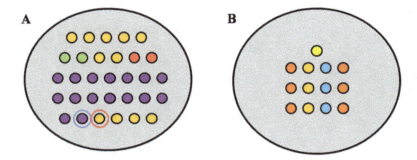

Fig. 1 The layout of bisection (**a**) and fusion (**b**) dishes. The media and solution legends are T2M (light orange), PRO-CS (green), T20M (red), CBTP (purple), PHA solution (yellow), T10M (orange), and CFM (light blue). Two droplets are outlined by blue and red markings. Check the text for details

4. Wash COCs in TL-HEPES and wash them twice in IVMM. Transfer 50 COCs per well of the IVMM dish and incubate at 38.5 °C in 5% CO_2, 5% O_2, and 90% N_2 atmosphere and maximum humidity for 41–42 h.

3.3 Oocyte Denudation and Zona Pellucida Digestion

1. Prepare bisection dish (Fig. 1a).
2. Collect COCs from wells after 41–42 h of IVM with minimum volume of IVM. Place COCs into the hyaluronidase solution. Add 3–4 µL mineral oil (*see* **Note 3**). Pipette gently. Avoid the contact between the pipette tip and the bottom of the tube. Vortex the tube for 1 min.
3. Transfer the content of the tube with oocytes into an empty 35 mm Petri dish, rinse the tube with T2M, and vortex for 5 s. Transfer the content to the Petri dish.
4. Search for denuded oocytes and distribute them evenly among the upper T2M droplets in the bisection dish.
5. Transfer 20 oocytes to a PRO-CS droplet and observe until definite *zona pellucida* deformation occurs.
6. Wash zona-free oocytes quickly in T2M and T20M droplets and distribute them between three CBTP droplets (i.e., 6–7 oocytes/drop). Search for oocytes with a visible polar body (PB) or extrusion cone (a small round extrusion of the membrane containing the chromatin). Line up oocytes (from north to south) with the PB or EC at 12 o'clock in a clock's face. Do this in all three CBTP droplets.

3.4 Somatic Cell Nuclear Transfer

3.4.1 Oocyte Enucleation by Bisection

1. Use a pipette with a medium inner diameter (~200 µm). Perform oriented manual bisection with splitting blades by cutting ~30% of the oocyte cytoplasm adjacent to the PB (*see* **Note 4**). After these oocyte bisections, put a new batch of 20 oocytes into a PRO-CS droplet.

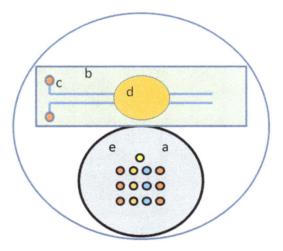

Fig. 2 The layout for fusion: (a) lid of a 60 mm Petri dish; (b) fusion chamber; (c) electrode connection; (d) fusion medium; (e) fusion dish

2. Tap the side of the dish with a fingernail to detach bisected oocytes from the plastic. Collect cytoplasts (i.e., enucleated halves) and transfer them to the T2M droplet with the red circle. Collect and transfer bisected oocyte halves with PB to the CBTP with the blue circle. Repeat the procedure. It should take less than 10 min to bisect 20 oocytes.

3.4.2 Cell Fusion

1. Place the fusion chamber and the lid of a 35 mm dish at the center of the lid of a 60 mm Petri dish (Fig. 2). Prepare fusion dish with 20 µL droplets in the lid of the 35 mm dish as shown in Fig. 1b.

2. Adjust the fusion machine as follows: direct current (DC) at 100 V, alternate current (AC; enable) at 0 V, and a single pulse of 9 µs. Attach the wires of the fusion machine to the fusion chamber.

3. Add 5–20 µL of the donor cell suspension to one or two T2M droplets, remove half of the cytoplasts (preferably the larger ones) from the bisection dish, and add them to the fusion dish, T10M droplets on the left. Leave the rest in the bisection dish, on the heated stage, and protected from light.

4. Put 500 µL CFM covering the two wires and aside.

5. With small diameter pipettes (~150 µm), transfer five cytoplasts to PHA solution for 2 s, wash in T2M, and roll them one at a time over a single-donor somatic cell. Transfer cell couplets to CFM. Repeat the whole procedure with five cytoplasts to create an additional five couplets.

190 Gábor Vajta et al.

6. Transfer the first five cell couplets to a fusion chamber close to the north of the wires. Place cell couplets between the wires (one by one). Increase AC to 5 V.

7. Apply the DC, decrease AC to 0, pick up the cell couplet, and repeat it for the remaining cell couplets. Return the five cell couplets to the last T10M drop in line.

8. Repeat the fusion process for the remaining cytoplasts (*see* **steps 5–7**).

9. Repeat the pairing-fusion procedure (steps) with an additional 10 cytoplasts (*see* **steps 5–8**).

10. After fusing 20 cell couplets, check the first 10 for fusion (absence of the donor cell adjacent to the cytoplast) and transfer to another T10M droplet below (*see* **Note 5**).

11. Protect the lid dish from light and keep it on the warm stage at 38.5 °C for 1 h (*see* **Note 6**). The preparation of 50 fused couplets usually requires 20–30 min.

3.5 *Oocyte Activation*

1. Adjust fusion machine to DC at 43 V, AC at 4 V, and a single pulse of 80 μs. Replace fusion droplets in the fusion dish and over the wire with CFM. Transfer all cytoplasts to a T10M droplet placed on the left. Transfer 10 cell couplets and 10 cytoplasts to the first CFM and second CFM droplets, respectively. Wait until they reach the bottom of the dish.

2. Transfer cell couplets and cytoplasts to the fusion chamber, north and south of the wires, respectively. Move 10 cytoplasts to the northern wire first and place a single-cell couplet to each cytoplast. Increase AC to 10 V, apply the DC, and decrease AC to 0 (*see* **Note 7**). Put these reconstructed oocytes in a T10M droplet. Repeat with remaining reconstructed oocytes. The preparation of 50 reconstructed oocytes usually requires 5 min.

3. Prepare two Nunc four-well dishes by adding 400 μL PZM-3 medium and 400 μL oil to each well. Prepare the first dish for oocyte activation by adding 4.0 μL T2CB and 10 μg/mL cycloheximide to wells #1 and #2. Mix by gentle pipetting. Transfer reconstructed oocytes to wells #1 and #2 of the activation dish. Incubate reconstructed oocytes with 5% CO_2, saturated humidity at 38.5 °C for 4 h.

3.6 *Embryo Culture*

1. Make enough Wells-of-the-Wells (WOWs) to all fused embryos in well #4 of the second culture dish (*see* **Note 8**).

2. Wash embryos in wells #3 and #4 of the dish 4 h after activation. Wash in all four wells of the second (washing) dish and transfer to wells #1 and #2 of the culture dish. Transfer embryos next to, then inside of, the WOWs (*see* **Note 9**).

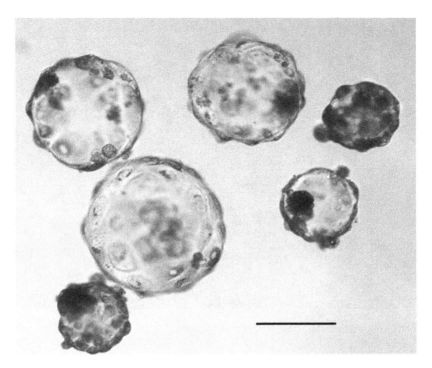

Fig. 3 Handmade-cloned pig blastocysts on day 5 of in vitro culture. Scale bar: 100 μm

3. Culture embryos with 5% CO_2 + 5% O_2, under saturated humidity at 38.5 °C, until day 5 (D5) post-activation (*see* **Note 10**).

4. Remove all embryos from the WOWs on D5 and determine blastocyst rates. Remove embryos from WOWs by filling the pipette with PZM-3 medium and gently flushing out embryos from WOWs. Forced aspiration may cause the blastocysts to collapse and lead to inaccurate evaluation (*see* **Note 11**). The selection of good-quality blastocysts (Fig. 3) by morphology grading allows full term development (Fig. 4).

3.7 Embryo Transfer

A pig farm providing appropriate conditions for surgical embryo transfer, recipient management, pregnancy monitoring, and caesarean section within a 3–4 h drive is an indispensable requirement. Collaboration with an experienced veterinarian is also required for surgical embryo transfer and caesarean section.

3.7.1 Recipient Preparation

Recipients were healthy sows on days 3–6 after estrus. No feeding since the afternoon of the day before surgery. In addition to the requirements of estrus time on the selection of receptors, there are also the following requirements:

1. Varieties: Acceptor varieties include large binary hybrid, long white, large white and three yuan hybrid, etc.

Fig. 4 Piglets born from vitrified handmade-cloned blastocysts

2. Age: 1.5-year-old sow.
3. Parity: Parturient sows should be selected as the main choice, and 1-parturient sows are preferred.
4. Body shape: medium body shape, not too fat, not too thin, and suitable for 150–180 kg.
5. Health status: Recipient sows should have no genetic diseases, no long-term records of non-estrus, abortion, dystocia, and no history of endometritis, mastitis, etc.
6. Reproductive performance: The recipient sows have excellent reproductive performance, such as large litter size, high survival rate at birth, sufficient milk production, and good motherhood.

3.7.2 Surgical Method for Embryo Transfer

1. Receptor anesthesia and stabilization with an intravenous injection of 15% ketamine:

    1.2–1.5 mL for body weight less than 150 kg, while 1.5 2.0 mL for body weight greater than 150 kg.

2. Place the recipient on the operating table and clean the surgical site with soap (between the penultimate 2–3 nipple), and the hair on the surgical site was shaved with a razor.
3. Maintain the respiratory maintenance anesthesia with a flow rate of oxygen and nitrous gas of 0.5 L/min with 5% isoflurane. Adjust the flow of oxygen and nitrous oxide appropriately according to the recipient's state of anesthesia.

4. Surgical preparation and disinfection: Prepare surgical instruments, gauze, surgical gowns, surgical gloves, masks, stitches, physiological saline, iodine, 75% alcohol, etc. The surgeon should wear masks, hats, surgical gowns, and surgical gloves. Pay attention to the rules and order of wearing, and avoid touching the outside of gloves during operation to prevent contamination. Use surgical tweezers to hold iodine tincture to disinfect the surgical site (from inside to outside, in a spiral shape), then spray 75% alcohol to deiodize, then rinse with normal saline, and finally wipe the surgical site with sterile gauze.

5. Abdominal surgery: Spread the surgical towel and fix it on the skin surface with four towel-holding forceps (avoid clipping the nipple and affecting later lactation). Between the second and third nipples, the skin and subcutaneous connective tissue are cut with a scalpel. Hemostatic gauze and hemostatic forceps are prepared for hemostasis of small vessels, nonabsorptive surgical sutures are prepared in advance, and ligation of large bleeding vessels is prepared at all times. The adipocytes are separated along the abdominal linea alba by hemostatic forceps or scalpel. Separate the fat layer from the peritoneum and cut the peritoneum. Remove the blood stains around the surgical site and fix two large gauze strips with tissue forceps on both sides of the midabdominal line, spread them out, and spray a certain amount of physiological saline on them to prevent uterine dehydration. The uterus is extracted from the abdominal cavity to check the ovulation of the ovary. In case of premature or late ovulation or severe uterine adhesion, the transplantation should be abandoned and the recipient replaced. Find the oviduct–uterine junction, use sterile blunt needle to pierce a small hole, the whole process should be timely to spray physiological saline into the uterus, and keep the uterus moist, not dehydration. Collect 50 good-quality blastocysts in TL-HEPES in a 35 mm dish and load them into a tomcat catheter. The embryo is slowly transferred into the oviduct–uterine junction, and then the hole is held with fingers or hemostatic forceps for a period of time to prevent bleeding. The same is done on the other side of the uterus.

6. Close abdomen: Remove the blood stains at the surgical site with gauze and restore the uterus into the body. Add a small amount of physiological saline to the abdominal cavity to clean the abdominal cavity and inject 12 mL dextran and two bottles of penicillin into the abdominal cavity. The peritoneum was sutured with absorbent surgical sutures using continuous suture method. When suturing, the needle spacing should not be too large because it will lead to hernia caused by intestinal exposure to the abdominal cavity, which should be less than

1 cm. During the suture process, the assistant shall use needle-holding forceps to assist the surgeon in needle insertion, and assist in removing blood stains at the surgical site, threading, leading, and cutting thread heads. The muscle layer and fat layer are sutured with the same method, and sulfa is used to reduce inflammation. The skin layer is sutured with nonabsorbable suture and nodular suture. After suturing, surgical tweezers are used to turn the skin at the sutured site outward to facilitate healing, and iodine tincture is used to disinfect the surgical site. Turn off the anesthesia machine. To ensure the safety of the anesthetized animal, oxygen can be continued and assisted breathing for a period of time.

4 Notes

1. It would be easier to make a final cc. stock, but the effect of the PHA may vary as it may need a more or less concentrated PHA solution. We found this arrangement to be more practical.

2. The PHA and pronase solutions may have some sediment. CB is collected only from the bottom.

3. Helps remove cumulus cells.

4. The separation of the two oocyte parts should happen before the *zona pellucida* is cut: the intruding half-digested, soft zona pushes away the two parts.

5. Low fusion rates may be due to low-quality somatic cells or too many cells attached to the oocyte surface, or alternatively due to too high AC. Selection of small, strongly double-refracting somatic cells may help.

6. Theoretically, this period is needed for reprogramming.

7. This second round of fusion will increase the size of the reconstructed oocyte (donor cell with two cytoplasts instead of one) and initiates oocyte activation.

8. For preparation of WOWs, support the bottom of well #4 of a Nunc four-well culture dish with a piece of glass (e.g., a light filter glass from a microscope) to avoid cracking during preparation. Under a stereomicroscope, make up to 7×7 WOWs with vertical steady pressure by using a BLS aggregation needle. The needle is sterilized shortly before use by holding it very briefly inside a gas flame.

9. Thorough washing with repeated pipetting in each well is required to get rid of potentially harmful chemicals used for activation.

10. Low oxygen concentration is essential to obtain appropriate blastocyst number and quality. Under the suggested conditions, no medium change is required during the culture period.

However, maximum humidity in the incubator is essential to avoid osmotic damage. We strongly recommend refraining from any embryo assessment (e.g., cleavage evaluation) during the 5 days of culture.

11. An established porcine HMC method can result in >50% blastocyst rate (based on the number of reconstructed oocytes). Approximately 80% of the blastocysts have well-defined inner cell masses and a trophectoderm of more than 10 cells. Trophectoderm cell counting requires a good microscope and considerable experience.

Acknowledgments

The authors are grateful to all members of the former Embryo Technology Laboratory, Department of Genetics and Biotechnology, Faculty of Agricultural Sciences, Aarhus University, Tjele, Denmark, and BGI Ark Biotechnology Co. LTD, Shenzhen, China, for support in establishing and optimizing the porcine handmade cloning method.

References

1. Vajta G, Gjerris M (2006) Science and technology of farm animal cloning: state of the art. Anim Reprod Sci 92:211–230

2. Kragh PM, Lade ÆA, Juan NÆ, Yutao LÆ, Lin DÆ, Holm ÆIE et al (2009) Hemizygous minipigs produced by random gene insertion and handmade cloning express the Alzheimer's disease-causing dominant mutation APPsw. Transgenic Res 18:545–558

3. Holm I, Alstrup A, Luo Y (2016) Genetically modified pig models for neurodegenerative disorders. J Pathol 238:267–287

4. Umeyama K, Watanabe M, Saito H, Kurome M, Tohi S, Matsunari H et al (2009) Dominant-negative mutant hepatocyte nuclear factor 1alpha induces diabetes in transgenic-cloned pigs. Transgenic Res 18:697–706

5. Al-Mashhadi RH, Sorensen CB, Kragh PM, Christoffersen C, Mortensen MB, Tolbod LP et al (2013) Familial hypercholesterolemia and atherosclerosis in cloned minipigs created by DNA transposition of a human PCSK9 gain-of-function mutant. Sci Transl Med 5:166ra1

6. Teboul L, Herault Y, Wells S, Qasim W, Pavlovic G (2020) Variability in genome editing outcomes: challenges for research reproducibility and clinical safety. Mol Ther 28:1422–1431

7. Vajta G, Lewis I, Tecirlioglu R (2006) Handmade somatic cell cloning in cattle. Methods Mol Biol 348:183–196

8. Vajta G, Lewis I, Trounson A, Purup S, Maddox-Hyttel P, Schmidt M, Pedersen H, Greve T, Callesen H (2003) Handmade somatic cell cloning in cattle: analysis of factors contributing to high efficiency in vitro. Biol Reprod 68:571–578

9. Vajta G (2007) Handmade cloning: the future way of nuclear transfer? Trends Biotechnol 25:250–253

10. Du Y, Li J, Kragh PM, Zhang Y, Schmidt M, Bøgh IB et al (2007) Piglets born from vitrified cloned blastocysts produced with a simplified method of delipation and nuclear transfer. Cloning Stem Cells 9:469–476

11. Vajta G, Callesen H (2012) Establishment of an efficient somatic cell nuclear transfer system for production of transgenic pigs. Theriogenology 77:1263–1274

12. Callesen H, Liu Y, Pedersen HS, Li R, Schmidt M (2014) Increasing efficiency in production of cloned piglets. Cell Reprogram 16:2–5

13. Yoshioka K, Suzuki C, Tanaka A, Anas Idris M-K, Iwamura S (2002) Birth of piglets derived from porcine zygotes cultured in a chemically defined medium. Biol Reprod 66:112–119

Chapter 10

Somatic Cell Nuclear Transfer in Pigs

Werner G. Glanzner, Vitor B. Rissi, and Vilceu Bordignon

Abstract

Somatic cell nuclear transfer (SCNT) has been successfully applied to clone animals of several species. Pigs are one of the main livestock species for food production and are also important for biomedical research due to their physiopathological similarities with humans. In the past 20 years, clones of several swine breeds have been produced for a variety of purposes, including biomedical and agricultural applications. In this chapter, we describe a protocol to produce cloned pigs by SCNT.

Key words Embryo transfer, Enucleation, Histone deacetylases, Nuclear transplantation, Nuclear reprogramming, SCNT, *Sus scrofa*, Swine, Transcriptional inhibition

1 Introduction

Mice were the first cloned mammals created by nuclear transfer (NT) from embryonic cells [1], while the first cloned animal from an adult somatic cell was produced more than two decades ago [2]. Since then, many animals of a variety of domestic and wild species have been cloned by somatic cell NT (SCNT) [3, 4]. This technology has been mainly applied to study cellular reprogramming and produce cloned animals. Cloned pigs were first produced in 1989 by NT from four-cell stage embryos [5], and from somatic cells in 2000 [6, 7]. A year later, SCNT was applied to produce the first transgenic pigs from cells that were genetically modified in vitro [8], which accelerated the interest and the use of this technology to create pigs with unique characteristics for use as models in biomedical research. Given their physiopathological similarities with humans [9–12], the main interest in swine cloning has been to produce genetic models of diseases [12, 13], and for research in xenotransplantation [14–16]. Despite much effort to improve SCNT efficiency, the proportion of SCNT embryos that develop to term and generate live cloned piglets is in general below 3% [9, 17–19], which limits the broader use of this technology.

Marcelo Tigre Moura (ed.), *Somatic Cell Nuclear Transfer Technology*, Methods in Molecular Biology, vol. 2647, https://doi.org/10.1007/978-1-0716-3064-8_10,
© The Author(s), under exclusive license to Springer Science+Business Media, LLC, part of Springer Nature 2023

SCNT is a multistep and complex technology, albeit incomplete nuclear reprogramming remains the main constraint affecting the efficiency to produce cloned animals [20]. Nuclear reprogramming in SCNT embryos involves the resetting of epigenetic marks such as histone acetylation and methylation and DNA methylation [21]. Aberrant DNA and histone methylation [22], and histone acetylation patterns have been observed in SCNT embryos compared with fertilized embryos [23], thus suggesting incomplete nuclear reprogramming.

Many studies have attempted to improve cell reprogramming in SCNT embryos by modulating epigenetic mechanisms [24, 25]. For example, histone deacetylase inhibitors (HDACi) were shown to improve SCNT efficiency in swine and in other species [26–30]. There is evidence that HDACi treatment promotes histone hyperacetylation and increases gene expression [31], alters DNA methylation patterns [32], and enhances DNA damage repair in SCNT embryos [33]. In addition, we observed that transcription inhibition during HDACi treatment improved gene expression patterns and cell numbers in SCNT embryos [34, 35].

Histone lysine methyltransferases (KMTs) and demethylases (KDMs) enzymes play key roles in nuclear reprogramming [36, 37], as well as in the regulation of gene expression [38], and embryo genome activation (EGA) during preimplantation development [39–42]. Moreover, expression of KDMs that act on histone 3 lysine 9 trimethylation (H3K9me3) improved nuclear reprogramming and development of SCNT embryos [36, 43–45]. These studies indicate that animal cloning efficiency may be further improved by protocols that enable better reprogramming of multiple epigenetic marks.

The main steps required to clone an animal by SCNT are oocyte enucleation, cell transfer, cell fusion, oocyte activation, embryo culture, and embryo transfer to a surrogate female. Oocytes for SCNT are usually in vitro matured oocytes obtained from abattoir-sourced ovaries. Matured oocytes are normally enucleated by using micropipettes attached to micromanipulators. After enucleation, a somatic cell is transferred to the perivitelline space and fused to the enucleated oocyte by an electric pulse. Donor cells for use in SCNT can be obtained from fetuses, newborn or adult animals, and in vitro cultured. Variations in cloning efficiency associated with donor cell types have been reported, but results are inconsistent. It is possible that cells from fetuses and newborn animals have less aging-associated epigenetic variations and may be easily reprogrammed following SCNT than cells from adult animals [46, 47]. Oocyte activation is performed by applying electrical pulses or by exposure to chemicals that induce calcium mobilization followed by a temporary inhibition of protein kinases activity or protein synthesis. After activation, SCNT embryos can be

Pig Cloning   199

cultured in vitro up to the blastocyst stage and are then transferred to synchronized surrogate females. In pigs, SCNT embryos are normally transferred surgically into the oviduct or uterus depending on their developmental stage at the time of transfer. This chapter provides a detailed description of the equipment, materials, and methods necessary to produce cloned pigs by SCNT.

2 Materials

2.1 Equipment

1. Stereomicroscope.

2. Inverted microscope.

3. Micromanipulation system (*see* **Note 1**): Inverted microscope equipped with micromanipulators, microinjectors, and holding and injection pipettes.

4. Holding pipettes with 80–120 μm external diameter and 20–30 μm internal diameter. Injection pipette with 18–25 μm internal diameter and a 45° bevel.

5. CO_2 incubator.

6. Electrofusion equipment and cell fusion chamber.

7. Cell-freezing container (e.g., Nalgene® Mr. Frosty).

8. Anesthetic machine.

9. Gas cylinders (CO_2, O_2).

10. Surgical instruments.

11. Ultrasound system.

2.2 Media and Solutions

All media and solutions are prepared using deionized ultra-pure water (18 MΩ.cm at 25 °C). Stock solutions are stored at −20 °C or −80 °C. Culture media are filtered (0.22 μm) and stored at 4 °C for up to 2 months, unless otherwise indicated.

1. Cell culture medium: Mix 89 mL DMEM-F12, 10 mL fetal bovine serum (FBS), and 1 mL antibiotics (10.000 units/mL penicillin and 10 mg/mL streptomycin). Store at 4 °C for 1 week.

2. Cell-freezing medium: Mix 9 mL of culture medium (DMEM + 10% FBS) with 1 mL dimethyl sulfoxide (DMSO). This medium is prepared immediately before use.

3. COC manipulation medium: Mix 98 mL TCM-199 supplemented with 25 mM HEPES, 5 mM sodium bicarbonate, 1 mL porcine follicular fluid (PFF), and 1 mL antibiotics (10.000 units/mL penicillin and 10 mg/mL streptomycin) (*see* **Note 2**).

4. IVM medium: TCM199 supplemented with 25 mM HEPES, 26 mM sodium bicarbonate, 20% PFF, 100 µg/mL cysteine, 0.91 mM sodium pyruvate, 3.05 mM D-glucose, 10 ng/mL epidermal growth factor (EGF), 20 µg/mL gentamicin, 10 µg/mL follicle-stimulating hormone (FSH), 5 UI/mL human chorionic gonadotropin (hCG), and 1 mM dibutyryl cyclic adenosine monophosphate (cAMP) (*see* **Note 3**).

5. Oocyte manipulation medium: 50 mL TCM199 supplemented with 25 mM HEPES, 26 mM sodium bicarbonate, 2 mg/mL bovine serum albumin (BSA), and 20 µg/mL gentamicin.

6. Oocyte manipulation medium with demecolcine and sucrose (*see* **Note 4**): TCM199 medium with 25 mM HEPES, 26 mM sodium bicarbonate, 2 mg/mL BSA, 400 ng/mL demecolcine, and 17 mg/mL sucrose.

7. PZM3 medium: 108 mM NaCl, 10 mM KCl, 0.35 mM KH_2PO_4, 0.4 mM $MgSO_4 \cdot 7H_2O$, 25.07 mM $NaHCO_3$, 0.2 mM Na-pyruvate, 2 mM Ca-$(lactate)_2 \cdot 5H_2O$, 1 mM L-glutamine, 5 mM hypotaurine, 20 ml/L Basal Medium Eagle amino acids (Sigma B6766), 10 ml/L Minimum Essential Medium nonessential amino acids (Sigma M7145), and 20 µg/mL gentamicin, and 3 mg/mL BSA [48].

8. Hyaluronidase solution: Dilute 1 mg/mL hyaluronidase in TCM199 medium supplemented with 25 mM HEPES and 5 mM sodium bicarbonate. Prepare 500 µL aliquots and store at −20 °C.

9. Cysteine solution: Dilute 100 mg cysteine in 1 mL TCM-199, prepare 20 µL aliquots, and store at −80 °C. Add 1 µL of the stock solution per mL of IVM medium.

10. Sodium pyruvate: Dilute 10 mg sodium pyruvate in 1 mL TCM-199, prepare 200 µL aliquots, and store at −80 °C. Add 10 µL of the stock solution per mL of IVM medium.

11. EGF: Dilute 10 µg EGF in 1 mL TCM-199, prepare 20 µL aliquots, and store at −80 °C. Add 1 µL of the stock solution per mL of IVM medium.

12. FSH: Dilute 12.5 mg FSH (standard Armour) in 5 mL TCM199, prepare 80 µL aliquots, and store at −80 °C. Add 4 µL of stock per mL of IVM medium.

13. hCG: Dilute 1 IU/µL hCG in TCM 199, prepare 100 µL aliquots, and store at −80 °C. Add 5 µL of the stock solution per mL of IVM medium.

14. cAMP: Dilute 9.8 mg cAMP in 1 mL TCM-199, prepare 100 µL aliquots, and store at −80 °C. Add 50 µL of the stock solution per mL of IVM medium.

15. Cytochalasin B solution (7.5 μg/mL): Dilute 7.5 mg in 1 mL DMSO and store 1 μL aliquots at −80 °C. Working solution: Dilute 1 μL stock solution in 1 mL oocyte manipulation medium.

16. Electrofusion solution (0.28 mM D-mannitol): Dilute 5.46 g D-mannitol, 100 μL stock solution (1000×) $CaCl_2$, 100 μL stock solution (1000×) $MgSO_4$, 50 mg BSA, and 11.8 mg HEPES in 100 mL ultrapure water. Stir, correct pH to 7.3–7.4, and sterile filter. Store at 4 °C for 4 months or at −20 °C for 12 months. For $CaCl_2$ 1000× stock solution: Dilute 73.5 mg in 10 mL of ultrapure water. For $MgSO_4$ 1000× stock solution: Dilute 246.6 mg in 10 mL of ultrapure water. Stir and filter both solutions, and store at 4 °C for 4 months.

17. Ionomycin solution: Dilute 1 mg ionomycin in 89 μL DMSO, prepare 1 μL aliquots, and store at −80 °C. Working solution (15 μM): Dilute 1 μL stock solution in 1 mL of TCM199 supplemented with 25 mM HEPES, 26 mM sodium bicarbonate, and 2 mg/mL BSA.

18. N,N,N′,N′-Tetrakis(2-pyridylmethyl)ethylenediamine (TPEN): Dilute 10.62 mg TPEN in 500 μL DMSO, prepare 2 μL aliquots, and store at −20 °C. Working solution (200 μM): Dilute 2 μL stock solution in 500 μL porcine zygote medium 3 (PZM3).

19. Cycloheximide (CHX) solution: Dilute 1 mg CHX in 100 μL ethanol (100%) or DMSO, prepare 1 μL aliquots, and store at −80 °C. Working solution (10 μg/mL): Dilute 1 μL stock solution in 1 mL PZM3.

20. Strontium chloride (Sr^2) solution: Dilute 317.06 mg Sr^2 in 1 mL ultrapure water, prepare 5 μL aliquots, and store at −80 °C. Working solution (10 mM): Add 5 μL stock solution per mL of PZM3.

21. Scriptaid solution: Dilute 0.1 mg Scriptaid in 612 μL DMSO, prepare 1 μL aliquots, and store at −80 °C. Working solution (500 μM): Add 1 μL stock solution per mL of PZM3 (see **Note 5**).

22. DRB (5,6-dichloro-1-beta-D-ribofuranosylbenzimidazole) solution: Dilute 3 mg DRB in 94 μL DMSO, prepare 1 μL aliquots, and store at −20 °C. Working solution (100 μM): Add 1 μL stock solution per mL of PZM3.

3 Methods

Oocyte and embryo handling, and SCNT are performed using stereoscopes or inverted microscopes equipped with warm stages with temperature set at 37 °C.

202 Werner G. Glanzner et al.

3.1 Tissue Collection and Culture of Somatic Cells

1. Disinfect the skin with 70% ethanol (*see* **Note 6**).

2. Collect one or more 0.5–1 cm^3 skin biopsies and transport at 4 °C to the laboratory in PBS containing antibiotics.

3. Fragment the biopsies in a 100 mm cell culture dish using a scalpel blade and place explants in culture or digest using 0.25% trypsin for 20 min. After tissue digestion, centrifuge cells at 500 g for 5 min, discard the supernatant, resuspend in 10 mL cell culture medium, and then transfer to 25 or 75 cm^2 culture flasks.

4. Culture cells until reaching ~90% confluence (Passage 0). The cells are then trypsinized for cryostorage, culture expansion, or use in SCNT. For trypsinization, cells are incubated with 0.25% trypsin for 5 min at 38.5 °C, washed in culture medium, centrifuged at 500 g for 5 min, and resuspended in culture medium. For freezing, resuspend cells in freezing culture medium, transfer to the freezing container overnight at −80 °C, and store at −196 °C indefinitely.

3.2 Cell Cycle Synchronization of Donor Cells

1. Porcine cloning is normally performed with oocytes at metaphase II stage and cells at G0 or G1 stage of the cell cycle. To obtain a higher proportion of cells at G1 stage, cells are seeded and maintained in culture for 2–3 days after they reach >90% confluency.

2. To prepare donor cells for SCNT, wash confluent cell cultures with 5 mL of warm PBS and incubate with 5 mL 0.25% trypsin solution for 5 min at 37 °C. Rinse and resuspend detached cells in 1–2 mL culture medium and add approximately 10 μL of cell suspension to each droplet of the micromanipulation dish.

3.3 Oocyte Collection and IVM

1. Collect porcine ovaries from gilts at a slaughterhouse and transport them to the laboratory in saline solution at 32–37 °C using a thermos bottle. In the laboratory, wash the ovaries with saline and keep them at 35 °C for follicle aspiration.

2. Aspirate 2–8 mm follicles using 10 mL syringes and 21G needles. Deposit the follicular content in 50 mL conical tubes. After aspiration, let the follicular content sediment by centrifugation at 20 g for 3 min, remove the supernatant fluid, and wash the pellet containing the cumulus-oocyte complexes (COCs) three times with 25 mL of COC manipulation medium.

3. Transfer the fluid to 100 mm plates and retrieve the COCs by searching under a stereoscope. Select the COCs having at least three layers of cumulus cells and oocytes with homogenous and agranular cytoplasm.

4. Wash groups of 30 COCs (Fig. 1a) in IVM medium and place them in 90 μL droplets of the same medium covered with

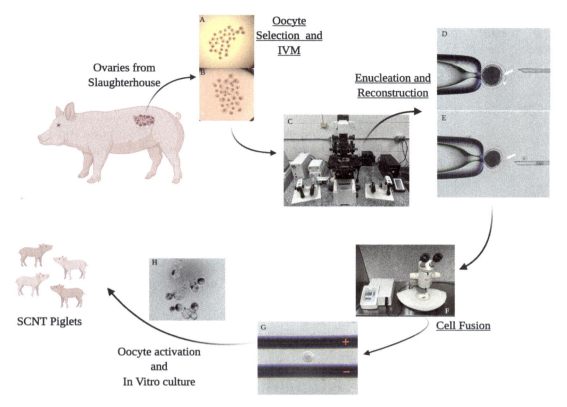

Fig. 1 Schematic representation of somatic cell nuclear transfer (SCNT) in pigs. Cumulus-oocyte complexes are selected and subject to in vitro maturation (IVM) for 44 h. (**a**) Immature (germinal vesicle stage) oocytes; (**b**) mature oocytes. Using a micromanipulation system (**c**), oocytes are enucleated (**d**, arrow indicates the polar body and arrowhead indicates the pseudo-second-polar body), and a nuclear donor cell is transferred to the perivitelline space (**e**, arrow indicates donor cell in the perivitelline space). Cell fusion is performed by using a cell electroporator (**f**) once the cell is aligned between the electrodes of the fusion chamber (**g**). Oocytes are then activated and cultured for up to 7 days to reach blastocyst stage (**h**) and are transferred to surrogate sows. IVM in vitro maturation. (This figure was created using Biorender)

mineral oil in 35 mm or 60 mm plates. Maturation comprises two periods of 22 h each, the first in the presence of gonadotropins (FSH and hCG) and cAMP, and the second in the absence of these compounds (*see* **Note 3**).

3.4 Oocyte Denuding

1. Remove cumulus cells after IVM by repetitive pipetting for 3 min in 500 μL hyaluronidase solution in a well of a four-well plate. Wash denuded oocytes twice in oocyte manipulation medium and select for the presence of the first polar body and normal morphology for use in SCNT (*see* **Note 7**).

2. Incubate selected oocytes in oocyte manipulation medium supplemented with demecolcine and sucrose for 45–60 min in the incubator (*see* **Notes 4** and **8**). After this treatment, oocytes should display a pseudo-second-polar body containing the oocyte spindle and chromatin (Fig. 1d).

204 Werner G. Glanzner et al.

3.5 Somatic Cell Nuclear Transfer

3.5.1 Oocyte Enucleation

1. Transfer ~25 oocytes from the demecolcine and sucrose treatment to a 50 μL drop of oocyte manipulation medium supplemented with 7.5 μg/mL cytochalasin B in a micromanipulation plate (*see* **Note 9**).

2. Grasp one oocyte with the holding pipette, position the pseudo-second-polar body to the 4 o'clock position in a clock's face, and aspirate it along with the first polar body and a small portion of surrounding cytoplasm using the injection pipette (Fig. 1c). After enucleation, keep the oocyte attached to the holding pipette and perform the reconstruction by injecting a nuclear donor cell in the perivitelline space.

3.5.2 Oocyte Reconstruction

1. Aspirate one cell into the injection pipette and maintain it close to the tip of the pipette.

2. Insert the pipette through the slit of the *zona pellucida* created during the enucleation and place the nuclear donor cell in the perivitelline space (Fig. 1e). Repeat the procedure of enucleation and cell injection for all oocytes of the same batch.

3. Transfer the reconstructed oocytes to oocyte manipulation medium at 38.5 °C and 5% CO_2 for 30–60 min before electrofusion.

3.5.3 Cell Fusion

1. Wash the reconstructed oocytes in a solution containing 500 μL of oocyte manipulation medium and 500 μL of electrofusion solution under the stereoscope (Fig. 1f), and then in electrofusion solution alone (*see* **Note 10**).

2. Fill the fusion chamber with electrofusion solution and align each oocyte between the two electrodes (Fig. 1g). Apply a single direct current (DC) pulse of 1.6 KV/cm and 70 μs duration to induce membrane fusion. After the DC pulse, keep the oocytes in oocyte manipulation medium for 45–60 min at 38.5 °C and 5% CO_2 to allow complete membrane fusion before performing oocyte activation.

3.6 Oocyte Activation

1. Initiate the oocyte activation protocol based on ionomycin and TPEN (*see* **Note 11**). Wash the oocytes three times in oocyte manipulation medium and transfer them to 15 μM ionomycin for 5 min.

2. Wash the oocytes three times in oocyte manipulation medium and transfer them to the TPEN solution for 15 min.

3. Wash the oocytes three times in PZM3 medium and transfer them to the embryo culture medium.

3.7 Embryo Culture

1. Transfer batches of 20–30 activated oocytes to 60 μL PZM3 droplets supplemented with 500 nM Scriptaid and

100 μM 5,6-dichloro-1-beta-D-ribofuranosylbenzimidazole (DRB) (*see* **Note 12**), and culture them under mineral oil at 38.5 °C and 5% CO_2 for 15 h (*see* **Note 13**).

2. Wash the reconstructed oocytes three times in PZM3 and transfer them in groups of 20–30 to 60 μL PZM3 droplets for culture under mineral oil at 38.5 °C and 5% CO_2.

3. Evaluate cleavage rates at 48 h of culture, remove the uncleaved oocytes, and culture the cleaved embryos in PZM3 supplemented with 0.3% BSA until day 5.

4. After 5 days of culture, replace 30 μL (50% of the drop volume) in each culture drop with PZM3 supplemented with 20% FBS. Alternatively, transfer the embryos to new 60 μL drops of PZM3 supplemented with 10% FBS.

5. Evaluate embryonic development at day 7 and classify blastocyst stage embryos as early, expanded, or hatched blastocysts (Fig. 1h). Blastocysts classification for embryo transfer will depend on the stage of development when embryo transfer will take place. Normally, early and expanded blastocysts on D5 and D6 of culture are selected for transfer into the uterine horns. Alternatively, reconstructed oocytes can be transferred into the oviduct 15–24 h after oocyte activation.

3.8 Embryo Transfer

1. Feed gilts or sows 20 mg Altrenogest (Regumate®) per day for 12–14 days for estrus synchronization (*see* **Note 14**). Inject 1000 I.U. eCG (Novormon®) on the last day of Altrenogest supplementation, followed by an intramuscular injection of 500 I.U. hCG (Chorulon®) given 72–78 h later (*see* **Note 15**).

2. Prepare animals for embryo transfer by administering intramuscularly a mix of 2.2–4.4 mg/kg xylazine and 10–18 mg/kg ketamine, intubate the animals (Fig. 2a), and maintain anesthesia with isoflurane (*see* **Note 16**).

3. Perform an aseptic midventral laparotomy of approximately 10 cm to exteriorize oviducts/ovaries or uterine horns (Fig. 2b, c). Transfer the embryos to the oviduct (embryos on day 1 after SCNT) or to the uterus, approximately 2 inches (~5 cm) away from the uterotubal junction (embryos on days 5–7 after SCNT; Fig. 2c).

4. Use a Tomcat catheter connected to a 1 mL syringe (Fig. 2d) filled with PZM3-HEPES medium for embryo transfer. Load embryos into the catheter as follows: (1) medium (~20 μL); (2) air column (~0.5 cm); (3) medium with embryos (~10 μL); (4) air column (~0.5 cm); and (5) medium (~10 μL) (Fig. 2d). Introduce the catheter via the fimbria when transferring embryos into the oviduct. Introduce the catheter via the uterine wall when transferring blastocysts into the uterine horns.

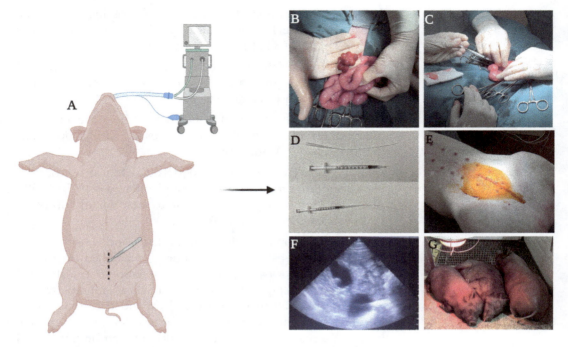

Fig. 2 Production of cloned piglets by somatic cell nuclear transfer (SCNT). Estrus synchronized gilts or sows are anesthetized and a midventral laparotomy is performed (**a**) to expose the ovaries/oviducts and uterine horns (**b**). Embryos are loaded into a Tomcat catheter attached to a syringe (**d**) and are transferred either into the oviduct or uterine horns (**c**) depending on the embryo developmental stage. The abdominal incision is closed using absorbable sutures (**e**). Pregnancy is confirmed by transabdominal ultrasonography at approximately 4 weeks after embryo transfer (**f**). Cloned piglets (**g**, 3-month-old Yucatan minipigs produced by SCNT) are normally delivered around 114 days of pregnancy. (This figure was created using Biorender)

The catheter insertion in the uterus is facilitated by making a small perforation with an 18G needle.

5. Use an absorbable suture (e.g., Vicryl) to close the incision (both abdominal wall and skin; Fig. 2e).

6. Administer post-operatory medication (analgesic and antibiotics), e.g., 3 mg/kg ketoprofen (ANAFEN®; analgesic), 20 mg/kg oxytetracycline (Oxyvet® 200 LA; antibiotic), both intramuscularly and monitor recovery.

7. Determine pregnancy rates by transabdominal ultrasound examination (Fig. 2f) ~4 weeks after embryo transfer. Ultrasound examinations can be repeated regularly (every 2 weeks) for monitoring fetal viability and development.

4 Notes

1. An alternative method named handmade cloning is also used for SCNT in pigs [49]. This method does not require the use of micromanipulators for enucleation and cell transfer, but

embryos need to be cultured individually until the blastocyst stage.

2. Porcine follicular fluid (PFF): Follicular fluid is collected from follicles of cyclic gilts (ovaries with the presence of corpora lutea), having 3–6 mm diameter from cycling gilts, centrifuged twice at 4700 g for 10 min to spin down cell debris, sterile filtered (0.22 μm), aliquoted (2 mL), and stored at −80 °C.

3. The use of cAMP during IVM is optional. Nonetheless, previous studies from our and other groups demonstrated beneficial effects of cAMP during IVM, especially in association with porcine follicular fluid [50–52].

4. Matured oocytes (with visible first polar body) are incubated in oocyte manipulation medium supplemented with demecolcine and sucrose, which results in the formation of a pseudo-second-polar body that contains the oocyte chromatin, which facilitates effective oocyte enucleation.

5. This inhibitor of histone deacetylases is used during the first 15 h of culture post activation.

6. Porcine skin biopsies are usually obtained from the tail or ears.

7. Oocyte denudation may be performed by vortexing. Transfer the COCs into a 2 mL tube containing 100 μL of 0.1% hyaluronidase solution and vortex for 2–3 min at maximum speed. Transfer the solution with the oocytes to a 35 mm plate, wash the tube twice with oocyte manipulation medium, and select oocytes with an extruded polar body and normal morphology.

8. Oocyte enucleation may be performed using DNA staining (e.g., Hoechst 33342), followed by a brief exposure to ultraviolet (UV) light to ensure enucleation. However, the protocol based on demecolcine and sucrose avoids detrimental consequences of UV radiation on oocyte organelles (e.g., mitochondrial DNA damage).

9. Because the pseudo-second-polar body tends to disappear if oocytes are maintained in the micromanipulation medium for a long time, transferring a small group each time (~25 oocytes) helps maximizing enucleation efficiency and oocyte usage.

10. Work with a small group of reconstructed oocytes (<20) each time. Wash oocytes in the mixed solution (50% oocyte manipulation medium and 50% fusion medium) to avoid osmotic stress before transferring them to the electrofusion solution. Oocytes should settle at the bottom of the plate during each wash before transferring them to the next solution. Align each oocyte between the electrodes of the fusion chamber to increase fusion efficiency. Keep the remaining oocytes away from electrodes to avoid excessive exposure to fusion electric pulses.

11. There are alternative protocols for chemical activation of pig oocytes. For many years, we used a protocol based on ionomycin, strontium chloride, cycloheximide, and cytochalasin B [53]. Our current protocol for oocyte activation relies on ionomycin and TPEN [54].

12. Culture reconstructed oocytes with inhibitors of histone deacetylases (e.g., Scriptaid) and transcription (DRB) for 15 h. We found that treatment with DRB improved the quality of both porcine and bovine SCNT embryos [34, 35].

13. Pig SCNT embryos can be cultured in a gas atmosphere of 5% CO_2 or 5% O_2, 5% CO_2 and 90% N_2, and temperature of 38.5 °C.

14. Experiments involving the production of SCNT animals must comply with national and institutional regulations. Normally, two gilts or sows are synchronized to be used as recipients for embryo transfer in each round of SCNT. At the time of embryo transfer, the ovaries are checked to confirm ovulation has occurred. Usually, 20–30 blastocysts are transferred to each uterine horn or 40–60 reconstructed oocytes are transferred to each oviduct.

15. Ovulation should occur approximately 42 h after hCG administration. Embryos are produced by SCNT 1 day before the estimated day of ovulation.

16. For intubation of animals for embryo transfer, topical anesthetic (2% lidocaine) is sprayed prior to insertion of the endotracheal tube.

Acknowledgments

The authors are thankful to the Brazilian Coordination for the Improvement of Higher Education Personnel (CAPES) for scholarships and the Natural Sciences and Engineering Research Council (NSERC) of Canada for the financial support.

References

1. Illmensee K, Hoppe PC (1981) Nuclear transplantation in Mus musculus: developmental potential of nuclei from preimplantation embryos. Cell 23:9–18

2. Wilmut I, Schnieke AE, McWhir J, Kind AJ, Campbell KH (1997) Viable offspring derived from fetal and adult mammalian cells. Nature 385:810–813

3. Keefer CL (2015) Artificial cloning of domestic animals. Proc Natl Acad Sci U S A 112:8874–8878

4. Matoba S, Zhang Y (2018) Somatic cell nuclear transfer reprogramming: mechanisms and applications. Cell Stem Cell 23:471–485

5. Prather RS, Sims MM, First NL (1989) Nuclear transplantation in early pig embryos. Biol Reprod 41:414–418

6. Polejaeva IA, Chen SH, Vaught TD, Page RL, Mullins J, Ball S et al (2000) Cloned pigs produced by nuclear transfer from adult somatic cells. Nature 407:86–90

7. Onishi A, Iwamoto M, Akita T, Mikawa S, Takeda K, Awata T et al (2000) Pig cloning by microinjection of fetal fibroblast nuclei. Science 289:1188–1190

8. Park KW, Cheong HT, Lai L, Im GS, Kuhholzer B, Bonk A et al (2001) Production of nuclear transfer-derived swine that express the enhanced green fluorescent protein. Anim Biotechnol 12:173–181

9. Vajta G, Zhang Y, Machaty Z (2007) Somatic cell nuclear transfer in pigs: recent achievements and future possibilities. Reprod Fertil Dev 19:403–423

10. Niemann H, Kues WA (2003) Application of transgenesis in livestock for agriculture and biomedicine. Anim Reprod Sci 79:291–317

11. Wernersson R, Schierup MH, Jorgensen FG, Gorodkin J, Panitz F, Staerfeldt HH et al (2005) Pigs in sequence space: a 0.66X coverage pig genome survey based on shotgun sequencing. BMC Genomics 6:70

12. Gutierrez K, Dicks N, Glanzner WG, Agellon LB, Bordignon V (2015) Efficacy of the porcine species in biomedical research. Front Genet 6:293

13. Wolf E, Braun-Reichhart C, Streckel E, Renner S (2014) Genetically engineered pig models for diabetes research. Transgenic Res 23:27–38

14. Yamada K, Sykes M, Sachs DH (2017) Tolerance in xenotransplantation. Curr Opin Organ Transplant 22:522–528

15. Sykes M, Sachs DH (2019) Transplanting organs from pigs to humans. Sci Immunol 4

16. Cooper DKC, Gaston R, Eckhoff D, Ladowski J, Yamamoto T, Wang L et al (2018) Xenotransplantation-the current status and prospects. Br Med Bull 125:5–14

17. Wolf DP, Mitalipov S, Norgren RB Jr (2001) Nuclear transfer technology in mammalian cloning. Arch Med Res 32:609–613

18. Prather RS (2007) Nuclear remodeling and nuclear reprogramming for making transgenic pigs by nuclear transfer. Adv Exp Med Biol 591:1–13

19. Lai L, Prather RS (2003) Creating genetically modified pigs by using nuclear transfer. Reprod Biol Endocrinol 1:82

20. Krishnakumar R, Blelloch RH (2013) Epigenetics of cellular reprogramming. Curr Opin Genet Dev 23:548–555

21. Morgan HD, Santos F, Green K, Dean W, Reik W (2005) Epigenetic reprogramming in mammals. Hum Mol Genet 14 Spec No 1:R47–R58

22. Kang YK, Koo DB, Park JS, Choi YH, Chung AS, Lee KK, Han YM (2001) Aberrant methylation of donor genome in cloned bovine embryos. Nat Genet 28:173–177

23. Zhao J, Whyte J, Prather RS (2010) Effect of epigenetic regulation during swine embryogenesis and on cloning by nuclear transfer. Cell Tissue Res 341:13–21

24. Glanzner WG, de Macedo MP, Gutierrez K, Bordignon V (2022) Enhancement of chromatin and epigenetic reprogramming in porcine SCNT embryos-progresses and perspectives. Front Cell Dev Biol 10:940197. https://doi.org/10.3389/fcell.2022.940197

25. de Macedo MP, Glanzner WG, Gutierrez K, Bordignon V (2022) Chromatin role in early programming of embryos. Anim Front 11(6): 57–65. https://doi.org/10.1093/af/vfab054

26. Mao J, Zhao MT, Whitworth KM, Spate LD, Walters EM, O'Gorman C et al (2015) Oxamflatin treatment enhances cloned porcine embryo development and nuclear reprogramming. Cell Reprogram 17:28–40

27. Wang LJ, Zhang H, Wang YS, Xu WB, Xiong XR, Li YY et al (2011) Scriptaid improves in vitro development and nuclear reprogramming of somatic cell nuclear transfer bovine embryos. Cell Reprogram 13:431–439

28. Bui HT, Wakayama S, Kishigami S, Park KK, Kim JH, Thuan NV et al (2010) Effect of trichostatin A on chromatin remodeling, histone modifications, DNA replication, and transcriptional activity in cloned mouse embryos. Biol Reprod 83:454–463

29. Rissi VB, Glanzner WG, Mujica LK, Antoniazzi AQ, Goncalves PB, Bordignon V (2016) Effect of cell cycle interactions and inhibition of histone deacetylases on development of porcine embryos produced by nuclear transfer. Cell Reprogram 18:8–16

30. Martinez-Diaz MA, Che L, Albornoz M, Seneda MM, Collis D, Coutinho AR et al (2010) Pre- and postimplantation development of swine-cloned embryos derived from fibroblasts and bone marrow cells after inhibition of histone deacetylases. Cell Reprogram 12:85–94

31. Kretsovali A, Hadjimichael C, Charmpilas N (2012) Histone deacetylase inhibitors in cell pluripotency, differentiation, and reprogramming. Stem Cells Int 2012:184154

32. Chen CH, Du F, Xu J, Chang WF, Liu CC, Su HY et al (2013) Synergistic effect of trichostatin A and scriptaid on the development of cloned rabbit embryos. Theriogenology 79: 1284–1293

33. Bohrer RC, Duggavathi R, Bordignon V (2014) Inhibition of histone deacetylases

enhances DNA damage repair in SCNT embryos. Cell Cycle 13:2138–2148

34. Rissi VB, Glanzner WG, de Macedo MP, Mujica LKS, Campagnolo K, Gutierrez K et al (2018) Inhibition of RNA synthesis during Scriptaid exposure enhances gene reprogramming in SCNT embryos. Reproduction 157: 123–133

35. de Macedo MP, Glanzner WG, Gutierrez K, Currin L, Guay V, Carrillo Herrera ME et al (2022) Simultaneous inhibition of histone deacetylases and RNA synthesis enables totipotency reprogramming in Pig SCNT Embryos. Int J Mol Sci 23(22):14142. https://doi.org/10.3390/ijms232214142

36. Liu X, Wang Y, Gao Y, Su J, Zhang J, Xing X et al (2018) H3K9 demethylase KDM4E is an epigenetic regulator for bovine embryonic development and a defective factor for nuclear reprogramming. Development 145(4)

37. Hormanseder E, Simeone A, Allen GE, Bradshaw CR, Figlmuller M, Gurdon J et al (2017) H3K4 methylation-dependent memory of somatic cell identity inhibits reprogramming and development of nuclear transfer embryos. Cell Stem Cell 21:135–143 e136

38. Hyun K, Jeon J, Park K, Kim J (2017) Writing, erasing and reading histone lysine methylations. Exp Mol Med 49:e324

39. Glanzner WG, Gutierrez K, Rissi VB, de Macedo MP, Lopez R, Currin L et al (2020) Histone lysine demethylases KDM5B and KDM5C modulate genome activation and stability in porcine embryos. Front Cell Dev Biol 8:151

40. Glanzner WG, Rissi VB, de Macedo MP, Mujica LKS, Gutierrez K, Bridi A et al (2018) Histone 3 lysine 4, 9 and 27 demethylases expression profile in fertilized and cloned bovine and porcine embryos. Biol Reprod 198:742–751

41. Liu X, Wang C, Liu W, Li J, Li C, Kou X et al (2016) Distinct features of H3K4me3 and H3K27me3 chromatin domains in pre-implantation embryos. Nature 537:558–562

42. Dahl JA, Jung I, Aanes H, Greggains GD, Manaf A, Lerdrup M et al (2016) Broad histone H3K4me3 domains in mouse oocytes modulate maternal-to-zygotic transition. Nature 537:548–552

43. Chung YG, Matoba S, Liu Y, Eum JH, Lu F, Jiang W et al (2015) Histone demethylase expression enhances human somatic cell

nuclear transfer efficiency and promotes derivation of pluripotent stem cells. Cell Stem Cell 17:758–766

44. Matoba S, Liu Y, Lu F, Iwabuchi KA, Shen L, Inoue A et al (2014) Embryonic development following somatic cell nuclear transfer impeded by persisting histone methylation. Cell 159: 884–895

45. Liu Z, Cai Y, Wang Y, Nie Y, Zhang C, Xu Y et al (2018) Cloning of macaque monkeys by somatic cell nuclear transfer. Cell 172(881–887):e887

46. Simoes R, Rodrigues Santos A Jr (2017) Factors and molecules that could impact cell differentiation in the embryo generated by nuclear transfer. Organogenesis 13:156–178

47. Meissner A, Jaenisch R (2006) Mammalian nuclear transfer. Dev Dyn 235:2460–2469

48. Yoshioka K, Suzuki C, Tanaka A, Anas IM, Iwamura S (2002) Birth of piglets derived from porcine zygotes cultured in a chemically defined medium. Biol Reprod 66:112–119

49. Vajta G, Lewis IM, Hyttel P, Thouas GA, Trounson AO (2001) Somatic cell cloning without micromanipulators. Cloning 3:89–95

50. Nascimento AB, Albornoz MS, Che L, Visintin JA, Bordignon V (2010) Synergistic effect of porcine follicular fluid and dibutyryl cyclic adenosine monophosphate on development of parthenogenetically activated oocytes from pre-pubertal gilts. Reprod Domest Anim 45: 851–859

51. Park SH, Yu IJ (2013) Effect of dibutyryl cyclic adenosine monophosphate on reactive oxygen species and glutathione of porcine oocytes, apoptosis of cumulus cells, and embryonic development. Zygote 21:305–313

52. Bagg MA, Nottle MB, Grupen CG, Armstrong DT (2006) Effect of dibutyryl cAMP on the cAMP content, meiotic progression, and developmental potential of in vitro matured pre-pubertal and adult pig oocytes. Mol Reprod Dev 73:1326–1332

53. Che L, Lalonde A, Bordignon V (2007) Chemical activation of parthenogenetic and nuclear transfer porcine oocytes using ionomycin and strontium chloride. Theriogenology 67:1297–1304

54. de Macedo MP, Glanzner WG, Rissi VB, Gutierrez K, Currin L, Baldassarre HB et al (2018) A fast and reliable protocol for activation of porcine oocytes. Theriogenology 123: 22–29

Chapter 11

Somatic Cell Nuclear Transfer Using Freeze-Dried Protaminized Donor Nuclei

Luca Palazzese, Marta Czernik, Kazutsugu Matsukawa, and Pasqualino Loi

Abstract

Somatic cell nuclear transfer (SCNT) is the only nuclear reprogramming method that allows rewinding an adult nucleus into a totipotent state. As such, it offers excellent opportunities for the multiplication of elite genotypes or endangered animals, whose number have shrunk to below the threshold of safe existence. Disappointingly, SCNT efficiency is still low. Hence, it would be wise to store somatic cells from threatened animals in biobanks. We were the first to show that freeze-dried cells allow generating blastocysts upon SCNT. Only a few papers have been published on the topic since then, and viable offspring have not been produced. On the other hand, lyophilization of mammalian spermatozoa has made considerable progress, partially due to the physical stability that protamines provide to the genome. In our previous work, we have demonstrated that a somatic cell could be made more amenable to the oocyte reprogramming by the exogenous expression of human *Protamine 1*. Given that the protamine also provides natural protection against dehydration stress, we have combined the cell protaminization and lyophilization protocols. This chapter comprehensively describes the protocol for somatic cell protaminization, lyophilization, and its application in SCNT. We are confident that our protocol will be relevant for establishing somatic cells stocks amenable to reprogramming at low cost.

Key words Cloning, Lyophilization, Protamine, Biobanking, Endangered species

1 Introduction

Somatic cell nuclear transfer (SCNT) empowers us to obtain almost unlimited numbers of genetically identical cells in a totipotent state. Thus, with the birth of Dolly the sheep, SCNT brought the revolution of asexual reproduction to mammalian species [1]. SCNT potential is clearly noticeable in fields such as animal breeding, transgenic animal production, and conservation of genetic resources. Thus, it is somehow frustrating to realize that the low cloning efficiency limits its applications.

Marcelo Tigre Moura (ed.), *Somatic Cell Nuclear Transfer Technology*, Methods in Molecular Biology, vol. 2647,
https://doi.org/10.1007/978-1-0716-3064-8_11,
© The Author(s), under exclusive license to Springer Science+Business Media, LLC, part of Springer Nature 2023

This is frustrating because we are certain that a reliable SCNT could positively impact the multiplication of elite genotypes, production of transgenic animals to produce biological peptides or act as animal models for human diseases, or even farm animals with low environmental impact [2]. Of relevance for the scope of this chapter, SCNT holds a remarkable potential to expand or restore animal populations threatened by extinction [3], a potential that regrettably remains theoretical for the above reasons. One consensus strategy to counteract species extinction is to establish biobanks in the form of nucleated cells from selected, endangered animals. Some biobanking organizations, including the Frozen Ark Consortium (www.frozenark.org) and the Frozen Zoo, are well advanced, albeit research groups have also stocked frozen cells of wild species. The Frozen Zoo, established in 1972, stores over 10,000 viable samples (somatic cells, oocytes, spermatozoa, and embryos) from almost 1,000 species and subspecies. The storage of genetic resources in biobanks is in liquid nitrogen or well under -100 °C. Deep freezing in liquid nitrogen is a robust and straightforward protocol applied worldwide, but it is expensive, with a heavy carbon footprint [4], and thus restricted to wealthy countries that have facilities for sustained liquid nitrogen production.

The demonstration that lyophilized cells can be reprogrammed and direct embryonic development upon SCNT opened a new and more affordable venue for biobanking [5–7]. As a widespread storage approach in unicellular eukaryotes, lyophilization remained unexplored in mammals until the discovery that non-motile lyophilized mouse spermatozoa support full-term development after intracytoplasmic sperm injection [8]. Since then, reports have demonstrated the feasibility of dry storage in spermatozoa, including the birth of a few large animals [9]. The physical stability of the spermatozoa on DNA is due to its packaging in tight toroid structures made of DNA and protamine(s), packaging that renders spermatozoa far more resistant to mechanical/physical stresses such as ionizing radiation than somatic cells [10]. In our continuous efforts to improve somatic cell nuclear reprogramming, we have recently demonstrated that exogenous expression of the human *Protamine 1* gene (*hPrm1*) in sheep fibroblasts remodels their nuclei into a spermatid-like structure [11]. Furthermore, injecting protaminized somatic nuclei into enucleated oocytes resulted in higher preimplantation development than histone-enriched cells [11–13]. Our next assumption was that if protamine confers spermatozoa with resistance to lyophilization, it would very likely show the same action in protaminized somatic cells. This chapter combines our latest breakthrough in SCNT nuclear programming with our expertise in spermatozoa lyophilization [14] to establish dry biobanks of protaminized cells. Dry biobanks, ideally with storage at room temperature (RT), would dramatically simplify the storage of cell lines collected from endangered animals.

2 Materials

2.1 Equipment

1. Stereomicroscope (Nikon).
2. Inverted microscope (Ti2-U, Nikon).
3. Benchtop centrifuge (Eppendorf).
4. Mister Frosty (cat. no. 432004, CoolCell LX, Corning).
5. Freeze-drier (VirTis 2.0 BenchTop, SP Scientific).
6. Deep freezer set to −80 °C.
7. Piezo-driven micropipette system (PiezoXpert, Eppendorf, Milan, Italy).
8. Microinjector for enucleation/injection micropipette (Cell-Tram Oil, Eppendorf).
9. Microinjector for holding micropipette (CellTram Air, Eppendorf).
10. Stereomicroscope (Nikon).
11. Microforge (cat. no. MF-900, Narishige, Tokyo, Japan).
12. Micropipette puller (model P-87 flaming/brown micropipette puller, Sutter Instrument Company).
13. Fluorescence unit (Intensilight C-HGFI, Nikon).
14. Microscope glass warming plate (Okolab).
15. Micromanipulation system (Narishige).

2.2 Tools and Consumables

1. Borosilicate glass capillaries: Outer diameter, 10 cm × 1 mm; inner diameter, 0.78 mm (cat. no. GC 1005-15, Harvard).
2. Embryo image capture software (OCTAX EyeWare imaging software, version 2.3.0.372, Octax Microscience GmbH, Altdorf, Germany).

2.3 Media and Solutions

All solutions should be prepared with cell culture-grade bi-distilled water and analytical-grade reagents. Chemicals were obtained from Sigma-Aldrich or as indicated below. Carry out all procedures at room temperature and in sterile condition, unless otherwise indicated.

1. HEPES-buffered oocyte holding medium (H-TCM-199): Mix 9.5% (w/v) TCM-199 (Gibco, Life Technologies, Milan, Italy), 2.0 mM L-glutamine, 2.2% (w/v) $NaHCO_3$, 5 mL/L gentamicin solution (Sigma-Aldrich, cat. no. G1397), 0.4% (w/v) bovine serum albumin (BSA), and 4.7% (w/v) HEPES. Adjust the osmolarity to 280 mOsm (*see* **Note 1**).
2. H-TCM-199 supplemented with heparin: Dilute 0.05% (w/v) heparin in H-TCM-199.

3. In vitro maturation (IVM) medium: Mix 9.5% (w/v) TCM-199 (Gibco), 2.0 mM (w/v) L-glutamine, 0.3 mM sodium pyruvate, 100 µM cysteamine, 10% fetal bovine serum (FBS; Gibco), 5.0 µg/mL follicle-stimulating hormone (FSH; Ovagen, ICP, Auckland, New Zealand), 5.0 µg/mL luteinizing hormone (LH), and 1.0 µg/mL 17 β-estradiol.

4. Cell culture medium (CCM): Dulbecco's Modified Eagle Medium (DMEM; Gibco, cat. no. 1320-033), 10% (v/v) FBS, and 5 mL/L gentamycin solution (Sigma-Aldrich, cat. no. G1397).

5. Freezing medium: Mix 60% (v/v) DMEM, 20% (v/v) FBS, and 20% (v/v) dimethyl sulfoxide (DMSO).

6. Serum starvation medium (SSM): DMEM (Gibco, cod. no. 11320-033), 0.5% FBS (v/v), and 5 mL/L gentamicin solution (Sigma-Aldrich, cat. no. G1397).

7. Lipofectamine solution: Prepare Solution A (containing lipofectamine) in a 1.8 mL microcentrifuge tube: Add 9.0 µL lipofectamine into 500 µL Opti-MEM. Mix gently up and down with a pipette and let it settle for 10 min at RT. Prepare Solution B (containing DNA) in a 1.8 mL tube: Add 4.0 µg phPrm1 and 500 µL Opti-MEM. Mix gently up and down with a pipette and let it settle for 5 min at RT. Spin briefly both solutions and transfer Solution B into Solution A. Vortex the final solution (B + A) and incubate for 30 min at RT. Spin the final solution and use it fresh.

8. TSA solution: DMEM, 10% (v/v) FBS, 50 nM Trichostatin A.

9. Freeze-drying medium (FDM): 1.0 mL 0.5 M Tris-HCL, 5.0 mL 0.5 M EGTA, and 2.5 mL 1 M NaCl in water.

10. Polyvinylpyrrolidone (PVP) solution: Dissolve 12% (w/v) PVP 360 KDa in PBS.

11. Hyaluronidase solution: Dissolve 0.3 mg/mL hyaluronidase in H-TCM-199 medium.

12. Hoechst solution: Dissolve 5.0 µg/mL Hoechst 33342 in H-TCM-199 medium.

13. Enucleation medium: Dissolve 7.5 µg/mL cytochalasin B (CB) in H-TCM-199 medium.

14. Ionomycin solution: Dissolve 5.0 µM ionomycin in H-TCM-199 medium (*see* **Note 2**).

15. Embryo in vitro culture medium (IVCM): BO-IVC (cat. no. 71005, IVF Bioscience).

16. 6-Dimethylaminopurine (6-DMAP) solution: Dissolve 2.0 mM 6-DMAP in IVC medium in incubator at 38.5 °C, 5% CO_2 in air.

17. DNA constructs pPrm1-GFP/RFP and GFP/RFPtag (pEGFPC2 vector, pERFPC2 vector; Clontech).
18. Ficoll (Ficoll-Paque PLUS, cat. no. 71101700-EK, GE Healthcare).

2.4 Micromanipulation Setup

1. Embryo culture dish: Prepare a 35 mm cell culture dish containing 7–8 IVCM droplets of 20 μL around the border (culture droplets) and three 20 μL IVC medium drops in the center of the Petri dish (wash droplets). Cover the droplets with mineral oil (Fig. 1a) and transfer the dish for equilibrating in the incubator for at least 2 h before use.

2. Micromanipulation pipettes: Use thin-wall borosilicate capillary glass without filaments (0.78 mm inner Ø). Pull all the above pipettes with the micropipette puller. Use the

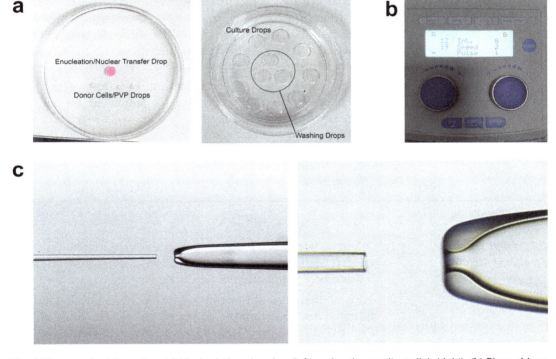

Fig. 1 Micromanipulator setup. (**a**) Manipulation chamber (left) and embryo culture dish (right). (**b**) Piezo-driven micropipette indicative settings display three parameters (Int.: pulse intensity; speed: pulse frequency; pulse: pulse number per foot press). The "A" settings are for entry through the *zona pellucida* and for breaking the donor cell membrane. The "B" settings are for donor nucleus injection into the oocyte. (**c**) Layout of holding and injection pipettes. 4× magnification (left) and 20× magnification (right)

216 Luca Palazzese et al.

Microforge to process the holding pipette with 100 μm outside Ø and bent at a 30° angle. Adjust the enucleation pipette to 20–22 μm outside Ø and bent at a 30° angle. Prepare the injection pipette with a 10 μm outside Ø and bent at a 25° angle. Load ~8–10 mm mercury (*see* **Note 3**) to enucleation and injection pipettes using a 10 μL microsyringe from the rear end and then fill with mineral oil to the uncut tip.

3. Enucleation dish: Use the lid of a 100 mm dish and place centrally ~150–200 μL enucleation medium droplet and 5–6 10 μL PVP droplets. Cover the droplets with mineral oil (Fig. 1b).

4. Set holding and injection pipettes to the proper positions (both pipettes should be positioned at the center of the field of view and parallel to the working plane Fig. 1c).

5. Wash the enucleation pipette with PVP solution by aspirating up and down a couple of times. Fill the enucleation pipette with a small amount of PVP solution and then with enucleation medium (indicatively, aspirate the PVP up to the pipette bending point). Set the piezo unit to high speed (>5) and power (>5) as shown below (Fig. 1b; *see* **Note 4**).

6. Reconstruction dish: For the nuclear injection of the enucleated oocytes with freeze-dried protaminized nuclei. This dish should be prepared as described above (*see* **item 3**), with the drop of enucleation medium replaced with a drop of H-TCM-199 (Fig. 1a). Set the injection pipette as was done for oocyte enucleation (Fig. 1c).

3 Methods

3.1 Adult Sheep Fibroblast Culture

1. Collect a biopsy of ~2.0 cm^2 from the selected tissue (tail, ear, or skin) of an adult sheep (ewe or ram) and wash it briefly in 75% medical-grade ethanol (*see* **Note 5**).

2. Work in sterile conditions in a tissue culture hood. Under a stereomicroscope, remove damaged tissue, fat, and/or hair from the skin biopsy. Wash the tissue twice in sterile warmed PBS incubate in 100 μL 0.25% trypsin-EDTA solution, cut it with a surgical scalpel into small pieces, smaller than 1–2 mm (*see* **Note 6**). Add 5 mL CCM to block the trypsin activity and transfer to a 15 mL conical tube. Centrifuge at 1,637 × g at room temperature (RT) for 5 min.

3. Remove the supernatant, add 2.0 mL CCM, and transfer the medium containing the tissue fragments into a treated 35 mm cell culture Petri dish. Incubate the tissue fragments in a humidified incubator set to 5% CO_2, saturated humidity, at 38.5 °C.

4. After 24 h, remove the CCM and all tissue fragments that did not attach to the dish surface. Gently add 2 mL of fresh CCM.

Cloning Freeze-Dried Protaminized Donor Nuclei   217

5. When cells reach 90% confluence, perform cell passaging as described below. Replenish CCM every 2 days thereafter.

6. Remove the CCM and gently wash cells twice with PBS. Add 1 mL 0.25% trypsin-EDTA solution and incubate for 2 min at 38.5 °C. Observe under the inverted microscope whether cells detached from the dish. If not, return cells to the incubator for 2 min. Add 5 mL CCM and transfer the content into a 15 mL conical tube. Centrifuge at $1,637 \times g$ at RT for 5 min.

7. Remove the supernatant and resuspend cells in a 1.0 mL CCM.

8. Estimate cell concentration with a Bürker chamber and cell viability Trypan blue staining. Plate 5.0×10^5 cells per 100 mm cell culture Petri dish.

9. Make sure that you stock (freeze) cells after at least three passages (subcultures).

10. Prepare cells for freezing by repeating previous passaging steps (*see* **steps 6–8**). Resuspend 1.0×10^6 cells in 500 µL CCM and transfer them into a 1.5 mL cryovial. Add 500 µL freezing medium very gently, drop-by-drop (*see* **Note 7**). Place the cryovial at −80 °C. After 24 h, store cells in liquid nitrogen (−196 °C).

11. Thaw cells by taking one cryovial from the liquid nitrogen tank. Work in sterile conditions under a tissue hood. Spray the surface of the cryovial with 75% medical ethanol. Add 0.5 mL warm (37 °C) CCM and transfer the cryovial into a water bath at 37 °C for 2–3 min. Transfer the cells from the cryovial into a 15 mL conical tube containing 9 mL CCM and centrifuge at $1,637 \times g$ at RT for 5 min.

12. Discard the supernatant, add 10 mL CCM, and plate cells in 100 mm cell culture dish.

13. Do not use thawed cells for transfection. Perform at least one passage before transfection, as described above (*see* **steps 5–8**).

3.2 Cell Transfection with phPrm1

1. Use cells that reached 60–80% confluence. Remove the CCM, wash twice with PBS, add 5 mL SSM to inhibit cell proliferation (cell synchronization at G0), and incubate with 5% CO_2, saturated humidity, at 38.5 °C for 24 h before transfection.

2. Remove the SSM 1 h before transfection, wash twice with PBS, add 1.0 mL Opti-MEM to a 100 mm Petri dish, and incubate in 5% CO_2, saturated humidity at 38.5 °C.

3. Prepare the lipofectamine solution and add dropwise to the cells in Opti-MEM medium. Gently rock the plate and incubate in 5% CO_2, with saturated humidity at 38.5 °C for 3.5–4.0 h (*see* **Note 8**).

4. Remove the Opti-MEM medium at the end of incubation and add 7 mL CCM containing 50 nM TSA. Incubate in 5% CO_2, with saturated humidity at 38.5 °C for 16–20 h. Remove the

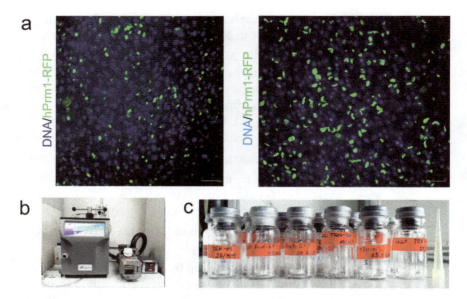

Fig. 2 (**a**) Spermatid-like cells isolated by Ficoll gradient centrifugation. Spermatid-like cells before the selection (left). Spermatid-like cells after the selection (right). (**b**) The freeze-drying device. (**c**) Freeze-dried protaminized-somatic cells stored in one-in-one vacuum-sealed glass vials (glass vial of Ø 8 mm in a glass vial of Ø 20 mm). Scale bar: 50 μm

CCM, wash twice with PBS, and add 7 mL of fresh CCM (without TSA).

5. Identify hPrm1-transfected cells under the fluorescence microscope 24 h post-transfection as cells with a single red (RFP) and green (GFP) nuclear spots. At 48 h post-transfection, protaminized cells are observed as spermatid-like cells. These cells detach from the Petri dish and remain nonadherent (Figs. 2a and 3a).

3.3 Lyophilization of Protaminized Somatic Cells

Freeze-dry protaminized cells with a spermatid-like morphology (Fig. 2a). To reach the proper conditions of the freeze-dryer (condenser at −58 °C and the freeze-drying chamber at −12 °C), the device must be on at least 3 h before use.

1. Aspirate the CCM with protaminized somatic cells at 48 h post-transfection (Fig. 2b; *see* **Note 9**) and transfer to a 15 mL conical tube with 2.0 mL Ficoll.

2. Centrifuge at 1,637 × *g* at RT for 5 min to isolate the spermatid-like cells, setting the centrifuge to activated up and down breaks (*see* **Note 10**). Discard the supernatant and resuspend the pellet in 200 μL CCM. Transfer cells into a 1.8 mL tube with 1.0 mL of a (1:1) mixture of PVP solution and Ficoll. Centrifuge at 1,637 × *g* at RT (centrifuge set to activate up and down breaks) for 5 min (*see* **Note 10**). Discard the supernatant and resuspend the pellet in 1.0 mL FDM. Using the Ficoll-PVP gradient technique, almost 80% of the collected

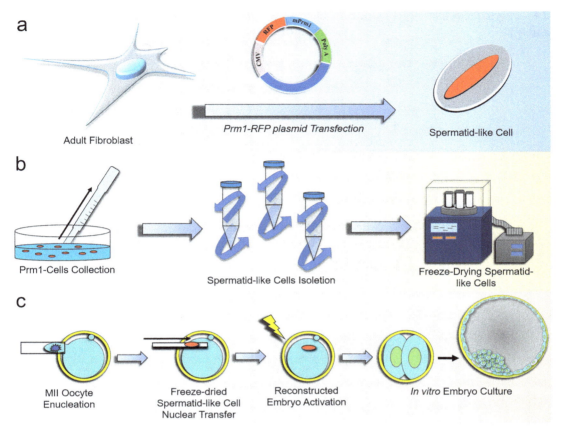

Fig. 3 Schematic representation of the experimental protocol. (**a**) Transfect adult sheep fibroblasts with plasmids containing the human *Prm1* gene tagged with Red Fluorescence Protein (phPrm1-RFP). At the end of protaminization, the remodeled somatic nucleus has a spermatid-like structure. (**b**) Collect protaminized somatic cells, subject them to differential gradient centrifugation, and freeze-dry for long-term storage. (**c**) Somatic cell nuclear transfer (SCNT). Aspiration of the metaphase II (MII) spindle of sheep oocytes. Injection of a somatic spermatid-like cell into the enucleated oocyte. The reconstructed oocyte subsequently undergoes chemical activation and in vitro embryo culture

protaminized cells (Fig. 2a). Count cells using Bürker chamber and dilute in FDM to obtain 1.0×10^6 cells/mL.

3. Transfer 100 μL of the cell suspension in FDM into glass vials (Ø 8 mm) and place these into Mister Frosty. Set the Mister Frosty to −80 °C for the duration required for the samples to reach −50 °C (cooling rate of −1 °C/min).

4. At the end of the freezing step, transfer the vials (Ø 8 mm) with the suspended cells into larger glass vials (Ø 20 mm) and place them in the freeze-drier (Fig. 2b). Place samples in the freeze-drier, activate the vacuum pump, and leave it working for 16 h until the pressure reaches 15 μbar. Seale the vials under vacuum (Fig. 2c).

5. Store the glass vials with the freeze-dried protaminized somatic cells at 4 °C in the dark until use.

220   Luca Palazzese et al.

3.4 Oocyte Collection and IVM

1. Collect sheep uteri at a local slaughterhouse and transport them to the laboratory within 1–2 h at 37–38 °C (*see* **Note 11**).

2. Cut ovaries from uteri with surgical scissors and place them in a beaker with warm PBS at 37 °C.

3. Wash the collected ovaries 2–3 times with warm PBS (*see* **Note 12**).

4. Aspirate cumulus-oocyte complexes (COCs) from 3 to 6 mm follicles using a 5 mL syringe coupled to 21G needle into H-TCM-199 with heparin. After aspirating 4–5 ovaries, transfer the syringe contents into a 100 mm dish.

5. Select COCs surrounded by at least two cumulus cell layers and with an evenly granulated oocyte cytoplasm. Place the COCs in a 35 mm dish containing 1.0 mL H-TCM-199. Work on a stereomicroscope under sterile conditions.

6. Add 500 μL IVM medium to each well of a four-well cell culture dish (*see* **Note 13**).

7. Wash the selected COCs twice in IVM medium and place ~35 COCs per well in the four-well dish.

8. Incubate COCs with 5% CO_2, saturated humidity at 38.5 °C for 22 h.

3.5 Somatic Cell Nuclear Transfer

The SCNT procedure contemplates oocyte enucleation and reconstruction with freeze-dried protaminized somatic nuclei (*see* **Note 14**). The microscope stage is equipped with a warming plate that should be set to 38.5 °C during oocyte enucleation. However, during the oocyte reconstruction step (i.e., nuclear injection), it is recommended not to use a warming plate or even cool it to 4–8 °C.

3.5.1 Oocyte Enucleation

1. Select COCs with uniformly expanded cumulus cell layers at 22 h post-IVM (*see* **Note 15**). Denude oocytes by gently pipetting of COCs in hyaluronidase solution (*see* **Note 16**). Select viable oocytes with a visible polar body (PB) and wash them 4–6 times in H-TCM-199.

2. Incubate oocytes in Hoechst 33342 solution for 10 min in the dark. Wash the oocytes in TCM-199 for 5 min and then place the first batch (~10 oocytes) into the central enucleation droplet in the manipulation dish. Incubate oocytes for ~3–5 min.

3. Grab an oocyte with the holding pipette and use the enucleation pipette to orientate the PB to an 11 o'clock position on a clock's face. Set the microscope focus on the PB, switch off the bright light of the microscope, and turn on the ultraviolet (UV) light for a few seconds (1–2 s maximum) to visualize the metaphase II (MII) spindle (*see* **Note 17**). Switch on the bright light, place the oocyte close to the enucleation pipette, and make a hole in the zona pellucida with piezo pulses (*see* **Note 4**).

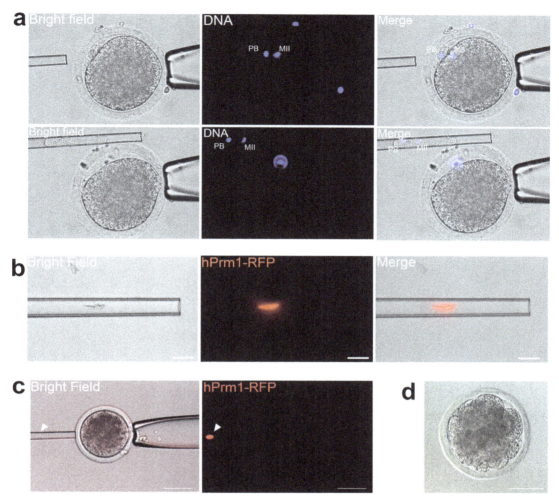

Fig. 4 Somatic cell nuclear transfer (SCNT). (**a**) Enucleation step. Mature sheep oocyte (top row) and enucleated oocyte (bottom row). (**b**) Selection of a spermatid-like cell with the injection pipette (Scale bar = 20 μm). (**c**) Nuclear injection step. A rehydrated somatic nucleus for injection into an enucleated sheep oocyte. Arrowheads point at the donor nucleus. (**d**) Embryo at 132 h of in vitro culture. Scale bar: 50 μm. MII metaphase II plate, PB polar body

4. Aspirate a small portion of cytoplasm containing the oocyte spindle. Check very briefly under UV light to ensure proper enucleation (Figs. 1c and 4a and *see* **Note 17**). Repeat the procedure with the remaining oocytes (*see* **Sect. 3.5.5** and **3.5.6**). After enucleating the oocyte batch, transfer them to H-TCM-199 in the incubator for 45 min before oocyte reconstruction.

3.5.2 Oocyte Reconstruction

1. Retrieve one cryovial containing freeze-dried protaminized somatic cells stored at 4 °C. Add 100 μL bi-distilled water with gentle pipetting. Take 5.0 μL of rehydrated protaminized

somatic cells and transfer them into a 1.8 mL tube containing 30 μL PVP solution. Pipet very gently. Take 5.0 μL of rehydrated cells in PVP and place them in one PVP droplet in the nuclear injection dish.

2. Place the first batch of 10 enucleated oocytes in the manipulation droplet.

3. Select a spermatid-like nucleus using the fluorescence light (GFP or red filter depending upon plasmid choice) and aspirate into the injection pipette (Fig. 4b).

4. Break the donor cell membrane with a few piezo pulses.

5. Hold the enucleated oocyte with the holding pipette. Drill the *zona pellucida* with light piezo pulses (Fig. 4c) or use the slit previously drilled during the enucleation step (*see* Subheading 3.5.1). Move one donor cell forward near the tip of the injection pipette.

6. Apply a piezo pulse to break the oolemma (indicated by a rapid relaxation of the oocyte membrane; *see* **Note 3**) and release the donor nucleus into the oocyte cytoplasm.

7. Withdraw the injection pipette from the oocyte very gently. Release the reconstructed oocytes by gently applying positive pressure within the holding pipette. Repeat the procedure with the other enucleated oocytes.

3.6 Oocyte Activation

1. Allow the reconstructed oocytes to recover in H-TCM-199 inside the incubator for 45 min.

2. Activate the reconstructed oocytes in ionomycin solution for 5 min. Wash reconstructed oocytes in H-TCM-199 4–6 times. Incubate the reconstructed oocytes in 6-DMAP solution for 3–5 h.

3.7 Embryo Culture

1. Wash activated oocytes twice in IVC medium for 5 min. Move the reconstructed embryos into culture drops on the in vitro embryo culture dish (place maximum of five embryos per droplet). Incubate embryos with 5% CO_2 7% O_2, under saturated humidity at 38.5 °C for 8 days. Check embryonic development at 24 h post-activation for cleavage rates (2-cellembryos) and on days 7–8 for blastocyst rates (Fig. 4d).

4 Notes

1. Store at 4 °C for up to 1 week.

2. Ionomycin is light sensitive, so handle it in the dark.

3. Mercury was extracted from old thermometers.

4. The piezo-driving system parameters shown in Fig. 1b are for illustration only and must be adapted at each manipulation session (and even during the same manipulation session). The fact remains that the "A" settings are "stronger" than the "B" settings, which need to be "softer" to allow punching the hole in the oolemma without oocyte lysis.

5. Prepare the cell line at least 3 weeks in advance before protaminization and/or an SCNT experiment. Carry out the tissue biopsy collection in accordance with relevant national and institutional regulations.

6. Warm the trypsin-EDTA solution at 37 °C before use. Do not keep the tissue in the trypsin solution for more than 10 min.

7. Equilibrate the freezing medium at −20 °C for at least 30 min before use.

8. Do not incubate for more than 4 h as the DNA/lipid complexed might be toxic to the cells.

9. The fully protaminized cells (spermatid-like cells) will float in the cell culture medium. Other cells (non-spermatid-like cells) will remain attached to the Petri dish.

10. To prevent the cells from mixing prematurely with the Ficoll, gently release the medium into the tube.

11. Keep the uteri at 37–38 °C to avoid thermal stress.

12. Perform sufficient washes to remove the blood.

13. Equilibrate the IVM medium for 2 h in the incubator (5% CO_2 at 38.5 °C) before use.

14. Perform micromanipulation on an inverted microscope equipped with a piezo-driven micropipette system connected to the enucleation/injection system. The left micromanipulation controls control a micropipette holder connected to an enucleation or injection pipette. The right micromanipulation controls control a micropipette holder connected to a holding pipette. The operator defines this setup, so it can also be reversed (holding system on the left and enucleation/injection system on the right).

15. Oocytes that have responded well to the gonadotropin treatment will display cumulus cell expansion.

16. Warm the hyaluronidase solution at 38.5 °C before use.

17. Expose oocytes to UV as short as possible.

Acknowledgments

This work has received funding from the European Union's Horizon 2020 Research and Innovation Programme under the Marie Skłodowska-Curie grant agreement no. 734434. Moreover, this study was carried out in the framework of the Project "Demetra" (Dipartimenti di Eccellenza 2018–2022, CUP_C46C18000530001), funded by the Italian Ministry for Education, University and Research. LP and MC acknowledge support from the National Science Centre, Poland, through grant no. 2016/21/D/NZ3/02610 (Sonata) and 2019/35/B/NZ3/02856 (Opus). Furthermore, the authors warmly thank Dr. Joseph Saragusty (University of Teramo, Italy) for his stylistic revision.

References

1. Wilmut I, Schnieke AE, McWhir J, Kind AJ, Campbell KH (1997) Viable offspring derived from fetal and adult mammalian cells. Nature 385:810–813

2. Tait-Burkard C, Doeschl-Wilson A, McGrew MJ, Archibald AL, Sang HM, Houston RD et al (2018) Livestock 2.0 – genome editing for fitter, healthier, and more productive farmed animals. Genome Biol 19:204

3. Saragusty J, Diecke S, Drukker M, Durrant B, Friedrich Ben-Nun I, Galli C et al (2016) Rewinding the process of mammalian extinction. Zoo Biol 35:280–292

4. Rockström J, Steffen W, Noone K, Persson A, Chapin FS 3rd, Lambin EF et al (2009) A safe operating space for humanity. Nature 461:472–475

5. Loi P, Matsukawa K, Ptak G, Clinton M, Fulka J Jr, Nathan Y et al (2008) Freeze-dried somatic cells direct embryonic development after nuclear transfer. PLoS One 3:e2978

6. Iuso D, Czernik M, Di Egidio F, Sampino S, Zacchini F, Bochenek M et al (2013) Genomic stability of lyophilized sheep somatic cells before and after nuclear transfer. PLoS One 8:e51317

7. Ono T, Mizutani E, Li C, Wakayama T (2008) Nuclear transfer preserves the nuclear genome of freeze-dried mouse cells. J Reprod Dev 54:486–491

8. Wakayama T, Yanagimachi R (1998) Development of normal mice from oocytes injected with freeze-dried spermatozoa. Nat Biotechnol 16:639–641

9. Saragusty J, Anzalone DA, Palazzese L, Arav A, Patrizio P, Gosálvez J et al (2020) Dry biobanking as a conservation tool in the Anthropocene. Theriogenology 150:130–138

10. Wakayama S, Kamada Y, Yamanaka K, Kohda T, Suzuki H, Shimazu T et al (2017) Healthy offspring from freeze-dried mouse spermatozoa held on the international Space Station for 9 months. Proc Natl Acad Sci U S A 114:5988–5993

11. Iuso D, Czernik M, Toschi P, Fidanza A, Zacchini F, Feil R et al (2015) Exogenous expression of human protamine 1 (hPrm1) remodels fibroblast nuclei into spermatid-like structures. Cell Rep 13:1765–1771

12. Czernik M, Iuso D, Toschi P, Khochbin S, Loi P (2016) Remodeling somatic nuclei via exogenous expression of protamine 1 to create spermatid-like structures for somatic nuclear transfer. Nat Protoc 11:2170–2188

13. Palazzese L, Czernik M, Iuso D, Toschi P, Loi P (2018) Nuclear quiescence and histone hyper-acetylation jointly improve protamine-mediated nuclear remodeling in sheep fibroblasts. PLoS One 13:e0193954

14. Palazzese L, Anzalone DA, Turri F, Faieta M, Donnadio A, Pizzi F et al (2020) Whole genome integrity and enhanced developmental potential in ram freeze-dried spermatozoa at mild sub-zero temperature. Sci Rep 10:18873

Chapter 12

Cattle Cloning by Somatic Cell Nuclear Transfer

Juliano Rodrigues Sangalli, Rafael Vilar Sampaio, Tiago Henrique Camara De Bem, Lawrence Charles Smith, and Flávio Vieira Meirelles

Abstract

Cloning by somatic cell Nuclear Transfer (SCNT) is a powerful technology capable of reprograming terminally differentiated cells to totipotency for generating whole animals or pluripotent stem cells for use in cell therapy, drug screening, and other biotechnological applications. However, the broad usage of SCNT remains limited due to its high cost and low efficiency in obtaining live and healthy offspring. In this chapter, we first briefly discuss the epigenetic constraints responsible for the low efficiency of SCNT and current attempts to overcome them. We then describe our bovine SCNT protocol for delivering live cloned calves and addressing basic questions about nuclear reprogramming. Other research groups can benefit from our basic protocol and build up on it to improve SCNT in the future. Strategies to correct or mitigate epigenetic errors (e.g., correcting imprinting loci, overexpression of demethylases, chromatin-modifying drugs) can integrate the protocol described here.

Key words Cattle, Nuclear transplantation, Reprogramming

---

1 Introduction

With the birth of the ewe Dolly in 1996, cloning by somatic cell nuclear transfer (SCNT) was demonstrated to be feasible in mammals [1], which was followed 2 years later by the announcement of the first cloned cattle [2]. The benefits of cloning by SCNT are shared by multiple fields, such as cell therapy, livestock production, and biotechnology [3, 4]. However, in spite of high blastocyst rates after SCNT, limited advances have been achieved in obtaining live offspring, which ranges around 1–5% of birth [5], causing limited usage due to high cost and low efficiency [3, 6]. While SCNT methodology has not advanced much technically, epigenetic

---

Juliano Rodrigues Sangalli, Rafael Vilar Sampaio and Tiago Henrique Camara De Bem contributed equally with all other contributors.

Marcelo Tigre Moura (ed.), *Somatic Cell Nuclear Transfer Technology*, Methods in Molecular Biology, vol. 2647, https://doi.org/10.1007/978-1-0716-3064-8_12, © The Author(s), under exclusive license to Springer Science+Business Media, LLC, part of Springer Nature 2023

mechanisms have been identified as the main culprits of low efficiency [7]. This finding has prompted laboratories around the world to adopt different methods to overcome barriers by using chemical probes targeting chromatin-modifying enzymes [8], gene overexpression, or knockdown [9, 10]. In this chapter, we will first briefly describe the epigenetic barriers that complicate faithful nuclear reprogramming after SCNT and attempts to overcome them to improve bovine cloning efficiency. Later, we will describe an SCNT protocol for cattle cloning routinely used for nuclear reprogramming experiments, which allows full-term development at reasonable efficiency (up to 10% of live calves from the total transferred blastocysts).

During SCNT, the oocyte cytoplasm often fails to completely erase the "identity" of the nuclear donor cell [11]. In fact, epigenetic marks such as DNA methylation, histone modification, and noncoding RNAs regulate many processes that determine cell-type identity [12]. The epigenetic memory inherited from the nuclear donor somatic cell retained in the embryo leads to abnormal gene expression and it is considered the main barrier to an efficient reprogramming process [13].

DNA methylation and histone modifications are the most studied among all the epigenetic marks. The lysine residue at position 9 on histone H3 (H3K9) is one of the most investigated sites. Special attention has been given to histone methylation since modifications on H3K9 are thought to be the main barrier affecting the SCNT reprogramming efficiency [10, 14]. Recent findings in mice have indicated that genomic areas refractory to reprogramming and classified as reprogramming-resistant regions (RRR) retain residual H3K9 trimethylation (H3K9me3) from the somatic nucleus and behave as the main barrier for efficient reprogramming, which supports the evidence that H3K9 methylation is critical for nuclear reprogramming [10]. Strikingly, silencing of histone methyltransferase enzymes by siRNA in the donor cells was able to reactivate genes in RRR, thus dramatically increasing embryonic development [10]. Another important barrier hampering the reprogramming efficiency is the H3K9 dimethylation (H3K9me2). Bovine SCNT embryos present abnormal H3K9me2 levels (i.e., increased compared with in vitro fertilized embryos), which may affect appropriate nuclear reprogramming [14, 15].

Different strategies have been attempted to transpose the barriers formed by these epigenetic marks. The most widely applied approach is treating donor somatic cells or early SCNT embryos with histone deacetylases or methyltransferases inhibitors [8, 16, 17]. However, these approaches remain controversial with promising results for cloning in mice [8] and pigs [18, 19], while no improvement for bovine embryonic development after SCNT [3, 20–22].

Attempting to modulate/overcome these barriers, we have shown recently that the catalytic inhibition of euchromatic histone-lysine N-methyltransferases 1 and 2 (*EHMT1* and *EHMT2,* also known as *GLP and G9a,* respectively), which are responsible for catalyzing H3K9me2, in both nuclear donor cells and cloned embryos reduced H3K9me2 and H3K9me3 levels at the blastocyst stage [16]. In addition, we observed increases in other epigenetic marks, such as 5mC and 5hmC. Although this approach reduced the levels of these histone marks, no improvement was observed in preimplantation development, thus suggesting that assessments at the blastocyst stage might not be the best developmental stage to measure the impact of H3K9me3 on cloning efficiency. In mice, the RRR are H3K9me3-enriched regions refractory to transcriptional activation at the two-cell stage in cloned embryos, the moment when the embryonic genome activation (EGA) occurs [10]. Likewise, the reduction in H3K9me3 levels during bovine EGA improved SCNT efficiency [14], indicating that 8-cell or 16-cell stages may be the most appropriate to examine the effects of H3K9me3 modulation in this species.

Liu and coworkers showed that overexpression of lysine demethylase 4E (*KDM4E*), an H3K9 lysine-specific demethylase, improved the cloning efficiency by not only improving blastocyst rate, but also live birth and postnatal survival rates in bovines [14]. Another report from the same group showed that H3K27me3 is also a barrier to efficient reprogramming of cattle [23]. Using the same strategy, these authors overexpressed lysine demethylase 6A (*KDM6A*), a specific demethylase for H3K9 modification, and obtained an improvement in blastocyst rate. Moreover, they found that reducing H3K27me3 levels enhanced the expression of genes involved in cell adhesion, cellular metabolism, and X-linked genes [23].

All these aforementioned studies paved the way to elucidate the main barriers hampering adequate nuclear reprogramming. From these ideas and strategies, we expect significant improvements to SCNT outcomes in the years to come.

2 Materials

Hereafter, we describe the basic SCNT protocol utilized by our lab in the last ~20 years that has propitiated the generation of several viable offspring and published research articles.

2.1 Equipment

1. Inverted microscope with Hoffman contrast optics and fluorescence illumination (ultraviolet light [UV], Fig. 1a).

2. Micromanipulation system containing a three-axis joystick system with oil hydraulic control (Fig. 1a).

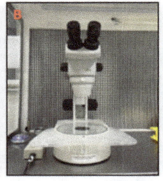

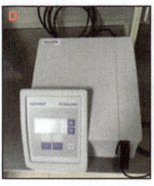
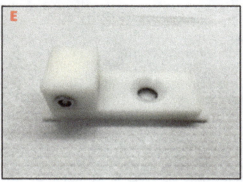

Fig. 1 Essential equipment required for somatic cell nuclear transfer. (**a**) Microscope and micromanipulation equipment. (**b**) Stereomicroscope. (**c**) Mouth pipette and Pasteur pipette. (**d**) Electrofusion generator. (**e**) Fusion chamber. (**f**) CO_2 incubator

3. Microinjectors: An oil-filled injector for holding pipette and air-filled injector for cell transfer pipette (Fig. 1a).
4. Stereomicroscope for oocyte and embryo handling (Fig. 1b).
5. Mouth pipettes (Fig. 1c), microdispensers, or Tomcat catheters for oocytes and embryo handling.
6. Electrofusion generator for cell fusion (Fig. 1d).
7. Fusion chamber with 0.2 mm gap between electrodes (Fig. 1e).
8. Incubator with atmosphere adjust to 5% CO_2, 5% O_2, and saturated humidity at 38.5 °C (Fig. 1f).

2.2 Media and Solutions

1. Oocyte-holding medium (OHM): Add 9 mL TCM-199 with Hank's salts (cat # 12350039), 1.0 mL fetal bovine serum (FBS), 50 μg/mL gentamicin, and 0.2 mM sodium pyruvate. Sterile filter (0.22 μm).
2. Oocyte maturation medium (OMM): Add 9.0 mL TCM-199 with Earle's salts (cat # 11150059), 1.0 mL FBS, 50 μg/mL gentamicin, 0.2 mM sodium pyruvate, 0.5 μg/mL

FSH, and 50 mg/mL hCG. Sterile filter (0.22 μm). The OHM and OMM can be prepared and stored at 4 °C for 2 weeks.

3. Fusion solution: Add 0.28 M mannitol, 0.1 mM MgSO$_4$, 0.5 mM HEPES sodium salt, and 0.05% bovine serum albumin (BSA) in ultrapure H$_2$O, pH 7.4. The fusion solution can be aliquoted and stored at −20 °C for several months.

4. Synthetic oviduct fluid supplemented with amino acids (SOFaa): 107.63 mM NaCl, 7.16 mM KCl, 1.19 mM KH$_2$PO$_4$, 1.51 mM MgSO$_4$, 1.78 mM CaCl$_2$.2H$_2$O, 5.35 mM sodium lactate, 25.00 mM NaHCO$_3$, 7.27 mM sodium pyruvate, 0.20 mM L-glutamine, 20 μL/mL BME amino acids, 10 μL/mL MEM amino acids, 0.34 mM Tri-sodium citrate, 2.77 mM Myo-inositol, 50 μg/mL gentamicin, and 10 μg/mL phenol red in ultrapure H$_2$O. Additionally, the SOFaa is supplemented with 5 mg/mL of BSA and 2.5% FBS. Sterile filter (0.22 μm). The SOFaa can be stored for over a month if kept in the refrigerator (4 °C) and protected from light.

2.3 Preparation of Dishes for Oocyte Handling

Before starting the oocyte denuding, make a culture dish to prepare and stain the oocytes for enucleation, to keep the oocyte batches for micromanipulation, and to return the enucleated oocytes, the reconstructed oocytes, and the fused cell couplets. We label this dish as the "working plate" since it is necessary for all steps during the micromanipulation procedure.

1. Prepare a microtube with 1 mL of SOFaa containing 7.5 μg/mL Cytochalasin B (CB) and 1.0 μg/mL Hoechst 33342.

2. Place two 60 μL droplets on the left side of a 60 mm Petri dish and use a marker pen to delimitate these droplets on the external surface of the bottom of the dish (Fig. 2d). These droplets are used to destabilize the actin cytoskeleton (CB) and stain the DNA (Hoechst 33342) of oocytes before enucleation.

3. On the same Petri dish, make at least eight additional droplets with SOFaa to place the oocytes from the other steps as described above. Allow the plate to equilibrate in the incubator for 2 h before starting the oocyte denuding.

2.4 Micro-manipulator Setup

On the day of SCNT, prepare the culture dishes for micromanipulation and oocyte handling, and assemble the micromanipulation microscope. To carry out SCNT, it is necessary to have a holding pipette and an injection pipette. High-quality pipettes are fundamental to minimize oocyte damage during the micromanipulation procedure. To ensure repeatability and for convenience, our laboratory prefers to use commercially available pipettes. In our experience, commercial pipettes possess high batch-to-batch production

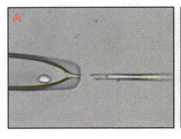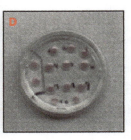

Fig. 2 Preparation of culture dishes and micromanipulator setup. (**a**) Holding pipette (on the left) and injection pipette (on the right). (**b**) Micromanipulation plate made using the lid of a 60 mm Petri dish containing manipulation medium droplets. (**c**) Manipulation plate placed at the center of the microscope stage. (**d**) Working plate for oocyte staining with the DNA-specific dye Hoechst 33342 to assist the oocyte enucleation step

consistency, are designed to optimize experiment outcome by minimal cell/oocyte trauma, and enable fine micromanipulation control. One option of commercial holding pipette possesses a 15 μm inner diameter and 100 μm outer diameter, and the tip is bent at a 35° angle to enable positioning parallel to the manipulation dish (Fig. 2a). The injection pipette has a 15 μm inner diameter and 20 μm outer diameter, while the tip is bent in a 20° angle to facilitate the enucleation or reconstruction (Fig. 2a). Nonetheless, micromanipulation pipettes can be made if the laboratory is equipped with micropipette puller, micro-forge, and micropipette grinder. A detailed protocol to prepare holding and transfer pipettes can be found elsewhere [24].

Carry out bovine SCNT in OHM droplets (*see* **Note 1**) prepared in the lid of 60 mm Petri dishes (*see* **Note 2**). We usually prepare both enucleation and reconstruction droplets in the same dish.

1. Make two droplets 200 μL OHM at the center of the dish (Fig. 2b). Add 7.5 ug/mL CB to one droplet destined for oocyte enucleation to destabilize the actin cytoskeleton (*see* **Note 3**).

2. Cover droplets with 9 mL mineral oil and transfer the plate to the center of the microscope stage (Fig. 2c).

3. Mount the holding pipette on the left micromanipulator, adjust the tip position parallel to the plate surface, and lower the pipette until visualizing the holding touching the bottom of the dish. Lift a few micrometers from the surface to avoid attrition. The holding pipette will be static during the whole micromanipulation procedure.

4. Next, connect the injection pipette to the hydraulic microinjector and mount it to the right side of the micromanipulator. As described above, we work with an air-filled microinjector for the injection pipette. To allow better micromanipulation

Bovine Somatic Cell Nuclear Transfer   231

control, we first lower the injection pipette into the oil to allow the oil to fill the tip of the pipette by capillarity. Then, we move the injection pipette to the center of the micromanipulation droplet facing the holding pipette.

After assembly of the micromanipulation setup, proceed to prepare the oocytes for nuclear transfer.

3 Methods

3.1 Preparation of Somatic Cells

Collect skin biopsies from the base of the tail due to the absence of large vessels, less exposure to UV radiation from the sunlight, and easy accessibility for veterinary handling. This protocol for fibroblast cell culture can be used to isolate cells from fetuses or deceased high-genetic merit animals.

1. Administer epidural anesthesia with 2% lidocaine chlorhydrate, trim, and carefully clean the base of the tail with 70% ethanol.

2. Perform an incision of 1.0 cm^2 at the base of the tail, retrieve the skin biopsy, and immediately add the biopsy to a 15 mL tube containing 5 mL of PBS + antibiotic-antimycotic (100×) (Gibco; catalog number: 15240096).

3. Transport the biopsy material on ice to the laboratory where it is minced using a sterile scalpel blade on a plastic Petri dish and incubate with 0.01% (w/v) type I collagenase diluted in Alpha Minimum Essential Medium (α-MEM) with antibiotic-antimycotic (100×) solution for 3 h at 38.5 °C.

4. Transfer the dissociated tissue to a microtube and centrifuge at 300 g for 5 min.

5. Discard the supernatant and resuspend the pellet in 2 mL of α-MEM supplemented with 10% FBS and antibiotic-antimycotic (100×) solution (cell culture medium [CCM]).

6. Plate the cell suspension in a 35 mm Petri dish and culture with 5% CO_2, under saturated humidity at 38.5 °C. Monitor cell culture dishes every day to assess growth rate.

7. Replenish CCM in 48 h intervals until primary fibroblast cultures reach ~90% confluence. Then, detach the cells by removing the CCM and adding 1 mL of trypsin. Return the cells to the incubator, wait for 5 min, and inactivate the trypsin by adding 1 mL of CCM. Recover the cells, transfer to 2 mL microcentrifuge tube, and centrifuge the cell suspension at 300 g for 5 min.

8. Remove the supernatant and resuspend the cells in 1.0 mL of cryopreservation medium (CCM supplemented with 10% dimethyl sulfoxide [DMSO]).

9. Count the cells and freeze at the density of 2.0×10^5 cells/per vial using a freezing container, Nalgene Mr. Frosty (Sigma C1562). After freezing, cell vials can be stored in liquid nitrogen tanks for unlimited time.

3.2 Fibroblast Cell Cycle Synchronization

The SCNT may be performed with several cell types (e.g., granulosa, fibroblasts, stem cells). Routinely in our laboratory we use fibroblasts since it is a cell line easy to obtain, grow, and maintain in vitro. Cells at different stages of the cell cycle (e.g., quiescent or mitotic) or physiological states (e.g., pluripotent, apoptotic) can also be used to produce cloned embryos [25–27]. There are several methods for cell cycle synchronization: cell cycle inhibitors (e.g., roscovitine), cell culture in 0.5% FBS (serum deprivation), and contact inhibition (100% confluence) [27, 28]. Synchronization of the cell cycle between the somatic donor cell and the oocyte is crucial to maintain the correct ploidy of the embryo. The full-term development of a viable offspring was only possible after a protocol to induce cell quiescence was established [29]. We describe the two methods used in our lab to arrest cells in G0 stage (noncycling cells): contact inhibition and serum deprivation (*see* **Note 4**).

1. Contact inhibition: Thaw a vial of fibroblasts in water bath at 37 °C. Transfer the cells to a 1.5 mL microtube and centrifuge for $300 \times g$ for 5 min (Fig. 3a). Discard the supernatant and resuspend the cells in 2 mL CCM. Plate the cells (2.0×10^5 cells) in a 35 mm Petri dish at least 4 days before the day of SCNT. Replenish CCM after 48 h and allow cells to grow for an additional period of 48 h. When the cells reach confluence (usually within 2 days in culture), most will stop in stage G0 of the cell cycle due to contact inhibition (Fig. 3b).

2. Serum deprivation: Thaw and plate fibroblasts as described above. Grow the cells in CCM for 24 h. Replace the CCM for α-MEM supplemented with only 0.5% FBS and antibiotic-antimycotic (100×) solution) and culture the cells for 72 h. The

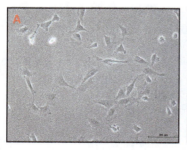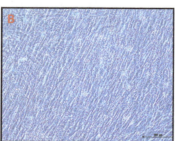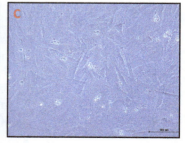

Fig. 3 Cell cycle synchronization of donor cells and oocyte reconstruction. (**a**) Bovine fibroblasts plated at low density in a Petri dish. (**b**) Bovine fibroblasts grown to full confluency (100%) to stop cell proliferation by contact inhibition. (**c**) Bovine fibroblasts under serum starvation for 72 h

Bovine Somatic Cell Nuclear Transfer 233

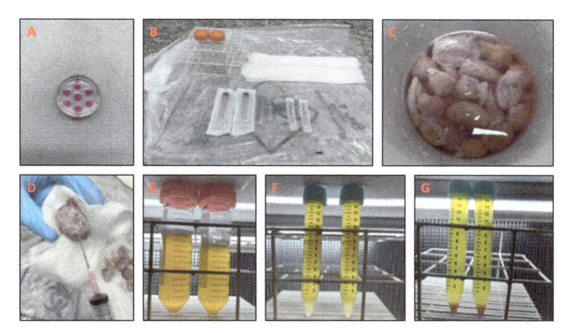

Fig. 4 Media preparation and oocyte aspiration. (**a**) Oocyte maturation dish. (**b**) Workstation to aspirate the ovaries. (**c**) Ovaries in an insulated container containing saline solution. (**d**) Ovary held with a gauze sponge in preparation for the aspiration. (**e**) Follicular fluid recovered from the ovaries decanting in the cell cabinet. (**f**) Supernatant recovered from the follicular fluid. (**g**) Supernatant recovered from the follicular fluid after centrifugation to remove blood and pieces of tissues and cells

serum deprivation will stimulate the cells to become quiescent and arrest at G0 (Fig. 3c). When the enucleated oocytes (ooplasts) are ready, dissociate the cells and proceed to the somatic cell transfer.

3.3 Oocyte Collection and In Vitro Maturation

Prepare OHM and OMM media in advance (see above) to allow them to equilibrate for at least 2 h in the CO_2 incubator.

1. Prepare oocyte maturation dishes by adding 100 μL OMM droplets in a 35 mm Petri dish (Fig. 4a), cover with 4 mL mineral oil, and equilibrate for at least 2 h in the incubator at 38.5 °C, humidity to saturation, and 5% CO_2 in air.

2. Set up a workstation to aspirate the ovaries after OHM and OMM media preparation (Fig. 4b). Using a clean working bench, position a biological waste bag nearby, a 50 mL sterile conical centrifuge tube, 10 mL syringes, 18G hypodermic needles, and gauze sponges (Fig. 4b).

3. Collect ovaries at the slaughterhouse (*see* **Note 5**) and place them in an insulated container containing 0.9% saline solution. At the laboratory, wash ovaries in warmed saline solution (~30 °C) to remove blood and debris. Transfer the ovaries to a new container with warmed saline solution (Fig. 4c).

4. Hold one ovary with gauze sponge and aspirate visible follicles (2–8 mm in diameter) using an 18G hypodermic needle connected to a 10 mL syringe (Fig. 4d).

5. Transfer the recovered follicular fluid to a 50 mL conical sterile polypropylene tube. Allow the follicular fluid to decant. After approximately 10 min, a pellet sediment will be visible (Fig. 4e) containing oocytes, small tissue fragments, and cells.

6. Recover the supernatant of the follicular fluid, transfer to a 15 mL conical sterile polypropylene centrifuge tube (Fig. 4f), and centrifuge 1000 × g for 3 min to produce a follicular fluid without cell debris (Fig. 4g).

7. Transfer 6 mL of follicular fluid to a gridded 100 mm Petri dish (Fig. 5a). The grid can be drawn with a marker pen on the external surface of the dish. Making strips at the bottom of the Petri dish will facilitate and guide the search for oocytes using a stereomicroscope.

8. Collect the sediment (containing oocytes and cell debris) using a disposable plastic Pasteur pipette, and mix and disperse with the follicular fluid previously prepared (Fig. 5b).

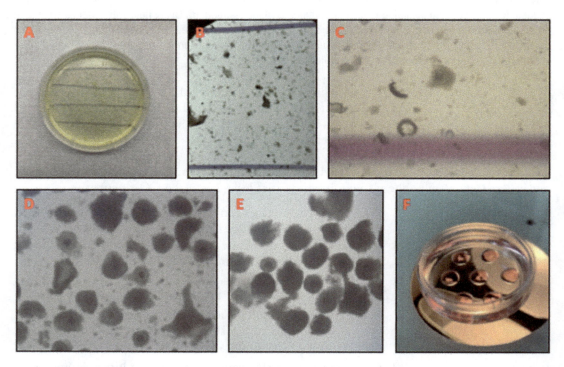

Fig. 5 Oocyte recovery, selection, and in vitro maturation (IVM). (**a**) Gridded 100 mm dish containing centrifuged follicular fluid. (**b**) Sediment containing cumulus-oocyte complexes (COCs) dispersed in the follicular fluid. (**c**) COCs under the stereomicroscope. (**d**) Recovered COCs. (**e**) COCs selected based on a homogeneously granulated cytoplasm and the number of cumulus cell layers. (**f**) COCs placed in medium droplets of the IVM dish

Bovine Somatic Cell Nuclear Transfer   235

9. Search and retrieve the cumulus-oocyte complexes (COCs) under a stereomicroscope using a mouth pipette, microdispenser, or pipette (Fig. 5c).

10. Transfer COCs to a 35 mm Petri dish containing 2 mL OHM (Fig. 5d). Select ~200 good-quality COCs taking into account the number of layers of cumulus cells (at least 3) and containing a light brown and homogeneous cytoplasm (Fig. 5e), which is associated with better developmental potential [30].

11. Wash the COCs three times in 200 μL OMM droplets and transfer pools of 15–20 COCs to each 100 μL OMM droplet (Fig. 5f). Place the Petri dish back in the incubator and allow the oocytes to undergo in vitro maturation for 18 h (*see* **Note 6**).

3.4 Oocyte Denuding and Staining

1. Remove COCs from the maturation dish using a P200 micropipette and wash three times in 200 μL OHM droplets to remove the residual mineral oil.

2. Transfer the COCs to a 400 μL trypsin (Tryple Express, Gibco) droplet and pipette gently for ~3 min to remove the cumulus cells (*see* **Note 7**). Wash denuded oocytes three times in OHM and transfer to an 800 μL OHM droplet to select oocytes with a visible polar body (PB), thus indicating a metaphase II (MII) oocyte (mature oocyte).

3. Examine the oocytes one by one searching for the first polar body (*see* **Note 8**). Transfer all mature oocytes to the top of the droplet and the oocytes without to the bottom. Calculate the maturation rate (number of matured oocytes divided by total COCs placed in IVM) and discard the remaining immature oocytes (without a visible PB).

4. Transfer ~30–40 oocytes to the working plate in the enucleation droplet (containing Hoechst 33342 and CB) and incubate for 15 min. Keep the remaining oocytes in SOFaa droplets until they are required for micromanipulation.

3.5 Somatic Cell Nuclear Transfer

3.5.1 Oocyte Enucleation

1. Place the stained oocytes on the top of the enucleation droplet. Move the microscope chariot and pull an oocyte with the holding pipette. Move back to the center of the droplet and adjust the microscope focus to the middle range of the oocyte, which enables a better of view of the *zona pellucida*. Using the enucleation pipette, rotate the oocyte to visualize and position the PB to the four o'clock position on a clock face (Fig. 6a).

2. Quickly expose the oocyte to the UV light to localize the metaphase plate and confirm if it is nearby the PB (Fig. 6b). Gently rotate the oocyte and make the required adjustments to ensure that both PB and metaphase plate are in the same focal plane.

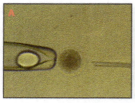

Fig. 6 Oocyte enucleation. (**a**) Oocyte with the first polar body (PB) positioned to 4 o'clock position on a clock's face. (**b**) Oocyte exposed to ultraviolet (UV) light to locate the metaphase II spindle. (**c**) Aspirated PB and surrounding cytoplasm. (**d**) Brief illumination of the pipette with UV light to ensure successful enucleation

3. Pierce the *zona pellucida* with the enucleation pipette, penetrate the perivitelline space and aspirate the PB. Next, aspirate ~3% of the oocyte cytoplasm (*see* **Note 9**) surrounding the PB (Fig. 6c).

4. Expose briefly the enucleation pipette to the UV light to confirm that the metaphase plate has been successfully removed (Fig. 6d).

5. Move the enucleated oocyte to the bottom of the drop to avoid mixing with the nonenucleated ones. Repeat the process until all oocytes in the batch are enucleated. It is advisable to place another batch of oocytes a few minutes before finishing to enucleate the previous batch (~15 min). In such a way, you can save time and speed up the process. In a typical day performing SCNT, we usually enucleate three batches of 30–40 oocytes, thereby totalizing ~100 enucleated oocytes.

6. Transfer the enucleated oocytes to a SOFaa droplet in the "working plate" after finishing the oocyte batch. Our laboratory uses this basic enucleation protocol, but there are alternative methods to enucleate oocytes utilizing oocytes in telophase stage [31] or special microscopes that dispense UV light exposure [32]. Oocytes enucleated at telophase do not require the use of G0/G1 donor cell synchronization to produce cloned blastocysts and viable offspring [31].

7. Proceed to oocyte reconstruction after finishing with the oocyte enucleation step.

3.6 Oocyte Reconstruction

1. Dissociate donor cells from the culture dish by removing the medium and add 1 mL of trypsin (Tryple Express, Gibco) and return the cell culture dish to the incubator for 5 min.

2. After detachment, transfer the cells to a 1.5 mL microtube and centrifuge at 300 × g for 5 min.

3. Remove the supernatant, resuspend the pellet in 100 μL OHM, and transfer ~5.0 μL of cell suspension to the oocyte reconstruction droplet. This cell suspension volume should provide enough cells to reconstruct all cytoplasts.

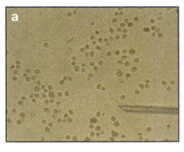 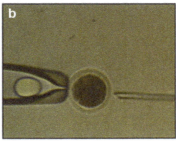 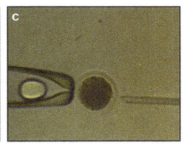

Fig. 7 (**a**) Dissociated fibroblasts for oocyte reconstruction and loading into the injection pipette. (**b**) Enucleated oocyte positioned for reconstruction. (**c**) Donor cell inserted into the perivitelline space and attached to the enucleated oocyte

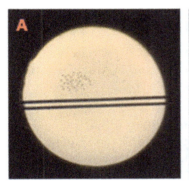 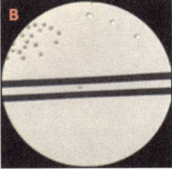 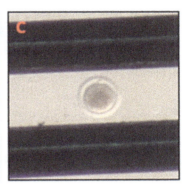

Fig. 8 (**a**) Cell couplets in the fusion chamber. (**b**) Cell couplet between the electrodes. (**c**) Cell couplet with the cell aligned at 12 o'clock to receive the electrical pulse

4. Transfer a group of 40–50 enucleated oocytes to the reconstruction drop. Load the injection pipette with a single small and round-shaped cell (Fig. 7a).

5. Pick one enucleated oocyte and hold it with the holding pipette. Pierce the zona pellucida in a region with the smallest space (*see* **Note 10**) between the *zona pellucida* and the oolema (Fig. 7b).

6. Repeat the procedure (*see* **Note 11**) for the remaining enucleated oocytes (Fig. 7c).

3.7 Cell Fusion

1. Fill the fusion chamber with 200 μL fusion solution and cover with 200 μL mineral oil to avoid evaporation.

2. Remove the oocytes from the incubator and wash three times in fusion solution to remove all traces of SOFaa medium. Place the couplets (ooplast + somatic cell) in the chamber on one side (e.g., top) of the electrodes (Fig. 8a).

3. Pick up one couplet using the mouth pipette and position it between the electrodes (Fig. 8b) by aligning the cell close and

perpendicular to the electrode axis at 6 or 12 o'clock position. Once the cell couplet was aligned, deliver two pulses of 1.75 kV/cm for 45 μs (Fig. 8c) (*see* **Note 12**).

4. Transfer the fused couplet to the bottom of the fusion chamber and repeat the process until all cell couplets were subject to cell fusion.

5. Wash fused cell couplets three times in SOFaa to remove residual mannitol solution and return them to the "working plate" in the incubator for 1 h to check the fusion rate.

6. Place all cell couplets subjected to fusion in an OHM droplet and confirm the fusion under a stereomicroscope by the absence of the somatic cells adjacent to the enucleated oocyte.

7. Discard all lysed and nonfused couplets and proceed to the oocyte chemical activation step.

3.8 Oocyte Chemical Activation

The artificial activation is carried out 26 h after the onset of oocyte IVM and usually 1–2 h after the cell fusion. We use the classical protocol described by Susko-Parrish [33], with minor modifications.

1. Prepare a culture dish with 60 μL SOFaa droplets containing 2.0 mM 6-dimethylaminopurine (6-DMAP) and equilibrate in the incubator for at least 2 h. This dish can be prepared together with the "working plate" and kept in the incubator.

2. To artificially activate the reconstructed couplets, select the successfully fused couplets and incubate in a 300 μL OHM droplet supplemented with 5 μM ionomycin calcium salt for 5 min.

3. Immediately after incubation, wash the oocytes at least three times in SOFaa + 2 mM 6-DMAP.

4. Incubate the activated oocytes in SOFaa + 2 mM 6-DMAP for 3 h.

3.9 Embryo Culture

1. Prepare a culture dish with multiple 100 μL SOFaa droplets under mineral oil and equilibrate in the incubator for at least 2 h before use. Prepare this dish during the incubation in SOFaa + 6-DMAP.

2. Remove activated oocytes from the activation dish and wash them extensively (i.e., at least five times) in fresh SOFaa droplets.

3. Transfer the presumptive embryos to the culture dish (20–30 embryos per SOFaa droplet) and incubate for 7 days in 5% CO_2, 5% O_2, and 90% N_2, under saturated humidity at 38.5 °C.

4. Determine embryonic developmental potential at 72 h post-activation (day 3) for cleavage rates and at 168 h post-activation (day 7) for cloned blastocyst development rates (Fig. 9a).

Fig. 9 (**a**) Cloned blastocysts at day 7 of in vitro culture. (**b**) Recipient cow used for transfer of cloned embryos. (**c**) Cloned calves obtained from the same cell line. (**d**) Neonate cloned calf stabilized after intensive care and oxygen supplementation. (**e**) Cloned calf fully recovered from post-delivery stabilization as demonstrated by suckling without assistance. (**f**) Cloned calves at ~3 months

3.10 Embryo Transfer

Examine the culture dishes for viable blastocysts on day 7 post-activation. At this point, they can be snap frozen in liquid nitrogen and stored at −80 °C for molecular analyses. To generate live offspring, proceed to embryo transfer to surrogate mothers as described below.

1. Select the high-grade cloned blastocysts from the culture dish (Fig. 9a) and load them into 0.25 mL sterile straws in HEPES-buffered SOFaa for embryo transfer.

2. Synchronize the recipient cows using a progesterone-releasing intravaginal device (PRID) plus 2 mg of estradiol benzoate (EB) at unknown stages of estrus cycle (day 0). On day 5, administrate 0.15 mg of D-cloprostenol (PGF) and 400 IU of eCG by intramuscular shots. Remove the PRID on day 8. Administrate 1 mg of EB on day 9. Day 10 is considered the day of estrus. Detailed protocols for recipient synchronization can be found elsewhere [34].

3. Transfer the embryos (blastocysts at day 7) transcervically to surrogate cows on day 17 into the uterine horn ipsilateral to the corpus luteum of previously synchronized recipient cows (Fig. 9b) (*see* **Note 13**).

240   Juliano Rodrigues Sangalli et al.

4. Monitor pregnancies by rectal ultrasound examination. Restrain the recipients in a cattle squeeze chute. Evacuate the feces from rectum. Lubricate the ultrasound transducer. Introduce into the rectum and position in close proximity to the dorsal surface of the uterine tract. Examine both horns for signs of pregnancy (e.g., presence of an echo-free fluid zone of varying size in the lumen of an echogenic uterine horn; presence of embryo or fetus).

5. Check the pregnancy every 30 days.

6. Expect the recipient to calving between 275 and 285 days after SCNT (*see* **Note 14**).

3.11 Delivery of Cloned Offspring and Neonatal Care

1. Start to closely monitor the recipient cows 1 week before the expected delivery date.

2. Prepare and sterilize the surgical instruments and clean the cattle squeeze chute.

3. Prepare a team of veterinarians to assist the recipient cow and the neonate.

4. Assist the recipient cow as soon as it starts showing signals of labor. In case it fails in naturally calving, proceed to caesarean section (*see* **Note 15**).

5. Perform a caesarean section and delivery the calf (Fig. 9c).

6. Check all vital parameters and estimate the calf viability.

7. Execute intensive care (Fig. 9d) immediately after birth (*see* **Note 16**).

8. Keep intensive care until vital parameters stabilize and the calf is able to stand up and suckle from the surrogate mother (Fig. 9e).

9. Monitor the calves during the neonatal period to avoid later complications such as diarrhea and umbilical infections.

10. Discharge the calves and transfer to a clean stall or pasture (Fig. 9f) (*see* **Note 17**).

4 Notes

1. We use OHM to micromanipulate because it has high buffering capacity preventing pH changes, and the FBS acts as a surfactant minimizing cell stickiness.

2. Micromanipulation dishes are prepared in the lid of 60 mm Petri dishes since the borders are lower, making it easier to assemble the holding and the injection pipettes.

3. The Cytochalasin B added to the micromanipulation drop will destabilize the actin cytoskeleton, which facilitates the micromanipulation procedure and prevents oocyte lysis.

4. In our experience, cellular synchronization by contact inhibition by cell confluence or serum deprivation propitiate the birth of viable offspring with similar efficiency.

5. We usually use oocytes from ovaries collected at local slaughterhouses. Oocytes can also be collected from animals by ultrasound-guided oocyte aspiration (ovum pick-up) when there is a desire for a specific cytoplasm donor (e.g., studies on mitochondrial inheritance).

6. Maturing the oocytes for 18 h propitiates at least 60% of oocytes to reach the metaphase II stage, and they are ready to be prepared for SCNT.

7. Partially denuded oocytes may be moved to another droplet with fresh trypsin to make sure all cumulus cells are removed. This will facilitate the selection of PB and also the enucleation step.

8. Stereomicroscopes with large zoom ranges facilitate the process.

9. During the enucleation, it is crucial to expose the oocytes to the UV light as quick as possible and to remove the minimal amount of cytoplasm necessary to enucleate it. These details avoid oocyte damage and help achieve higher fusion rate and embryonic development. When the metaphase plates are far from the PB, aspirate first the PB and then rotate the oocyte to find a better position to remove the metaphase plate.

10. Transferring the cell to the region with the smallest perivitelline space will help position the donor cell in close contact with the cytoplast and increase the fusion rates.

11. After finishing the first batch of cell transfer, if there are two researchers working, it is advisable to proceed immediately to cell fusion. In case you are working alone, return the cell couplets to the SOFaa drop in the "working plate" in the incubator and perform another round of oocyte reconstruction.

12. Herein, we described the fusion carried out in an Eppendorf Multiporator 4308 Electroporation System. The protocol can be adapted and tested in other electrofusion apparatus. The voltage to achieve satisfactory fusion rates with low cell lysis may vary among electrofusion apparatuses. Usually, best results are obtained with voltages between 1.75 and 2.25 kV/cm [24]. Moreover, the duration and number of pulses must be adjusted for each fusion apparatus.

242   Juliano Rodrigues Sangalli et al.

13. It is crucial to plan in advance and start to synchronize the recipient cows 10 days before the expected date to carry out the SCNT. Then, the recipient cow's uterus and the embryos will be synchronized to maximize the chance to get pregnant.

14. The cloning efficiency has ranged from 0% to 10% according to the cell batch being used. Therefore, we usually transfer ~50 embryos to ensure the production of at least one viable clone.

15. In our own experience, over 90% of all cloned pregnancies require a C-section since the recipients often fail to calve naturally.

16. The neonate clones often display comorbidities such as hypoglycemia, enlarged umbilical veins, and suffer from respiratory distress requiring oxygen supplementation (Fig. 9d). Intensive care immediately after birth (e.g., fluid drainage from the respiratory system, clamping the umbilicus) can help stabilize the calf's vital parameters, enabling it to stand up and suckle from the surrogate mother (Fig. 9e). The details of the management and intensive neonatal care can be found in a review article published by our group [6]. After the critical neonatal period, the surviving calves tend to develop into healthy adults (Fig. 9f) that reproduce normally.

17. Some donor cell lineages are more amenable to reprogramming and with a few numbers of transferred embryos enable the production of several viable offspring (Fig. 9c). Contrariwise, some lineages seem unclonable. On some occasions, we carried out 3–4 complete rounds of SCNT embryo transfers and consistently obtained only offspring that died a few minutes after birth, suggesting that sometimes the epigenetic errors acquired by the somatic donor cells cannot be reprogrammed.

Acknowledgments

The authors thank the staff and students at the Laboratory of Molecular Morphophysiology and Development. Juliano Rodrigues Sangalli is supported by São Paulo Research Foundation—FAPESP, grant number #2016/13416-9. Rafael Vilar Sampaio is supported by MITACS and L'Alliance Boviteq Inc. scholarship (FR39379). Tiago Henrique Camara De Bem is supported by São Paulo Research Foundation—FAPESP, grant number #2016/22790-1. Flávio Vieira Meirelles is supported by São Paulo Research Foundation—FAPESP, grant number #2013/08135-2, National Counsel of Technological and Scientific Development (CNPq) grant number 465539/2014-9 and CAPES. The funders had no role in study design, data collection and analysis, decision to publish, or preparation of the manuscript.

References

1. Wilmut I, Schnieke AE, McWhir J, Kind AJ, Campbell KH (1997) Viable offspring derived from fetal and adult mammalian cells. Nature 385:810–813

2. Kato Y, Tani T, Sotomaru Y, Kurokawa K, Kato JY, Doguchi H et al (1998) Eight calves cloned from somatic cells of a single adult. Science 282:2095–2098

3. Galli C, Lagutina I, Perota A, Colleoni S, Duchi R, Lucchini F et al (2012) Somatic cell nuclear transfer and transgenesis in large animals: current and future insights. Reprod Domest Anim 47:2–11

4. Yang X, Smith SL, Tian XC, Lewin HA, Renard J-P, Wakayama T (2007) Nuclear reprogramming of cloned embryos and its implications for therapeutic cloning. Nat Genet 39:295–302

5. Wilmut I, Beaujean N, de Sousa PA, Dinnyes A, King TJ, Paterson LA et al (2002) Somatic cell nuclear transfer. Nature 419:583–586

6. Meirelles FV, Birgel EH, Perecin F, Bertolini M, Traldi AS, Pimentel JRV et al (2010) Delivery of cloned offspring: experience in Zebu cattle (Bos indicus). Reprod Fertil Dev 22:88–97

7. Pasque V, Jullien J, Miyamoto K, Halley-Stott RP, Gurdon JB (2011) Epigenetic factors influencing resistance to nuclear reprogramming. Trends Genet 27:516–525

8. Kishigami S, Mizutani E, Ohta H, Hikichi T, Van Thuan N, Wakayama S et al (2006) Significant improvement of mouse cloning technique by treatment with trichostatin A after somatic nuclear transfer. Biochem Biophys Res Commun 340:183–189

9. Inoue K, Kohda T, Sugimoto M, Sado T, Ogonuki N, Matoba S et al (2010) Impeding Xist expression from the active X chromosome improves mouse somatic cell nuclear transfer. Science 330:496–499

10. Matoba S, Liu Y, Lu F, Iwabuchi KA, Shen L, Inoue A et al (2014) Embryonic development following somatic cell nuclear transfer impeded by persisting histone methylation. Cell 159:884–895

11. Halley-Stott RP, Gurdon JB (2013) Epigenetic memory in the context of nuclear reprogramming and cancer. Brief Funct Genomics 12:164–173

12. Burrill DR, Silver PA (2010) Making cellular memories. Cell 140:13–18

13. Ng RK, Gurdon JB (2005) Epigenetic memory of active gene transcription is inherited through somatic cell nuclear transfer. Proc Natl Acad Sci U S A 102:1957–1962

14. Liu X, Wang Y, Gao Y, Su J, Zhang J, Xing X et al (2018) H3K9 demethylase KDM4E is an epigenetic regulator for bovine embryonic development and a defective factor for nuclear reprogramming. Development 145:1–12

15. Santos F, Zakhartchenko V, Stojkovic M, Peters A, Jenuwein T, Wolf E et al (2003) Epigenetic marking correlates with developmental potential in cloned bovine preimplantation embryos. Curr Biol 13:1116–1121

16. Sampaio RV, Sangalli JR, De Bem THC, Ambrizi DR, del Collado M, Bridi A et al (2020) Catalytic inhibition of H3K9me2 writers disturbs epigenetic marks during bovine nuclear reprogramming. Sci Rep 10:1–13

17. Zhao J, Hao Y, Ross JW, Spate LD, Walters EM, Samuel MS et al (2010) Histone deacetylase inhibitors improve in vitro and in vivo developmental competence of somatic cell nuclear transfer porcine embryos. Cell Reprogram 12:75–83

18. Martinez-Diaz M, Che L, Albornoz M, Seneda M, Collis D, Coutinho A et al (2010) Pre- and postimplantation development of swine-cloned embryos derived from fibroblasts and bone marrow cells after inhibition of histone deacetylases. Cell Reprogram 12:85–94

19. Zhao J, Ross JW, Hao Y, Spate LD, Walters EM, Samuel MS et al (2009) Significant improvement in cloning efficiency of an inbred miniature pig by histone deacetylase inhibitor treatment after somatic cell nuclear transfer. Biol Reprod 81:525–530

20. Sangalli JR, Chiaratti MR, De Bem THC, de Araújo RR, Bressan FF, Sampaio RV et al (2014) Development to term of cloned cattle derived from donor cells treated with valproic acid. PLoS One 9:e101022

21. Sangalli JR, De Bem THC, Perecin F, Chiaratti MR, Oliveira LDJ, De Araújo RR et al (2012) Treatment of nuclear-donor cells or cloned zygotes with chromatin-modifying agents increases histone acetylation but does not improve full-term development of cloned cattle. Cell Reprogram 14:1–13

22. Hosseini SM, Dufort I, Nieminen J, Moulavi F, Ghanaei HR, Hajian M et al (2016) Epigenetic modification with trichostatin A does not correct specific errors of somatic cell nuclear transfer at the transcriptomic level; highlighting the non-random nature of oocyte-mediated reprogramming errors. BMC Genomics 17:1–21

23. Zhou C, Wang Y, Zhang J, Su J, An Q, Liu X et al (2019) H3K27me3 is an epigenetic barrier while KDM6A overexpression improves nuclear reprogramming efficiency. FASEB J 33:4638–4652

24. Ross PJ, Cibelli JB (2010) Bovine somatic cell nuclear transfer. Methods Mol Biol 636:155–177

25. Bogliotti YS, Wu J, Vilarino M, Okamura D, Soto DA, Zhong C et al (2018) Efficient derivation of stable primed pluripotent embryonic stem cells from bovine blastocysts. Proc Natl Acad Sci 115:2090–2095

26. Miranda MDS, Bressan FF, De Bem THC, Merighe GKF, Ohashi OM, King WA et al (2012) Nuclear transfer with apoptotic bovine fibroblasts: can programmed cell death be reprogrammed? Cell Reprogram 14:217–224

27. Wells DN, Laible G, Tucker FC, Miller AL, Oliver JE, Xiang T et al (2003) Coordination between donor cell type and cell cycle stage improves nuclear cloning efficiency in cattle. Theriogenology 59:45–59

28. Campbell KH (1999) Nuclear equivalence, nuclear transfer, and the cell cycle. Cloning 1:3–15

29. Campbell KHS, McWhir J, Ritchie WA, Wilmut I (1996) Sheep cloned by nuclear transfer from a cultured cell line. Nature 380:64–66

30. De Bem THC, Adona PR, Bressan FF, Mesquita LG, Chiaratti MR, Meirelles FV et al (2014) The influence of morphology, follicle size and Bcl-2 and bax transcripts on the developmental competence of bovine Oocytes. Reprod Domest Anim 49:576–583

31. Bordignon V, Smith LC (1998) Telophase enucleation: An improved method to prepare recipient cytoplasts for use in bovine nuclear transfer. Mol Reprod Dev 49:29–36

32. Kim EY, Park MJ, Park HY, Noh EJ, Noh EH, Park KS et al (2012) Improved cloning efficiency and developmental potential in bovine somatic cell nuclear transfer with the oosight imaging system. Cell Reprogram 14:305–311

33. Susko-Parrish JL, Leibfried-Rutledge ML, Northey DL, Schutzkus V, First NL (1994) Inhibition of protein kinases after an induced calcium transient causes transition of bovine oocytes to embryonic cycles without meiotic completion. Dev Biol 166:729–739

34. Nasser LF, Reis EL, Oliveira MA, Bó GA, Baruselli PS (2004) Comparison of four synchronization protocols for fixed-time bovine embryo transfer in Bos indicus x Bos taurus recipients. Theriogenology 62:1577–1584

Chapter 13

Production of Water Buffalo SCNT Embryos by Handmade Cloning

Prabhat Palta, Naresh L. Selokar, and Manmohan S. Chauhan

Abstract

Cloning by somatic cell nuclear transfer (SCNT) involves the transfer of a somatic nucleus into an enucleated oocyte followed by chemical activation and embryo culture. Further, handmade cloning (HMC) is a simple and efficient SCNT method for large-scale embryo production. HMC does not require micromanipulators for oocyte enucleation and reconstruction since these steps are carried out using a sharp blade controlled by hand under a stereomicroscope. In this chapter, we review the status of HMC in the water buffalo (*Bubalus bubalis*) and further describe a protocol for the production of buffalo-cloned embryos by HMC and assays to estimate their quality.

Key words Buffaloes, Cloning, Embryogenesis, Nuclear transplantation, SCNT

---

1 Introduction

The technique of cloning by somatic cell nuclear transfer (SCNT) involves the transfer of a somatic cell nucleus to an enucleated oocyte. The ooplasm of the enucleated oocyte converts the differentiated somatic cell nucleus to a totipotent state through a process called nuclear reprogramming. In the conventional SCNT method, both oocyte enucleation and reconstruction require micromanipulators [1], a fact which confined SCNT to a few laboratories that could afford expensive equipment and had skilled scientists to operate them. A simpler and less expensive SCNT method was developed and called handmade cloning (HMC), which does not require micromanipulators, and SCNT is done by hand using a sharp blade under a stereomicroscope [2]. Further, HMC led to a much wider use of SCNT and is suited to large-scale cloning programs, thus offering high embryo yields and improved live birth rates in cattle, sheep, and pigs [3–9].

Marcelo Tigre Moura (ed.), *Somatic Cell Nuclear Transfer Technology*, Methods in Molecular Biology, vol. 2647,
https://doi.org/10.1007/978-1-0716-3064-8_13,
© The Author(s), under exclusive license to Springer Science+Business Media, LLC, part of Springer Nature 2023

We used HMC for the production of the world's first cloned water buffalo (*Bubalus bubalis*) [10]. During the last decade, we made improvements in buffalo HMC, which included standardization of embryo culture [11, 12], improved cell fusion conditions [13], enhanced donor cell synchronization [14], selection of oocytes with greater developmental competence [15], and selection of early cleaved embryos for improved full-term development [16]. Furthermore, we tested modulations of the epigenetic status (donor cells, reconstructed embryos, or both) with histone deacetylase inhibitors [17–23] or inhibition of DNA methyltransferase activity [18, 19, 24]. Attempts to improve blastocyst yields by modulating small noncoding RNAs showed limited potential [25–27]. Collectively, these modifications in buffalo HMC improved blastocyst rates, thus making it comparable to in vitro fertilization [28, 29]. Recently, we have shown that treatment of reconstructed oocytes with Dickkopf-1, an inhibitor of canonical WNT signaling pathway, increased both conception and live birth rates to as high as 25% [30].

To prospect the most promising cell type for higher cloning efficiency (i.e., high blastocyst and live birth rates), we tested several cell types such as fetal fibroblasts [10], newborn fibroblasts [10], adult fibroblasts [31], milk-derived cells [31, 32], blood-derived cells [33], and from the trophectoderm [34]. Live births using buffalo HMC were derived from fetal fibroblasts [10], embryonic stem cell-like cells [35], adult fibroblasts [30, 31], seminal plasma-derived cells [36], and urine-derived cells [37]. Therefore, this chapter describes our HMC method used for producing clone calves from adult water buffaloes.

2 Materials

2.1 Equipment

The equipment required for production of embryos by HMC is illustrated below (Fig. 1).

1. Laminar flow hood.

2. Two stereomicroscopes.

3. Inverted microscope equipped with Hoffmann and epifluorescence.

4. Two CO_2 incubators: one with 5% CO_2 with high oxygen tension (20% O_2) and the other with 5% CO_2 and low oxygen tension (5% O_2).

5. Electrofusion machine (ECM 2001, BTX San Diego, CA, USA).

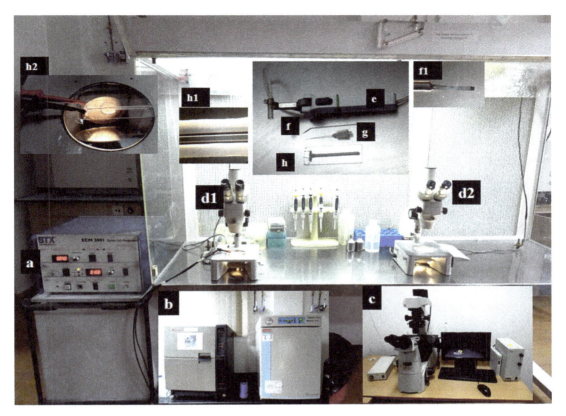

Fig. 1 Instruments for buffalo handmade cloning. (**a**) Electrofusion machine. (**b**) CO_2 incubators. (**c**) Inverted microscope with fluorescence. (**d**) Stereomicroscopes. (**e**) Oocyte handling pipette. (**f**) Microblade. (**f1**) Microblade tip. (**g**) Glass capillary attached to pipette holder. (**h**) Electrofusion chamber. (**h1**) Electrofusion chamber. (**h2**) Fusion medium droplet placed on the fusion chamber that is attached to red (positive) and black (negative) electrodes

2.2 Tools and Consumables

1. Oocyte handling pipettes (Unopette, Becton Dickinson & Co., USA).
2. Microblades (MTB-05; Micromanipulator Microscope Co. Inc., Carson City, USA).
3. Microslide 0.5 mm gap, model 450 (BTX, San Diego, USA).

2.3 Media and Solutions

Use only high-grade reagents (*see* **Note 1**). Store the stock solutions at −20 °C until use, unless specified otherwise (*see* **Note 2**).

1. Ovary washing medium (OWM): Mix 1 L distilled water, 9.0 g sodium chloride, 100,000 IU penicillin, and 100 mg streptomycin. Filter-sterilize (0.22 μm) and store at 4 °C (*see* **Note 3**).
2. Aspiration medium (APM): Mix 50 mL TCM-199 medium (Sigma, M7528), 0.3% (w/v) bovine serum albumin (BSA), 0.68 mM L-glutamine, and 50 μg/mL gentamicin.

3. Washing medium (WSM): Mix 36 mL TCM-199, 4.0 mL fetal bovine serum (FBS), 3.6 mg sodium pyruvate, 4.0 mg L-glutamine, and 2.0 mg gentamicin.

4. Brilliant cresyl blue medium (BCBM): Dissolve 0.4045 mg BCB (Sigma, B5388) in 20 mL PBS supplemented with 0.4% BSA.

5. In vitro maturation medium (IVMM): Mix 10 mL WSM with 5.0 μg/mL porcine FSH and 1.0 μg/mL 17-β estradiol (*see* **Note 4**).

6. Handling media (HM): Prepare HM0 with 30 mL TCM-199 and 0.84 mg sodium pyruvate, 3.0 mg L-glutamine, and 1.5 mg gentamicin. Prepare HM2 with 13.72 mL HM0 supplemented with 280 μL FBS. Prepare HM20 with 12.8 mL HM0 supplemented with 3.2 mL FBS.

7. Hyaluronidase solution: Mix 200 mL TCM-199 with 2.0% (v/v) FBS and 100 mg hyaluronidase. Store 500 μL aliquots in 1.5 mL tubes.

8. Pronase solution: Dissolve 100 mg pronase E in 50 mL TCM-199 medium supplemented with 10% FBS. Store 500 μL aliquots in 1.5 mL tubes. Spin down for 1 min before use.

9. Phytohemagglutinin (PHA) solution: Dissolve 50 mg PHA in 10 mL TCM-199 medium. Store 50 μL aliquots in 1.5 mL tubes. Prepare the working solution by thawing a 50 μL stock aliquot and add 450 μL HM2. Spin down for 1 min and aspirate 400 μL of the supernatant.

10. DPBS: PBS supplemented with 50 μg/mL gentamicin. DPBS-BSA: DPBS supplemented with 0.4% BSA.

11. Cell fusion medium (CFM): Mix 50 mL ultrapure water with 2.73 g D-mannitol, 1.0 mg $MgCl_2.6H_2O$, and 0.368 mg $CaCl_2.2\ H_2O$. Add 50.0 mg polyvinyl alcohol and homogenize. Incubate overnight at 4 °C. Store in 1.0 mL aliquots in 1.5 mL tubes.

12. Embryo in vitro culture medium (IVCM): 3.0 mL KRVCL medium (Cook®, Australia, K-RVCL-50) with 30 mg fatty acid-free BSA (Sigma, A8806).

13. Calcium ionophore solution: Dissolve 1.0 mg calcium ionophore in 1.0 mL dimethyl sulfoxide (DMSO). Store 5.0 μL aliquots in 0.2 mL tubes protected from light (it is light sensitive). Spin down for 30 s before use.

14. 6-Dimethylaminopurine (6-DMAP): Dissolve 100 mg 6-DMAP in 6.13 mL PBS. Shake the solution in a water bath at 55 °C until the solution becomes clear. Store 10 μL aliquots in 0.5 mL tubes.

15. Cell culture media (CCM20): Mix 20 mL FBS in 80 mL of DMEM-F12 (Sigma, D8437) supplemented with 50 µg/mL gentamicin. CCM10: Mix 10 mL FBS in 90 mL of DMEM supplemented with 50 µg/mL gentamicin.

16. Cell freezing media (CFZM): Mix 2 mL FBS, 1 mL DMSO in 7 mL of DMEM-F12 (Sigma, D8437) supplemented with 50 µg/mL gentamicin.

3 Methods

Experiments involving live animals must comply with institutional and national regulations. Carry out all the procedures at room temperature unless specified otherwise. Perform all handling and manipulation of oocytes under a stereomicroscope (*see* **Note 5**). The CO_2 incubator used for most steps should have 5% CO_2 and 20% O_2 (*see* **Note 6**), while embryo culture under 5% CO_2 and 5% O_2 supports higher embryonic developmental rates.

3.1 Preparation of Donor Cells

1. Restrain a water buffalo, wash the ventral part of the tail (just above the anal region) with soap, and disinfect with 70% ethanol. Wipe it dry with sterile cotton and collect biopsy of skin (approximately 2.0 cm^2) with a notcher. Transfer the biopsy to a sterile 15 mL tube containing 10 mL DPBS. Alternatively, skin biopsy may be taken from ear pinna tissue.

2. Cut the tissue biopsy into small fragments (~1.0 mm^2/each) immersed in 3.0 mL DPBS containing 35 mm dish. Transfer the tissue fragments to another 35 mm dish containing 3.0 mL DPBS and wash them 2–3 times with 3.0 mL CCM10. Usually, 15–20 fragments can be made from each tissue biopsy.

3. Place 3–5 tissue fragments in 10 µL CCM20 droplets (one tissue fragment per droplet) to a 25 cm^2 culture flask and incubate with 5% CO_2, saturated humidity at 37 °C 10–15 h. Replenish 4.0 mL CCM20 at 15 h and replenish 3.0 mL CCM20 every third day during the culture period. Cell outgrowths from the explants usually become visible after 5 days of culture (Fig. 2).

4. Remove the explants and dissociate cell outgrowths with 0.25% trypsin after 10–12 days of culture. Add 3.0 mL CCM10 and transfer the cell suspension to a sterile 15 mL tube. Centrifuge the tube at 800 g for 5 min and suspend the pellet in 3.0 mL CCM10. Divide the cell suspension into 3–4 aliquots (split ratio of 1:3–1:4 depending upon the total number of cells). Passage cells (subculture) 3–5 times in 25 cm^2 culture flasks in CCM10 with 5% CO_2, saturated humidity at 37 °C to establish fibroblast cultures without epithelial cells.

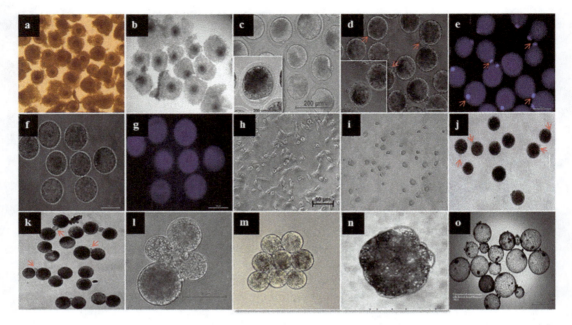

Fig. 2 (**a**) Immature cumulus-oocyte complexes (COCs). (**b**) COCs showing cumulus expansion. (**c**) Denuded oocytes. (**d**) *Zona pellucida*-free oocytes displaying protrusion cones (arrows). (**e**) DNA staining of zona-free oocytes in which the arrows indicate the metaphase-II plates. (**f**) Enucleated oocytes (demicytoplasts) under bright light. (**g**) Demicytoplasts stained with DNA dye H33342 under ultraviolet light. (**h**) Fibroblast cell line. (**i**) Donor cells after trypsinization. (**j**) Cell couplets made with donor cells (arrows) attached to demicytoplasts. (**k**) Cell couplets made with somatic cells (arrows) in between two demicytoplasts. (**l**) Four-cell embryo. (**m**) Eight-cell embryo. (**n**) Compact morula. (**o**) Blastocysts

5. Cryopreserve cells at each passage by loading cryovials with 1.0×10^6 to 1.0×10^7 cells in 1.0 mL CFZM and subject it to slow freezing by placing in −80 °C overnight. Next day, transfer cryovials for store in N_2 at −196 °C.

6. Before use, thaw a Cryovial in a water bath at 37 °C and wash cells 2–3 times with 5 mL CCM20, centrifugation at 800 g, and remove the supernatant. Place cells 400 μL CCM10 per well of a four-well dish and incubate with 5% CO_2, saturated humidity, at 37 °C for 5–7 days.

7. Cells grow for 5–7 days to make them full confluence before SCNT. Trypsinize cells immediately before use and resuspend the cell pellet in CCM10 at a concentration of 1.0×10^5 cells/mL.

3.2 Oocyte Collection and In Vitro Maturation

1. Collect ovaries (preferably >100 per collection) at a slaughterhouse and wash 3–4 times with OWM at 30 °C. Transport ovaries to the laboratory as soon as possible (less than 6 h) in a thermos flask containing OWM at 30 °C. In the laboratory, wash ovaries thoroughly with OWM at 30 °C and remove any extra tissue around them. Transfer the ovaries to a beaker containing warm OWM at 30 °C.

Handmade Cloning in Water Buffaloes   251

2. Hold one ovary with a sterile filter paper and aspirate all the visible antral follicles with an 18-gauge needle attached to a 10 mL syringe containing APM. Repeat the process with all the ovaries. Whenever the syringe gets filled, transfer the aspirated content to a 15 mL sterile tube. Keep all tubes in a dry bath at 30 °C and wait for cumulus-oocyte complexes (COCs) to settle for 15 min.

3. Remove the top portion of the APM of each tube with a sterile Pasteur pipette and transfer the pellet to a 100 mm dish with a 13 mm grid. Add 5 mL APM and shake the dish gently to distribute its contents evenly.

4. Collect COCs from the dish under a stereomicroscope. Use a sterile glass Pasteur pipette with mouth diameter of approximately 300–400 μM.

5. Transfer the COCs to a 35 mm dish containing 3.0 mL WSM. Select COCs with two or more compact cumulus cell layers and evenly granular cytoplasm (Fig. 2a). Transfer the selected COCs to another dish containing 3.0 mL DPBS-BSA. Discard the remaining oocytes.

6. Wash COCs twice in DPBS-BSA. Incubate COCs in BCBM in the CO_2 incubator at 38.5 °C for 90 min. Wash COCs once with DPBS-BSA and examine under a stereomicroscope. Collect COCs with discernible degree of blue coloration (BCB+ oocytes). These are oocytes of high developmental competence (*see* **Note 7**).

7. Transfer the BCB+ oocytes to another 35 mm dish containing 3.0 mL WSM and wash them three times with IVMM. Prepare 100 μL IVMM droplets in a cell culture grade 35 mm dish and cover with sterile mineral oil (*see* **Note 8**). Transfer 15–20 COCs to each IVMM droplet. Perform IVM of COCs with 5% CO_2, saturated humidity, at 38.5 °C for 21 h.

3.3 Oocyte Preparation for Enucleation

1. Examine the COCs after IVM under an inverted microscope. Select COCs with adequate cumulus expansion (Fig. 2b). Transfer COCs to hyaluronidase solution while avoiding transferring them with IVMM. Incubate COCs in hyaluronidase at 38.5 °C for 1 min. Pipette gently to assist oocyte denuding. Vortex gently if oocytes are not completely denuded.

2. Transfer the content of the tube to 35 mm dish containing 3.0 mL HM2. Pick all denuded oocytes (Fig. 2c) and wash twice in HM2 to remove residual cumulus cells (*see* **Note 9**).

3. Prepare a 400 μL pronase droplet in a 35 mm dish, transfer 200–250 denuded oocytes, and incubate with 5% CO_2, saturated humidity at 38.5 °C for 8–10 min. Swirl the dish gently a few times.

252   Prabhat Palta et al.

4. Examine the oocytes under a stereomicroscope. Transfer the oocytes with completely digested *zona pellucida* to another 35 mm dish containing 3.0 mL HM20. Wash twice with HM20 and incubate in a 35 mm dish containing 3.0 mL HM20 in a CO_2 incubator at 38.5 °C for 15–20 min.

3.4 Somatic Cell Nuclear Transfer

3.4.1 Oocyte Enucleation

1. Examine the oocytes for a "protrusion cone" under an inverted microscope (Fig. 2d, e). Transfer pools of 8–10 protrusion-bearing oocytes (each time) to a 35 mm dish in 4.0 mL HM20 containing 2.5 µg/mL cytochalasin B. Position them in a row for manual bisection.

2. Slice oocytes (i.e., one at a time) using the microblade into two parts, although intending to leave the protrusion cone in the smaller part (*see* **Note 10**). The larger oocyte part lacks the metaphase plate (oocyte spindle) and is thereafter called "demi-cytoplast" (Fig. 2f). Enucleate the remaining protrusion-bearing oocytes (*see* **Note 11**).

3. Transfer demicytoplasts to a 35 mm dish containing 3.0 mL HM20 and incubate with 5% CO_2, saturated humidity at 38.5 °C for 10–15 min, to enable them to regain spherical shape.

4. Incubate demicytoplasts in a 200 µL droplet of HM20 with 10 µg/mL Hoechst 33342 under 5% CO_2, saturated humidity at 38.5 °C for 10 min.

5. Subject demicytoplasts to a brief ultraviolet light exposure (i.e., few seconds) to confirm enucleation (Fig. 2g). Discard none-nucleated demicytoplasts and wash enucleated demicytoplasts 2–3 times in HM20.

3.4.2 Oocyte Reconstruction

1. Prepare a four-well dish with the following solutions: Well #1 with 400 µL HM20, well #2 with 400 µL PHA solution, well #3 with 400 µL HM20, and well #4 with 400 µL CFM.

2. Fill the space between wells of the four-well dish with 4.0 mL HM2 and add 8–10 µL of the donor cell suspension to it.

3. Transfer demicytoplasts to well #1. Pick 5–8 demicytoplasts from well #1 and immerse them in well #2 for 3–4 s.

4. Prepare cell couplets by gently rolling the demicytoplast over a single donor cell, thus keeping tight membrane contact (Fig. 2i). Repeat the process until half of the demicytoplasts are paired with donor cells (Fig. 2j). Electrofused couplets are incubated in well #3 for 4 h.

5. Transfer cell couplets to well #4 and incubate the dish in the CO_2 incubator at 38.5 °C for 10 min.

Handmade Cloning in Water Buffaloes   253

3.5 Cell Fusion

1. Turn on the electrofusion machine set with the following parameters: alternate current (AC) at 4 V (one pulse for 4 ms), direct current (DC) at 160 V (one pulse for 6 μs), and post-AC pulse at 0 V.

2. Place the fusion chamber on the stage of another stereomicroscope in the laminar hood. Fix the fusion chamber to the microscope stage with sticky tape. Attach electrode wires of the electrofusion machine to electrode tips of the fusion chamber.

3. Cover the electrodes with a 1.0–2.0 mL CFM droplet at the center of the fusion chamber and transfer cell couplets from well #4 to the northern part of the fusion chamber (*see* **Note 12**).

4. Transfer the remaining demicytoplasts to well #4 of the four-well dish and incubate the dish at room temperature for 5 min. Then, transfer the demicytoplasts to the southern part of the fusion chamber (distant from the platinum wires).

5. Transfer 1–2 cell couplets to the space between platinum wires and gently steer them with the glass capillary (Unopette) until the fibroblast of the cell couplet is placed between the demicytoplasts and faces the negative electrode.

6. Trigger the electrofusion. This will result in movement of the couplets and the demicytoplasts toward each other in such a way that the somatic cell gets sandwiched between the two demicytoplasts (Fig. 2k).

7. Collect gently all fused couplets and transfer them to well #3 of the four-well dish to recover from cell fusion. Continue the electrofusion until all cell couplets and demicytoplasts are fused to form reconstructed oocytes (*see* **Note 13**) and incubate them in well #3 in 5% CO_2, saturated humidity at 38.5 °C for 4 h.

3.6 Oocyte Activation

1. Prepare a four-well dish as follows: Well #1 with 5 μM calcium ionophore in 400 μL HM20, well #2 with 400 μL HM20, well #3 with 400 μL HM20, and well #4 with 2.0 mM 6-DMAP in 400 μL HM20.

2. Transfer reconstructed oocytes to well #1 and incubate in the CO_2 incubator at 38.5 °C for 5 min. Wash reconstructed oocytes in well #2 and well #3. Transfer reconstructed oocytes to well #4 and incubate with 5% CO_2, saturate humidity at 38.5 °C for 4 h (*see* **Note 14**). Wash reconstructed oocytes three times in HM20 after incubation with 6-DMAP-containing medium.

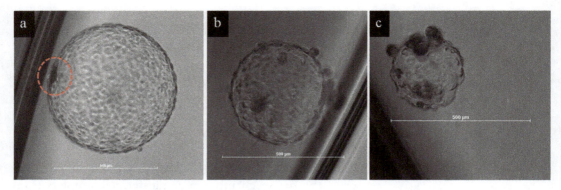

Fig. 3 Buffalo-cloned blastocysts produced by handmade cloning. (**a**) Grade A blastocyst with tightly packed inner cell masses (circles). (**b**) Grade B blastocyst. (**c**) Grade C blastocyst

3.7 Embryo Culture

1. Add 400 μL KRVCL medium per well of a four-well dish. Cover the medium in each well with 300 μL sterile mineral oil. Place gently 15–18 embryos per well along the periphery of the well at some distance from each other to avoid aggregation.

2. Transfer the dish very gently to the CO_2 incubator with 5% CO_2 and 5% O_2. Avoid shaking the dish (*see* **Note 15**). Replace gently 4.0 μL of the culture medium with 4 μL DKK1 stock on day 5 post-activation. Resume embryo culture until day 8 post-activation and record the embryo developmental rate (Fig. 2l–o) and evaluate the blastocyst quality (*see* **Note 16**).

3.8 Embryo Evaluation

3.8.1 Blastocyst Morphology

Capture an image of the blastocyst under a phase-contrast inverted microscope at 100× magnification. Use the following criteria for evaluation of the quality of blastocysts (Fig. 3).

Grade A: Blastocyst size is greater than 350 μm with tightly packed inner cell mass (ICM) cells and trophectoderm (TE) cells forming an organized cell layer. Grade B: Blastocyst size is 200–350 μm with both ICM and TE cells loosely packed. Grade C: Blastocyst size is less than 200 μm, with no distinct ICM and very few TE cells. For embryo transfer, use only grade A blastocysts.

3.8.2 Blastocyst Total Cell Number

1. Incubate blastocysts in a 50 μL DPBS droplet containing 10 μg/mL Hoechst 33342 and keep on a warm plate at 38.5 °C for 10 min.

2. Wash the blastocysts 2–3 times with DPBS.

3. Place blastocysts in a 10 μL DPBS droplet on a glass slide and cover it with a coverslip.

4. Capture the image of each blastocyst using a confocal fluorescence microscope at 100× magnification.

5. Count the number of nuclei visualized as blue dots (total cell number). Grade A blastocysts usually have 250–400 nuclei.

4 Notes

1. Success of HMC depends heavily on the quality of reagents and consumables. If possible, purchase all media in ready-to-use liquid form. Use chemicals of cell culture grade or wherever possible embryo culture grade. Use cell culture-grade plastic ware.

2. Prepare solutions in ultrapure water.

3. Prepare media or solutions and filter-sterilize using 0.22 μm membranes immediately before use.

4. There is a significant variation among batches of animal-derived products (e.g., FBS, BSA). Test the quality of each new batch before wide use. This can be done by producing parthenogenetic embryos comparing old and new batches simultaneously.

5. Keep media and solutions at 38.5 °C throughout oocyte and embryo handling outside the CO_2 incubator. Dishes with oocytes or embryos should be placed on a heating plate at 38.5 °C.

6. Oocytes and embryos must not be kept out of the CO_2 incubator any longer than necessary. Avoid the exposure of oocytes and embryos to light during microscopic examinations.

7. Select only those COCs with a clearly discernible blue coloration (i.e., BCB+) to ensure higher oocyte developmental competence.

8. The mineral oil deteriorates over time. Do not store it for long periods of time or exposed to intense sunlight (causes peroxidation). Use embryo-tested mineral oil.

9. The inner capillary diameter should be 250–300 μm for handling zona-enclosed oocytes or reconstructed embryos, and 150–200 μm for handling zona-free oocytes. The opening of the capillaries should be smoothened by flame polishing to avoid oocytes or embryo damage.

10. The metaphase II plate (oocyte spindle) lies below the protrusion cone. For enucleation, place the cutting edge of the blade on the oocyte and lower the blade slowly to bisect each oocyte into two parts. Care should be taken to ensure that a minimal amount of cytoplasm (no more than 30–40%) is lost during removal of the protrusion cone.

11. A skilled worker should be able to obtain more than 90% survival rate after enucleation of oocytes.

256 Prabhat Palta et al.

12. Errors in preparing CFM cause embryo sticking to the surface of the fusion chamber or osmotic stress, thus causing cell lysis. The osmolarity of CFM should be checked immediately before use. It should be 250–350 mOsm.

13. The number of reconstructed oocytes per session depends on experience and skill. Usually, one worker can produce 45–60 reconstructed oocytes per session.

14. Place reconstructed oocytes distant from one another during activation since embryos become sticky during incubation with 6-DMAP. Wash reconstructed oocytes 2–3 times with HM20 to ensure that they do not aggregate.

15. Place incubators on a nonvibrating table and avoid unnecessary opening of CO_2 incubators during embryo culture. Zona-free embryos are very fragile and may disintegrate by any disturbance. Allocate one CO_2 incubator strictly for embryo culture.

16. A video of our HMC method is available in another publication [28].

References

1. Wilmut I, Schnieke AE, McWhir J, Kind AJ, Campbell KH (1997) Viable offspring derived from fetal and adult mammalian cells. Nature 385:810–813

2. Vajta G, Lewis IM, Hyttel P, Thouas GA, Trounson AO (2001) Somatic cell cloning without micromanipulators. Cloning 3:89–95

3. Vajta G (2007) Handmade cloning: the future way of nuclear transfer? Trends Biotechnol 25:250–253

4. Tecirlioglu RT, Cooney MA, Lewis IM, Korfiatis NA, Hodgson R, Ruddock NT et al (2005) Comparison of two approaches to nuclear transfer in the bovine: hand-made cloning with modifications and the conventional nuclear transfer technique. Reprod Fertil Dev 17:573–585

5. Vajta G, Callesen H (2012) Establishment of an efficient somatic cell nuclear transfer system for production of transgenic pigs. Theriogenology 77:1263–1474

6. Vajta G, Lewis IM, Tecirlioglu RT (2006) Handmade somatic cell cloning in cattle. Methods Mol Biol 348:183–196

7. Vajta G, Kragh PM, Mtango NR, Callesen H (2005) Hand-made cloning approach: potentials and limitations. Reprod Fertil Dev 17:97–112

8. Vajta G, Maddox-Hyttel P, Skou CT, Tecirlioglu RT, Peura TT, Lai L et al (2005) Highly efficient and reliable chemically assisted enucleation method for handmade cloning in cattle. Reprod Fertil Dev 17:791–797

9. Vajta G, Lewis IM, Trounson AO, Purup S, Maddox-Hyttel P, Schmidt M et al (2003) Handmade somatic cell cloning in cattle: analysis of factors contributing to high efficiency in vitro. Biol Reprod 68:571–578

10. Shah RA, George A, Singh MK, Kumar D, Anand T, Chauhan MS et al (2009) Pregnancies established from handmade cloned blastocysts reconstructed using skin fibroblasts in buffalo (*Bubalus bubalis*). Theriogenology 71:1215–1219

11. Shah RA, George A, Singh MK, Kumar D, Chauhan MS, Manik R et al (2008) Handmade cloned buffalo (*Bubalus bubalis*) embryos: comparison of different media and culture systems. Cloning Stem Cell 10:435–442

12. Saini M, Selokar NL, Agrawal H, Singla SK, Chauhan MS, Manik RS et al (2015) Low oxygen tension improves developmental competence and reduces apoptosis in hand-made cloned buffalo (*Bubalus bubalis*) embryos. Livest Sci 172:106–109

13. Selokar NL, Shah RA, Saha AP, Muzaffar M, Saini M, Chauhan MS et al (2012) Effect of post-fusion holding time, orientation and position of somatic cell-cytoplasts during electrofusion on the development of handmade cloned embryos in buffalo (*Bubalus bubalis*). Theriogenology 78:930–936

14. Selokar NL, Saini M, Muzaffer M, Krishnakanth G, Saha AP, Chauhan MS et al (2012) Roscovitine treatment improves

synchronization of donor cell cycle in G0/G1 stage and in vitro development of handmade cloned buffalo (*Bubalus bubalis*) embryos. Cell Reprogram 14:146–154

15. Mohapatra SK, Sandhu A, Neerukattu VS, Singh KP, Selokar NL, Singla SK et al (2015) Buffalo embryos produced by hand-made cloning from oocytes selected using BCB staining have better developmental competence and quality, and are closer to embryos produced by in vitro fertilization in terms of their epigenetic status and gene expression pattern. Cell Reprogram 17:141–150

16. Kaith S, Saini M, Raja AK, Sahare AA, Jyotsana B, Madheshiya P et al (2015) Early cleavage of hand-made cloned buffalo (*Bubalus bubalis*) embryos is an indicator of their developmental competence and quality. Reprod Domest Anim 50:214–220

17. Panda SK, George A, Saha AP, Sharma R, Singh AK, Manik RS et al (2012) Effect of scriptaid, a histone deacetylase inhibitor, on the developmental competence of handmade cloned buffalo (*Bubalus bubalis*) embryos. Theriogenology 77:195–200

18. Saini M, Selokar NL, Agrawal H, Singla SK, Chauhan MS, Manik RS et al (2017) Treatment of donor cells and reconstructed embryos with a combination of trichostatin-A and 5-aza-2′-deoxycytidine improves the developmental competence and quality of buffalo embryos produced by handmade cloning and alters their epigenetic status and gene expression. Cell Reprogram 19:208–215

19. Saini M, Selokar NL, Agrawal H, Singla SK, Chauhan MS, Manik RS et al (2016) Treatment of buffalo (*Bubalus bubalis*) donor cells with trichostatin and 5-aza-2′-deoxycytidine alters their growth characteristic, gene expression and epigenetic status and improves the in-vitro developmental competence, quality and epigenetic status of cloned embryos. Reprod Fertil Dev 28:824–837

20. Selokar NL, Saini M, Agrawal H, Palta P, Chauhan MS, Manik R et al (2016) Buffalo (*Bubalus bubalis*) SCNT embryos produced from somatic cells isolated from frozen-thawed semen: effect of trichostatin A on the in vitro and in vivo developmental potential, quality and epigenetic status. Zygote 24:549–553

21. Selokar NL, Saini M, Agrawal H, Palta P, Chauhan MS, Manik R et al (2017) Valproic acid increases histone acetylation and alters gene expression in the donor cells but does not improve the in vitro developmental competence of buffalo (*Bubalus bubalis*) embryos produced by hand-made cloning. Cell Reprogram 19:10–18

22. Agrawal H, Selokar NL, Saini M, Singh MK, Chauhan MS, Palta P et al (2018) Epigenetic alteration of donor cells with histone deacetylase inhibitor m-carboxycinnamic acid bishydroxymide improves the in vitro developmental competence of buffalo (*Bubalus bubalis*) cloned embryos. Cell Reprogram 20:76–88

23. Agrawal H, Selokar NL, Saini M, Singh MK, Chauhan MS, Palta P et al (2018) m-carboxycinnamic acid bishydroxamide improves developmental competence, reduces apoptosis and alters epigenetic status and gene expression pattern in cloned buffalo (*Bubalus bubalis*) embryos. Reprod Domest Anim 53:986–996

24. Selokar NL, Saini M, Agrawal H, Palta P, Chauhan MS, Manik R et al (2015) Downregulation of DNA methyltransferase 1 in zona-free cloned buffalo (*Bubalus bubalis*) embryos by small interefering RNA improves in vitro development but does not alter DNA methylation level. Cell Reprogram 17:89–94

25. Rashmi, Sah S, Shyam S, Singh MK, Palta P (2019) Treatment of buffalo (*Bubalus bubalis*) SCNT embryos with microRNA-21 mimic improves their quality and alters gene expression but does not affect their developmental competence. Theriogenology 126:8–16

26. Singh S, Shyam S, Sah S, Singh MK, Palta P (2019) Treatment of buffalo (*Bubalus bubalis*) somatic cell nuclear transfer embryos with microRNA-29b mimic improves their quality, reduces DNA methylation, and changes gene expression without affecting their developmental competence. Cell Reprogram 21:210–219

27. Sah S, Sharma AK, Singla SK, Singh MK, Chauhan MS, Manik RS et al (2020) Effects of treatment with a microRNA mimic or inhibitor on the developmental competence, quality, epigenetic status and gene expression of buffalo (*Bubalus bubalis*) somatic cell nuclear transfer embryos. Reprod Fertil Dev 32:508–521

28. Saini M, Selokar NL, Palta P, Chauhan MS, Manik RS, Singla SK (2018) An update: reproductive handmade cloning of water buffalo (*Bubalus bubalis*). Anim Reprod Sci 197:1–9

29. Selokar NL, Saini M, Palta P, Chauhan MS, Manik RS, Singla SK (2018) Cloning of buffalo, a highly valued livestock species of south and Southeast Asia: any achievements? Cell Reprogram 20:89–98

30. Shyam S, Goel P, Kumar D, Malpotra S, Singh MK, Lathwal SS et al (2020) Effect of Dickkopf-1 and colony stimulating factor-2 on the developmental competence, quality, gene expression and live birth rate of buffalo

(*Bubalus bubalis*) embryos produced by handmade cloning. Theriogenology 157:254–262

31. Jyotsana B, Sahare AA, Raja AK, Singh KP, Singla SK, Chauhan MS et al (2015) Handmade cloned buffalo (*Bubalus bubalis*) embryos produced from somatic cells isolated from milk and ear skin differ in their developmental competence, epigenetic status, and gene expression. Cell Reprogram 17:393–403

32. Golla K, Selokar NL, Saini M, Chauhan MS, Manik RS, Palta P et al (2012) Production of nuclear transfer embryos by using somatic cells isolated from milk in buffalo (*Bubalus bubalis*). Reprod Domest Anim 47:842–848

33. Jyotsana B, Sahare AA, Raja AK, Singh KP, Nala N, Singla SK et al (2016) Use of peripheral blood for production of buffalo (*Bubalus bubalis*) embryos by handmade cloning. Theriogenology 86:1318–1324

34. Mohapatra SK, Sandhu A, Singh KP, Singla SK, Chauhan MS, Manik R et al (2015) Establishment of trophectoderm cell lines from buffalo

(*Bubalus bubalis*) embryos of different sources and examination of in vitro developmental competence, quality, epigenetic status and gene expression in cloned embryos derived from them. PLoS One 10:e0129235

35. George A, Sharma R, Singh KP, Panda SK, Singla SK, Palta P et al (2011) Production of cloned and transgenic embryos using buffalo (*Bubalus bubalis*) embryonic stem cell-like cells isolated from in vitro fertilized and cloned blastocysts. Cell Reprogram 13:263–272

36. Selokar NL, Saini M, Palta P, Chauhan MS, Manik R, Singla SK (2014) Hope for restoration of dead valuable bulls through cloning using donor somatic cells isolated from cryopreserved semen. PLoS One 9:e90755

37. Madheshiya PK, Sahare AA, Jyotsana B, Singh KP, Saini M, Raja AK et al (2015) Production of a cloned buffalo (*Bubalus bubalis*) calf from somatic cells isolated from urine. Cell Reprogram 17:160–169

Chapter 14

Bovid Interspecies Somatic Cell Nuclear Transfer with Ooplasm Transfer

L. Antonio González-Grajales and Gabriela F. Mastromonaco

Abstract

Interspecies somatic cell nuclear transfer (iSCNT) contributes to the preservation of endangered species, albeit nuclear–mitochondrial incompatibilities constrain its application. iSCNT, coupled with ooplasm transfer (iSCNT-OT), has the potential to overcome the challenges associated with species- and genus-specific differences in nuclear–mitochondrial communication. Our iSCNT-OT protocol combines the transfer of both bison (*Bison bison bison*) somatic cell and oocyte ooplasm by a two-step electrofusion into bovine (*Bos taurus*) enucleated oocytes. The procedures described herein could be used in further studies to determine the effects of crosstalk between nuclear and ooplasmic components in embryos carrying genomes from different species.

Key words Cloning, Cross-species, Cybrid, Heteroplasmy, Interspecies, Micromanipulation, Nuclear transplantation, Cytoplasm transfer

---

1 Introduction

Scarcity and technical challenges in acquiring viable gametes for use in the propagation of genetically valuable individuals of threatened species fueled an interest in alternative technologies for embryo production in genetic conservation programs. Interspecies somatic cell nuclear transfer (iSCNT) offers the possibility for cloning infertile, reproductively senescent or even deceased individuals, as well as reproductively healthy ones, thereby increasing contributions to the gene pool [1]. Studies have shown that the extent of evolutionary relatedness (i.e., taxonomic distance) between the species chosen for donating donor somatic cells and recipient oocytes in the context of iSCNT has significant effects on embryo developmental potential, which has been linked to mitochondrial and nucleus compatibility [2].

Marcelo Tigre Moura (ed.), *Somatic Cell Nuclear Transfer Technology*, Methods in Molecular Biology, vol. 2647,
https://doi.org/10.1007/978-1-0716-3064-8_14,
© The Author(s), under exclusive license to Springer Science+Business Media, LLC, part of Springer Nature 2023

Ooplasm transfer (OT) following iSCNT (iSCNT-OT) has been attempted to enhance nuclear–mitochondrial communication in the reconstructed embryos, but with contradictory results to date. While some studies reported significant benefits to transferring conspecific ooplasm (oocyte cytoplasm) along with the donor cell [3], others reported no improvements in embryonic development after embryonic genome activation [4–6]. Further studies are needed to better understand the implications of adding conspecific ooplasm or mitochondria to the reconstructed iSCNT embryo, including defining the components present in the transferred ooplasm and injecting isolated organelles (e.g., mitochondria) independently of other factors to evaluate specific outcomes. Meanwhile, the challenges associated with iSCNT-OT must be overcome. Limited availability and access to specimens of rare breeds and endangered wildlife species restrict the availability of oocytes and recipient females for iSCNT. Furthermore, the technical demands involved with increased handling and micromanipulation steps may put added stress on iSCNT embryos. This chapter describes our method to supplement ooplasm during reconstruction of iSCNT embryos (iSCNT-OT) to investigate the effects of ooplasmic factors during bison (*Bison bison bison*) embryonic development and may be adapted to other species.

2 Materials

2.1 Equipment

1. Stereomicroscope.
2. Inverted microscope.
3. Micromanipulators.
4. Electro Square Porator ECM 830 (BTX, Harvard Apparatus, Holliston, MA, USA).
5. CO_2 incubator.
6. Enucleation pipette (inner diameter of 10 μm), injection pipette (inner diameter of 15 μm), and holding pipette (inner diameter of 120 μm).
7. Fusion chamber (0.5 mm, Harvard Apparatus, Holliston, MA, USA).

2.2 Media and Solutions

1. 0.9% sodium chloride solution (Baxter Corporation, ON, Canada).
2. Oocyte collection medium: 9.8 g Ham's F-10 powder, 10 mL of 1 M HEPES buffer stock, 20 mL of 2% steer serum, 5000 IU/mL penicillin and 5000 μg/mL streptomycin, 2000 IU/L heparin, and 1.2 g $NaHCO_3$ in 1 L ultra-pure water (*see* **Note 1**).

3. IVM medium: TCM-199 (Fisher Scientific, Mississauga, ON, Canada), 25 mM HEPES, and 2% steer serum (Cansera). For IVM + H medium, add 2.0 μg/mL FSH, 14 IU/mL hCG, and 1.0 μg/mL estradiol to the IVM medium (*see* **Note 2**).

4. HEPES-TALP: 100 mL HEPES-TALP salt stock, 0.011 g sodium pyruvate, 0.21 g $NaHCO_3$, 0.31 mL sodium lactate syrup, 5 mg/mL gentamicin, 0.1 mL HEPES buffer stock, and 0.632 g BSA (*see* **Note 3**).

5. Hyaluronidase solution: Dissolve 1 mg/mL in HEPES-TALP, aliquot in 15 mL centrifuge tubes (2 mL/tube), and store at − 20 °C for 12 months.

6. Mannitol solution (0.28 M): Add 5.1 g D-mannitol, 100 mL ultrapure water, 50 μL of 1 M HEPES buffer stock, and 3.33 μL of 30% BSA solution (*see* **Note 4**).

7. Calcium chloride (10 mM): Dissolve 0.0147 g calcium chloride-$2H_2O$ in 10 mL ultrapure water. Sterile filter (0.22 μm) and store at 4 °C up to 6 months.

8. Magnesium chloride (10 mM): Dissolve 0.0203 g magnesium chloride-$6H_2O$ in 10 mL ultrapure water. Sterile filter (0.22 μm) and store at 4 °C up to 6 months.

9. Hi BSA (30 mg/mL): Dissolve 1 mL of 30% BSA solution in 9 mL HEPES-TALP. Mix well and store at 4 °C up to 4 weeks.

10. Hoechst 33342 (1 mg/mL): Prepare the 10 mg/mL primary stock by adding 25 mg Hoechst 33342 into 2.5 mL ultrapure water. Mix well to dissolve. Prepare the 1 mg/mL working stock by adding 100 μL of primary stock in 900 μL ultrapure water. Mix well and make 5 μL aliquots in 0.5 mL microtubes and store at −20 °C.

11. Cytochalasin B (1 mg/mL): Add 1 mL DMSO to a vial containing 1 mg cytochalasin B. Mix well to dissolve. Make 20 μL aliquots in 0.5 mL microtubes. To prepare the micromanipulation droplets, add 5 μL cytochalasin B to 1 mL HEPES-TALP (final concentration of 5 μg/mL).

12. Ionomycin (500 μM): Add 2.677 mL of DMSO to a vial containing 1 mg ionomycin. Mix well to dissolve. Make 50 μL aliquots in 0.5 mL microtubes. Store at –20 °C.

13. 6-Dimethylaminopurine (6-DMAP; 32.4 μg/mL): Add 50 mL DMSO to 0.8160 g 6-DMAP. Mix well to dissolve. Make 20 μL aliquots in 1.5 mL microtubes. Store at –20 °C.

14. Synthetic oviductal fluid (SOF): Mix 10 mL SOF medium (Chemicon-Millipore, Billerica, MA, USA) and add 50 μL sodium pyruvate, 200 μL of 100× nonessential amino acids solution, 100 μL of 50× essential amino acids solution, 5 μL gentamicin, 560 μL of 15% essentially fatty acid-free BSA (dissolved in SOF), and 200 μL steer serum in a 15 mL centrifuge tube (*see* **Note 5**).

262   L. Antonio González-Grajales and Gabriela F. Mastromonaco

3   Methods

The methods described here focus on the details for performing iSCNT+OT protocol. Further instructions on SCNT, in general, including preparation of somatic cells, can be found in previous publications [4, 7]. All media and solutions are used warm at 38.5 °C for 2 h unless otherwise specified.

3.1 Oocyte Collection and IVM

1. Collect domestic cattle (*Bos taurus*) and plains bison (*Bison bison bison*) ovaries from slaughterhouses and transport them in a sealed container (e.g., thermos) in 0.9% sodium chloride (i.e., saline solution) at 37 °C within a maximum of 3 h after slaughter.

2. Rinse ovaries with sterile saline solution in a sieve to remove blood and other debris. Transfer ovaries to a beaker of clean, warm saline, and keep at 37 °C until ready to process.

3. Collect cumulus-oocyte complexes (COCs) by follicular aspiration using a vacuum pump or hand-held syringe with 18-gauge needle into a 50 mL tube containing 5–10 mL of oocyte collection medium (*see* **Note 1**).

4. Allow the cellular content to settle to the bottom of the 50 mL tube, remove the supernatant, and resuspend the settled cells in 10 mL of oocyte collection medium. Transfer the cellular solution into a 100 mm Petri dish.

5. Search for the COCs using a stereomicroscope and select COCs with optimal morphological characteristics for IVM according to de Loos et al. [8].

6. Wash COCs in a four-well dish containing 0.5 mL per well twice in IVM medium and once in IVM + H medium (*see* **Note 2**).

7. After the series of washes, transfer COCs into 50 μL droplets of equilibrated IVM + H medium covered by 3.5 mL of conditioned silicone oil in a 35 mm Petri dish (maximum of 10 oocytes/droplet).

8. Place dishes in an atmosphere of 5% CO_2 in air, saturated humidity at 38.5 °C for 18 h.

3.2 Oocyte Denuding and Staining

1. At approximately 17.5 h of IVM, strip oocytes by manual pipetting in 500 μL droplets of hyaluronidase solution for 3 min at 37 °C.

2. Wash denuded oocytes 2× in a four-well dish containing 0.5 mL HEPES-TALP per well.

3. Select mature oocytes [Metaphase II (MII)] with visible extruded polar body (PB) and a homogeneously granulated cytoplasm under a stereomicroscope.

Ooplasm Transfer in Interspecies SCNT 263

4. Transfer MII oocytes back into new droplets containing IVM medium. Repeat the above steps until all cattle and bison oocytes are denuded.

5. Prior to micromanipulation, stain denuded oocytes with 5.0 µg/mL Hoechst 33342 and incubate in an atmosphere of 5% CO_2 in air, saturated humidity at 38.5 °C for 3 min. Wash the oocytes once in a four-well dish containing 0.5 mL HEPES-TALP per well.

3.3 Somatic Cell Nuclear Transfer with Ooplasm Transfer

1. Perform all micromanipulations, including enucleation and reconstruction (transfer of somatic cells and ooplasm), on an inverted microscope with attached micromanipulators.

3.3.1 Oocyte Enucleation

2. In a 60 mm Petri dish, prepare 40 µL droplets of HEPES-TALP supplemented with 5.0 µg/mL cytochalasin B for all micromanipulations and cover with conditioned silicone oil.

3. Restrain an oocyte using the holding pipette and place the PB at 4–5 o'clock.

4. Pierce the *zona pellucida* with the enucleation pipette and slowly aspirate the PB and a small amount of ooplasm.

5. Quickly expose the oocyte to UV light to confirm successful enucleation (PB and metaphase plate should be seen fluorescing in the pipette and not within the oocyte).

6. Place the enucleated oocytes back in new droplets containing IVM medium. Repeat the above steps until all cattle and bison oocytes are enucleated.

3.3.2 Oocyte Reconstruction and Ooplasm Transfer

1. Dissociate bison somatic cells from a 35 mm Petri dish using 1 mL of trypsin/EDTA solution (*see* **Note 6**) followed by washing with 4 mL HEPES-TALP using centrifugation at $120 \times g$ for 5 min.

2. Resuspend the pellet in 2 mL HEPES-TALP.

3. Place between 100 and 200 cells (~5 µL) from the cell suspension into a separate 40 µL micromanipulation droplet (HEPES-TALP droplets covered by conditioned silicone oil).

4. Add a maximum of 10 enucleated cattle oocytes and 1–2 enucleated bison oocytes in a 40 µL micromanipulation droplet of HEPES-TALP + cytochalasin B covered by conditioned silicone oil. Keep oocytes from both species as far apart as possible. For instance, within the same micromanipulation droplet, keep cattle, bison, and discarded oocytes at 9, 3, and 6 o'clock positions, respectively.

5. Move the injection pipette to the droplet with somatic cells and aspirate five good-quality (smooth membrane, small-medium sized sphere) bison fibroblasts into the end of the pipette. Make sure the cells are close to each other within the pipette to

localize them easily. Thereafter, move the injection pipette into the micromanipulation droplet containing the ooplasm-donor oocytes (bison) and enucleated recipient oocytes (cattle) mentioned above.

6. Restrain a bison oocyte with the holding pipette and roll it around with the injection pipette to locate the previous slit (from enucleation) within the *zona pellucida*. If the site is not easily detected, proceed to slowly introduce the pipette at a new location through the *zona pellucida*. When some ooplasm content extrudes out of the oocyte from the piercing previously performed during enucleation, it is recommended to stop the procedure and relocate the transfer pipette near the site of extrusion to facilitate entry of the pipette. Aspirate 10–15% bison ooplasm in the same pipette containing the bison somatic cells.

7. Release the bison oocyte at the 3 o'clock position and proceed to pick up a cattle oocyte at the 9 o'clock position with the holding pipette and roll it around to locate the previous piercing site within the *zona pellucida* whenever possible.

8. Slowly expel the bison ooplasm followed immediately by one bison somatic cell into the perivitelline space. Avoid expelling too much medium along with the cell. Place the reconstructed oocyte at the 12 o'clock position (*see* **Note 7**).

9. Pick up the bison oocyte (donor ooplasm) again with the holding pipette and repeat **steps 6–8**. Continue until all cattle oocytes have a bison cell and bison ooplasm transferred into them.

10. After removal of more than 70% of ooplasm from the bison oocyte, select a new one as ooplasm donor. Place oocytes transferred with a somatic cell and ooplasm back into the incubator in new droplets containing IVM medium.

3.4 Electrofusion of Reconstructed Embryos

1. Set up two stereomicroscopes side by side for the fusion procedure. One is used for attaching the fusion chamber and the other to manipulate the reconstructed embryos before electrofusion.

2. Tape the chamber onto the stereomicroscope stage and attach the positive electrode to the top rod and the negative electrode to the bottom rod.

3. Add 100 μL $CaCl_2$ solution and 100 μL $MgCl_2$ solution to the tube containing mannitol solution and mix well.

4. Add 1.0 mL of mannitol solution in the form of a large bubble to the center of the fusion chamber avoiding formation of small air bubbles.

5. Set the parameters required to fuse the transferred ooplasm to the reconstructed embryo. We recommend using low voltage and short exposure time for the first DC pulse at 1.5 kV/cm for 20 μs in a 0.5 mm gap electrofusion chamber.

6. On the free stereomicroscope, transfer five reconstructed embryos to a 35 mm Petri dish containing 3 mL HEPES-TALP, and immediately thereafter to a 35 mm Petri dish containing 3 mL mannitol solution at room temperature. Then, move the couplets to the fusion chamber on the outside of the rods (on the bottom side).

7. Start the fusion of each couplet by moving one reconstructed embryo at a time between the rods and lining it up so that the transferred ooplasm is at 12 o'clock and in the same plane as the oocyte.

8. Press the fusion button using the foot pedal. Rotate the reconstructed embryo to ensure proper positioning within the rods. Finally, move the fused couplet to the outside of the rods (on the top side).

9. Repeat **steps 7** and **8** until fusion of all five couplets is done, and then transfer them back to the dish of HEPES-TALP. Once the batch of oocytes is fused, place them back in a new droplet of IVM medium covered by conditioned silicone oil for 30 min in the incubator (*see* **Note 8**).

10. Determine fusion rates of the transferred ooplasm and proceed to apply a second DC pulse to fuse the somatic cell into the ooplasm this time. To accomplish this step, repeat the steps explained above (**steps 1–9**).

11. Set values for the second pulse at 2.1 kV/cm for 32 μs. After each pulse, reconstructed oocytes are subsequently washed in 3.0 mL HEPES-TALP and placed in IVM medium in an atmosphere of 5% CO_2 in air, saturated humidity at 38.5 °C. Assess fusion rates for the somatic cell 30 min after completion of the second pulse. The micromanipulation and fusion steps conducted on cattle and bison oocytes are described below (Fig. 1).

3.5 Chemical Activation and In Vitro Culture (IVC)

1. Initiate oocyte activation 24 h post-IVM.

2. Transfer only fused couplets (somatic cell + ooplasm) to a 35 mm Petri dish containing HEPES-TALP.

3. Make a four-well dish as follows: Add 10 μL ionomycin to 1.0 mL of HEPES TALP for a final concentration of 5.0 μM to well #1, 1.0 mL HEPES TALP to well #2, 1.0 mL Hi BSA solution to well #3, and 20 μL of 2.0 mM 6-dimethylaminopurine (6-DMAP) diluted in 980 μL synthetic oviductal fluid (SOF) to well #4 keeping at 38.5 °C in 5% CO_2 with saturated humidity.

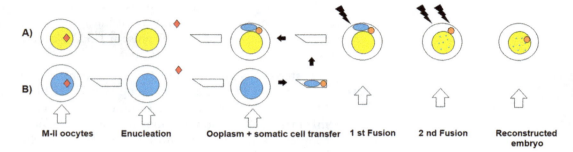

Fig. 1 Schematic representation of interspecies somatic cell nuclear transfer and ooplasm transfer (iSCNT-OT). (**a**) Cattle oocytes used for enucleation and as recipients of bison somatic cells and ooplasm. (**b**) Bison oocytes used as ooplasm donors. Red diamond and small orange circles represent the oocyte metaphase II spindle and bison somatic cell, respectively. Blue oval-shaped structures represent the ooplasm removed from a bison oocyte for iSCNT-OT. Reconstructed oocytes are subject to in vitro culture

4. Transfer the fused couplets to well #1 (ionomycin solution) and incubate for 5 min at 37 °C.

5. Thereafter, transfer the fused couplets to well #2 (HEPES-TALP) and wash stringently.

6. Transfer the reconstructed embryos to well #3 (Hi BSA solution) and incubate at 37 °C for 5 min.

7. Wash the reconstructed embryos in well #4 (6-DMAP) and culture embryos in SOF + 6-DMAP in a new well using the same concentration described above for four additional hours at 38.5 °C in 5% CO_2 with saturated humidity.

8. After 4 h, wash the reconstructed embryos twice in SOF medium and culture them at 38.5 °C in 30 µL droplets of SOF covered by conditioned silicone oil previously equilibrated in a humidified atmosphere of 5% CO_2, 5% O_2, and 90% N_2 for up to 10 days.

4 Notes

1. Stir well to dissolve, sterile filter (0.22 µm), and store at 4 °C for up to 3 weeks. We use steer serum from Cansera (Rexdale, ON, Canada). HEPES buffer stock (1 M): 23.8 g HEPES (acid form) in 100 mL ultrapure water. Stir well to dissolve, sterile filter (0.22 µm), and store at 4 °C for up to 6 months.

2. Make up fresh for every experiment. We use FSH (National Hormone & Peptide Program, Torrance, CA, USA), human chorionic gonadotropin (Chorulon, Intervet Canada, Kirkland, QC, Canada), estradiol (Sigma-Aldrich, Oakville, ON, Canada), and conditioned silicone oil (Paisley Products, Scarborough, ON, Canada).

3. HEPES-TALP salts stock (1 L): Add 28.5 g NaCl, 1.18 g KCl, 0.24 g $NaH_2PO_4.H_2O$, 1.47 g $CaCl_2$. $2H_2O$, 1.01 $MgCl_2$. $6H_2O$, and ultrapure water. Stir well to dissolve and sterile filter (0.22 μm). Store at 4 ° C for up to 6 months. Dissolve well by stirring in 75% of final volume for approximately 1 h. Add pellets of NaOH while stirring and adjust pH to 7.3–7.4. Make up to final volume. Check osmolarity: 1600–1800 mOsm. Check pH again. Sterile filter and store at 4 °C for up to 6 months. 60% syrup (density 1.32 g/mL). HEPES-TALP: Add all components except BSA, which should be added last. Stir well to dissolve. Adjust pH to 7.4. Check osmolarity = 290–305 mOsm. Add BSA once pH and Osm have been adjusted. Stir well to dissolve. Sterile filter (0.22 μm) and store at 4 °C for up to 2 weeks.

4. Dissolve the mannitol in water. Add the other components in order. Adjust pH to 7.2–7.4 and osmolarity to 300 Osm. Sterile filter (0.22 μm) and aliquot in 15 mL tubes. Store at 4 °C up to 1 month.

5. Add BSA last and dissolve all components. Sterile filter (0.22 μm) and make up fresh.

6. Trypsin/EDTA solution: Dissolve 50 mL Hanks balanced salt solution 10× (w/o Ca^{++} or Mg^{++}) in 450 mL Milli-Q water. Continue adding the following: 0.6 g $NaHCO_3$, 100 μL phenol red (0.5% solution), 1.25 g trypsin (from porcine pancreas 1:250), 0.224 g EDTA-Na salt, and 5 mL HEPES (1 M solution). Mix all components and adjust pH to 7.4. Filter sterilize and aliquot in 15 mL centrifuge tubes (10 mL/tube). Store at −20 °C up to 12 months.

7. Discontinue the procedure if the bovine or bison ooplasms lyse.

8. If the fusion parameters are working properly, expect more than 90% fusion rates after this time.

Acknowledgments

The development of this protocol was supported by the Natural Sciences and Engineering Research Council of Canada Discovery Grant (GFM).

References

1. Mastromonaco GF, King WA (2007) Cloning in companion animal, non-domestic and endangered species: can the technology become a practical reality? Reprod Fertil Dev 19:748–761

2. Loi P, Modlinski JA, Ptak G (2011) Interspecies somatic cell nuclear transfer: a salvage tool seeking first aid. Theriognology 76:217–228

3. Yao L, Wang P, Liu J, Chen J, Tang H, Sha H (2014) Ooplast transfer of triploid pronucleus zygote improve reconstructed human-goat embryonic development. Int J Clin Exp Med 7:3678–3686

4. Gonzalez-Grajales LA, Favetta LA, King WA, Mastromonaco GF (2016) Lack of effects of ooplasm transfer on early development of interspecies somatic cell nuclear transfer bison embryos. BMC Dev Biol 16:36

5. Sansinena MJ, Lynn J, Bondioli KR, Denniston RS, Godke RA (2010) Ooplasm transfer and interspecies somatic cell nuclear transfer: heteroplasmy, pattern of mitochondrial migration and effect on embryo development. Zygote 19:147–156

6. Lee WJ, Lee JH, Jean RH, Jang SJ, Lee SC, Park JS et al (2017) Supplement of autologous ooplasm into porcine somatic cell nuclear transfer embryos does not alter embryo development. Reprod Dom Anim 52:437–445

7. Mastromonaco GF, Favetta LA, Smith LC, Filion F, King WA (2007) The influence of nuclear content on developmental competence of gaur x cattle hybrid in vitro fertilized and somatic cell nuclear transfer embryos. Biol Reprod 76:514–523

8. de Loos F, van Vliet C, van Maurik P, Kruip TAM (1989) Morphology of immature bovine oocytes. Gamete Res 24:197–220

Chapter 15

Horse Somatic Cell Nuclear Transfer Using Zona Pellucida-Enclosed and Zona-Free Oocytes

Daniel Salamone and Marc Maserati

Abstract

Horse cloning by somatic cell nuclear transfer (SCNT) is an attractive scientific and commercial endeavor. Moreover, SCNT allows generating genetically identical animals from elite, aged, castrated, or deceased equine donors. Several variations in the horse SCNT method have been described, which may be useful for specific applications. This chapter describes a detailed protocol for horse cloning, thus including SCNT protocols using zona pellucida (ZP)-enclosed or ZP-free oocytes for enucleation. These SCNT protocols are under routine use for commercial equine cloning.

Key words *Equus caballus*, Somatic cell nuclear transplantation, Cloning, Micromanipulation

---

1 Introduction

Cloning by somatic cell nuclear transfer (SCNT) was developed in equids for propagating elite horses due to the low efficiency of other techniques such as superovulation and conventional in vitro fertilization (IVF). Further, SCNT cloning remains as the sole approach to rescue genetics of orchiectomized or recently deceased animals. Since initial reports of equine cloning [1, 2], numerous authors have produced cloned horses [3–8]. Briefly, SCNT involves the enucleation of a mature oocyte followed by its reconstruction with a donor cell by fusion or injection. This exposure of the donor cell to the oocyte cytoplasm reprograms the somatic epigenetic state to an undifferentiated state compatible with full-term development [1–8].

The production of cloned offspring by SCNT has been successfully attained in most of the species tested so far. Nonetheless, horses are one of the few species that SCNT continues to be of commercial interest. Failures in superovulation and conventional

Marcelo Tigre Moura (ed.), *Somatic Cell Nuclear Transfer Technology*, Methods in Molecular Biology, vol. 2647,
https://doi.org/10.1007/978-1-0716-3064-8_15,
© The Author(s), under exclusive license to Springer Science+Business Media, LLC, part of Springer Nature 2023

270 Daniel Salamone and Marc Maserati

in vitro fertilization (IVF) have been also reasons for the use of this technology in horses. Besides, the high cost of valuable individuals has justified the use of this expensive technology.

In horses, three different techniques have been used for cloning. The first to be used was enucleation with zona pellucida (ZP) followed by transfer of the donor cell into the perivitelline space, followed by cell fusion using an electrical pulse [1, 2]. One of the problems with this technique is fusion failure by the lack of good contact of the donor cell with the oolemma. The second way of performing SCNT in horses is by introducing the somatic nucleus in the ZP-enclosed oocyte by microinjection, most frequently using a piezoelectric equipment to facilitate the process [3–6]. The mechanical damage caused to the oocyte after injecting the donor cell is greater and can lead to oocyte lysis during microinjection. The great advantage of this protocol is that cells with damaged plasma membrane can be used in SCNT. The third way of performing SCNT is by applying the ZP-free procedure, which allows working with a blunt enucleation pipette and a single micromanipulation arm attached to the inverted microscope, followed by making cell–cell contact with phytohemagglutinin under a stereoscopic microscope, and then conducting cell fusion as in first protocol [7, 8]. The fusion rates increase significantly, albeit it requires a microwell system to keep blastomeres together during embryo culture. This methodology improves fusion rates, possibly due to a greater cell–cell by phytohemagglutinin. Other advantage of this protocol is the easier embryo aggregation to increase blastomere numbers while making epigenetic compensation since it uses more than one SCNT embryo.

Another aspect to take into account is the degree of synchrony of the mitotic stage of the donor cell with the meiotic stage of the oocyte. Numerous publications showed that it is essential that oocytes have high levels of the maturation-promoting factor. The donor cells can be either G1 or G0, but they could be in G2 or metaphase. In the latter two cases, it is important to allow releasing the extra chromosome set.

In this chapter, we will describe a detailed protocol for horse cloning using both zona pellucida (ZP)-enclosed and ZP-free oocytes.

2 Materials

2.1 Equipment

1. Two humidified CO_2 incubators set at 38.5 °C (Fig. 1). One incubator with high oxygen tension (20%) and another with low oxygen tension (5%), while both incubators with 5% CO_2. Eventually, the last one can be replaced by a modular incubator chamber (MIC-101, Billups-Rothenberg Inc.) or common

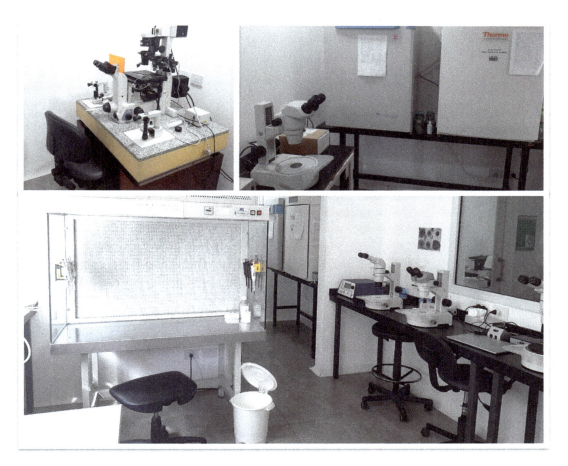

Fig. 1 Laboratory facilities for horse SCNT. Micromanipulation room with inverted microscope on anti-vibratory table. Wear a jacked incubator and stereomicroscope. Tissue culture hood with electrofusion machine, stereomicroscope, and thermic plate

household plastic storage container (i.e., a commercially purchased premade chamber), roughly 35.1 × 25.4 × 18.6 cm, with adaptor and tubing with a clamp to introduce a gas mixture and purge the air (5% O_2, 5% CO_2, and 90% N_2).

2. Tissue culture hood.
3. Micropipette puller (Sutter, P-97 Flaming).
4. Microforge (Narishige).
5. Microgrinder (Narishige).
6. Stereomicroscope.
7. Heating plate.
8. Centrifuge.
9. Vortex mixer.

272   Daniel Salamone and Marc Maserati

10. Inverted microscope (Nikon, Eclipse TE-300 microscope) with fluorescence system (Fig. 1).

11. Micromanipulator set (Narishige).

12. Electrofusion machine (BTX Electro-Cell Manipulator 830).

2.2 Tools and Consumables

1. Holding pipette for ZP-free oocytes: Prepare the holding pipette by pulling a glass capillary on a micropipette puller. Make a blunt tip with an approximate outer diameter of 120 μm with the microforge. Close the tip slightly by exposing it to a flame.

2. Holding pipette for ZP enclosed oocytes: All the steps were similar to previous pipette until pipette tip closing (**step 1**). Here, use the microforge to reduce the pipette opening to 25–30 μm.

3. Injection pipette for ZP-free oocytes: Pull a glass capillary with a micropipette puller. Make a blunt tip with an approximate outer diameter of 25–30 μm using the microforge. Prepare 2–3 pipettes of each type before SCNT session.

4. Injection pipette for ZP enclosed oocytes: All the steps are similar to the previous pipette (**step 3**). However, after the beveled tip (*see* **Note 1**) of 45° is made with a microgrinder. Prepare the beveled pipette tip using the microgrinder. Wash the pipette extensively with alcohol 70%. Make a spike in the pipette tip using the microforge.

5. Microslides for electrofusion chambers: Use slides with two stainless steel tubes mounted on a glass slide at 10 mm gaps to provide homogenous fields (BTX™, Cat# 15447250).

2.3 Media and Solutions

Unless otherwise stated, all chemicals were obtained from Sigma-Aldrich. Ready-to-use media after the addition of all the supplements are subject to sterile filtering (0.22 μm; Corning; Cat #431219 – cellulose acetate membrane/surfactant-free) and stored at 4 °C for a week. Store stock solution at −20 °C indefinitely.

1. Media or solution used for handling and transporting the skin biopsy.

2. Cell culture medium (CCM): 1:1 mixture of Dulbecco's modified Eagle's medium and Ham's F12 media (DMEM/F12; Thermo Fisher Scientific, 11320-033) supplemented with 10% (v/v) fetal bovine serum (FBS; GE Healthcare Life Sciences, HyClone), 1.0 μL/mL insulin–transferrin–selenium (ITS; Thermo Fisher Scientific, Cat# 51300-044), and 1.0% (v/v) penicillin and streptomycin solution (ATB; Cat# P4458).

3. Cell freezing solution (CFS): DMEM/F12 supplemented with 10% (v/v) FBS and 10% (v/v) dimethyl sulfoxide (DMSO).

4. Aspiration medium (APM): HEPES-buffered Tyrode's medium containing albumin, lactate, and pyruvate (H-TALP). Use this media for all manipulations outside the incubator. Add 6.62 g NaCl, 0.239 g KCl, 0.294 g CaCl$_2$-2H$_2$O (Sigma; Cat# C-7902), 0.102 g MgCl$_2$-6H$_2$O, 0.168 g NaHCO$_3$, and 2.38 g HEPES to 981.86 mL sterile embryo tested water (Sigma-Aldrich; Cat# W1503). Swirl and add 10 mL antibiotic–antimycotic (ATB; Thermo Fisher; Cat# 15240-096), 1.0 mL phenol red (Sigma-Aldrich; Cat# P0290), 1.44 mL sodium lactate (Sigma-Aldrich; Cat# L1375), 1.0 mL of 100 mM sodium pyruvate stock (Sigma-Aldrich; Cat# P2256), and 3.0 g bovine serum albumin (Sigma-Aldrich; Cat# A7906). Adjust the pH to 7.2–7.3 and the osmolality to 275 ± 10 mOsm. Aliquot in 50 mL tubes and store at 4 °C for a month.

5. In vitro maturation medium (IVMM): Bicarbonate-buffered Tissue Culture Medium 199 (TCM-199; Thermo Fisher Scientific, Cat# 11150–059) supplemented with 10% (v/v) FBS, 1.0 μL/mL ITS, 1.0 mM sodium pyruvate, 100 mM cysteamine, 10 μg/mL follicle-stimulating hormone (FSH; Bioniche, Folltropin, NIH-FSH-P1, Ontario, Canada), and 1.0% (v/v) ATB.

6. Hyaluronidase solution: Dissolve 5.0 mg hyaluronidase (Cat# H3506) in 5.0 mL H-TALP. Make ready-to-use 50 μL aliquots.

7. Pronase solution: Dissolve 1.5 mg pronase (Sigma-Aldrich; Cat# P8811) in 1.0 mL H-TALP. Make 5 μL aliquots and add 100 μL H-TALP before use.

8. Cytochalasin B (Sigma-Aldrich; Cat# C6762): Prepare 0.5 μg/μL stock in DMSO. Make 5 μL aliquots and store at −20 °C. Dilute 2.5 μL stock/mL in culture medium before use.

9. Hoechst 33342 (Sigma-Aldrich; Cat# B2261): Dissolve 1.0 mg/mL in H-TALP. Make 5 μL aliquots and 1:100 dilutions in culture medium.

10. Demecolcine (Sigma-Aldrich; Cat# D1925): Demecolcine (D1925) 4 μM in SOF for 20 min to induce protrusion of the chromosome plate, for facilitating the enucleation.

11. Cell fusion media (CFM): 0.3 M mannitol (Sigma-Aldrich; Cat# M9647), 0.1 mM MgSO$_4$, 0.05 mM CaCl$_2$, and 1.0 mg/mL polyvinyl alcohol (PVA) in ultrapure water. Store at −20 °C.

12. Ionomycin (Thermo Fisher Scientific; Cat# I24222): Prepare 5.0 mM stock by dissolving 1.0 mg ionomycin in 267.6 μL DMSO. Make 5.0 μL aliquots and store at −20 °C. Prepare working aliquots by diluting 1.74 μL stock per mL H-TALP.

274   Daniel Salamone and Marc Maserati

13. 6-Dimethylaminopurine (6-DMAP; Sigma-Aldrich; Cat# D2629): Dissolve 1.0 mg 6-DMAP in 30 mL PBS (Gibco; Cat# 21600–051). Place the 6-DMAP solution in boiling water to facilitate dissolving. Store at −20 °C. Make 2.0 mM working solution by diluting 0.5 μL stock per 100 μL culture medium.

14. Cycloheximide solution: Prepare stock with 1.0 mg/mL in PBS. Make a 10 μg/mL working solution in culture medium.

15. Oocyte activation medium (OAM): DMEM/F12 supplemented with 5.0% (v/v) FBS, 2.0 mM 6-DMAP, 10 μg/mL cycloheximide, and 1.0% (v/v) ATB.

16. Embryo culture medium (ECM): 50% (v/v) DMEM/F12 and 50% Global Total (Life Global, Cat# LGGT-030), and 1.0% ATB.

17. Embryo feeding medium (EFM): DMEM/F-12 supplemented with 10% (v/v) FBS and 1.0% ATB.

2.4 Micromanipulation Setup

Prepare micromanipulation pipettes in advance (*see* Subheading 2.2). Connect pipettes to the micromanipulation system with holding and injection pipettes placed on opposite sides within the microscope field. Aspirate and eject media using both holding and injection pipettes to ensure adequate control of micromanipulation pipettes (*see* **Note 2**).

2.4.1 Dishes for Handling, Culture, and Micromanipulation

1. Micromanipulation dish: This procedure is typically performed by using the lid of a 100 mm Petri dish with 100 μL H-TALP droplets under mineral oil.

2. Before micromanipulation, oocytes are maintained for 15 min in a culture medium containing 1.0 μg/mL Hoechst 33342, 1.0 μg/mL cytochalasin B, and 4 μL in 100 μL demecolcine.

3. Cell fusion dish: BTX™ Microslides for Electrofusion Chambers, Fisher Scientific.

4. Oocyte activation dish: Activation can be in culture media, but in ZP-free SCNT should be in 5 μL droplets.

5. Embryo culture dish: Group of at least 10 embryos should be cultured together but in ZP-free SCNT every embryo should be located in small microwells to avoid dispersion. These are produced previously using a heated glass capillary slightly pressed to the bottom of a 35 mm diameter Petri dish. Microwells can be produced following the procedure described elsewhere [9] and are covered with 50–100 μL ECM droplets.

3 Methods

3.1 Preparation of Donor Somatic Cells

1. Restrain the animal chosen as cell donor for SCNT (*see* **Note 3**). Apply local sedation before biopsy collection.

2. Shave and clean with alcohol 70% of the surface area to be biopsied.

3. Collect with a skin biopsy using a punch between 5 and 8 mm of diameter from the neck or base of the tail of the donor animal.

4. Transfer the biopsy to a 15 mL tube a solution at room temperature or 4 °C and transport it to the laboratory.

5. Mince the biopsy into small fragments with a sharp blade (*see* **Note 4**).

6. Plate in small fragments tissue in 30 mm or smaller Petri dish covered with media but preventing them from floating. Culture with 5% CO_2, saturated humidity at 38.5 °C.

7. Replenish the CCM at 48 h intervals until reaching full (100%) confluence (*see* **Note 5**).

8. Passage (subculture) the confluent dish by removing CCM and adding 5 mL of 0.25% trypsin. Place the dish in a thermic plate controlling microscopically the detachment (usually takes 5–10 min).

9. Add 5–10 mL CCM to inactivate the trypsin. Transfer the cell suspension to a 15 mL tube and centrifuge at 400 g for 10 min.

10. Discard the supernatant and resuspend the cell pellet with 10 mL CCM. Passage the fibroblast culture in 4- to 6-day intervals.

11. Cryopreserve cells by resuspending the cell pellet (**step 10**) in 1–2 mL CFS. Transfer the cell suspension to cryovials. Place cryovials at −70 °C in a freezing container (Nalgene® Mr. Frosty) for 24 h and transfer to liquid nitrogen (−196 °C) for long-term storage.

12. Thaw and prepare the donor cell line a few weeks before its use for SCNT. Make sure to cryopreserve cell stocks as a backup.

3.2 Cell Cycle Synchronization of Donor Cells

There are several protocols for cell cycle synchronization in G0 or G1. For zona-free oocytes, we used (a) cell confluence and serum deprivation (G0) and (b) culture after cell confluence for 24 h (G1).

For zona enclosed enucleation is with a telophase II (TII) oocytes [10], a procedure using roscovitine (G2/M of the cell cycle synchronizer). Incubate fibroblast culture at less than 80% of confluence for 16–24 h 25–50 µM roscovitine diluted in CCM with 5% CO_2, saturated humidity at 38.5 °C for 24 h.

1. Remove the media for cell cycle synchronization and dissociate cells with 5 mL 0.25% trypsin for 5–10 min in the incubator. Add 5–10 mL CCM to inactivate the trypsin.

2. Transfer the cell suspension to a 15 mL tube and centrifuge at 400 g for 10 min.

3. Remove the supernatant and resuspend the cell pellet in 1–3 mL H-TALP for SCNT.

3.3 Oocyte Collection

1. Collect equine ovaries at slaughterhouses and transfer them to a recipient bottle with saline solution (0.9% NaCl) at 27–29 °C (see **Note 6**). Transport ovaries to the laboratory within 4–7 h after slaughter.

2. Wash ovaries in the laboratory and transfer to a Becker with saline solution in a water bath at 30 °C.

3. Aspirate all visible antral follicles by applying scraping and washing with an 18-gauge needle attached to a syringe with 10 mL APM.

4. Transfer the aspirated content to a 50 mL tube and let it settle for at least 10 min in the water bath at 35 °C.

5. Aspirate the pellet of the 50 mL tube and transfer to a 100 mm dish with 10 mL H-TALP.

6. Recover cumulus-oocyte complexes (COCs) under the stereomicroscope.

3.4 Oocyte In Vitro Maturation and Denuding

1. Select COCs with at least three compact layers of cumulus cells and a homogeneously granulated cytoplasm. Wash COCs three times in IVMM.

2. Transfer COCs to each 100 μL IVMM droplet under mineral oil and for IVM with 5% CO_2, saturated humidity at 38.5 °C for 24–26 h.

3. Denude COCs by brief exposure to 0.05% trypsin-EDTA and gentle pipetting in a solution of at 100 μL hyaluronidase solution for 2 min.

4. Wash denuded oocytes twice in H-TALP.

5. Select mature oocytes [i.e., metaphase II (MII) oocytes] for SCNT with homogeneously granulated cytoplasm and a visible polar body (PB) under a stereomicroscope (Fig. 2).

3.5 Zona Pellucida Removal

This step is strict to the ZP-free SCNT protocol.

1. Incubate MII oocytes in a 100 μL pronase solution droplet at 35–38 °C for 3–6 min. Monitor oocytes under a stereomicroscope until the ZP begins thinning. Collect oocytes as soon as the ZP disappears.

Horse Somatic Cell Nuclear Transfer 277

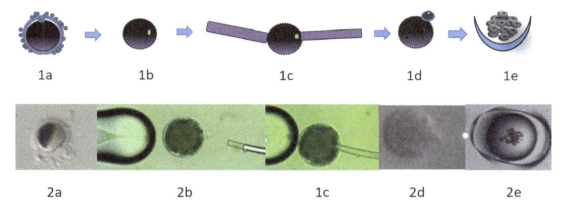

Fig. 2 Enucleation and reconstruction of zona pellucida (ZP)-free horse oocytes. (**a–c**) Schematic representation of oocyte enucleation and reconstruction using ZP-free oocytes. (**a–e**) Microscope or stereomicroscope view of micromanipulation using ZP-free oocytes

2. Wash ZP-free oocytes extensively in H-TALP.
3. Place ZP-free oocytes in maturation or culture medium until enucleation (*see* **Note 7**).

3.6 Somatic Cell Nuclear Transfer (SCNT)

Most horse cloning relies on SCNT protocols using ZP-enclosed oocytes [1, 2, 5, 6], albeit may display minor variations in experimental conditions (e.g., oocyte IVM, embryo culture).

3.6.1 Enucleation and Reconstruction of ZP-Enclosed Oocytes

1. Incubate 10–20 oocytes per CB droplet in the micromanipulation dish. Place 1–5 μL donor cell suspension in an H-TALP droplet.
2. Aspirate 1–5 small-donor cells into the injection pipette.
3. Grab an oocyte with the holding pipette. Make sure the pressure is sufficient to hold the oocyte by the ZP but allows its rotation with the injection pipette.
4. Rotate the oocyte with the injection pipette and position the PB in the three o'clock position in a clock's face (*see* **Note 8**).
5. Use the beveled tip of the inject pipette to penetrate the ZP. Apply a subtle aspiration with the injection pipette to remove the PB with 5–10% of adjacent cytoplasm.
6. Remove the injection pipette from the perivitelline space and expose the PB and aspirated cytoplasm to the ultraviolet light (UV) to ensure enucleation and discard the aspirated content (*see* **Note 9**).
7. Use the same ZP hole made during enucleation to introduce the injection pipette and place the donor cell into the perivitelline space. Push the ZP with the injection pipette to ensure the attachment of donor cells on enucleated oocytes, which is essential for cell fusion (*see* **Note 10**).

3.6.2 Enucleation and Reconstruction of ZP-Free Oocytes

ZP-free SCNT was developed initially for the handmade cloning protocol [9], in which oocyte enucleation relies on manual bisection with a sharp blade under a stereomicroscope. This method was adapted to equine cloning using micromanipulators and ZP-free TII oocytes [7, 8]. ZP-free embryos require a specific well-of-the-well (WOW) culture system, which keeps developing blastomeres in proximity and avoids embryo aggregation.

1. Subject MII oocytes to chemical activation for obtaining TII oocytes (*see* Subheading 3.8.1).

2. Place 10–20 oocytes per CB droplet of the micromanipulation dish for 20 min. Add 1–5 µL donor cell suspension in an H-TALP droplet.

3. Use the injection pipette to aspirate 5–7 large fibroblasts, which are at the G2/M stage of the cell cycle (*see* **Note 11**). Keep donor cells far from the injection pipette tip, such that they are not exposed to the UV light during enucleation.

4. Move pipettes to a droplet with activated oocytes. Place an oocyte adjacent to the holding pipette and apply the UV light.

5. Select activated oocytes with a cytoplasm protrusion, which indicates the location of the oocyte spindle.

6. Rotate the oocyte with the injection pipette and place the protrusion in the three o'clock position in a clock's face. Aspirate the protrusion with 5–10% of adjacent cytoplasm using the blunt injection pipette under exposure to the UV.

7. Stop the oocyte exposure to the UV. Discard the aspirated oocyte cytoplasm and its protrusion. Move one donor cell to the tip of the injection pipette and stick it to the enucleated oocyte (*see* **Note 12**).

3.7 Cell Fusion

Place a fusion microslide on a 100 mm dish and attach it with vacuum grease. Place the dish on the stereomicroscope. Attach fusion cables to the microslide and turn on the electrofusion machine. Cell fusion is done with two 1.2 kV/cm pulses [direct current (DC)], in which each pulse lasted for 30 µs and 0.1 s apart.

1. Place cell couplets in CFS for 2–3 min. Add 2.0 mL CFS at the center of the fusion microslide.

2. Wash cell couplets quickly in CFS and transfer to the microslide (up to four cell couplets).

3. Turn on the AC and align cell couplets on one electrode (cell couplets will stick to the electrode with AC on). Place the surface of contact between the donor cell and the enucleated oocyte in parallel to electrodes.

Horse Somatic Cell Nuclear Transfer   279

4. Apply the DC pulses and place cell couplets back to H-TALP droplets. Transfer cell couplets subject to fusion to individual 10 μL of culture media droplets and incubated with 5% CO_2, saturated humidity at 38.5 °C for 30 min.

5. Check for fused cell couplets 30–50 min after fusion (i.e., absence of the donor cell on the oocyte surface). Non-fused cell couplets are subject to the second round of cell fusion.

6. ZP-enclosed reconstructed oocytes are subject to chemical activation (*see* Subheading 3.8.1) and ZP-free undergo embryo culture (*see* Subheading 3.9).

3.8   Oocyte Activation

3.8.1   Chemical Activation for SCNT Using ZP-Free Oocytes

1. Incubate reconstructed oocytes in 10 μM ionomycin for 5 min.

2. Rinse and incubate reconstructed oocytes BSA containing media for 10 min.

3. Place reconstructed oocytes in ECM until the emergence of the second PB.

3.8.2   Chemical Activation for SCNT Using ZP-Enclosed Oocytes

Perform oocyte activation 2 h after the completion of cell fusion.

1. Wash reconstructed oocytes once in ionomycin solution.

2. Culture reconstructed oocytes in 10 mM ionomycin solution for 4 min.

3. Wash reconstructed oocytes once in activation solution.

4. Culture reconstructed oocytes individually in single 10 μL activation droplets for 4 h.

3.9   Embryo Culture and Grading

1. Wash reconstructed embryos three times in ECM droplets.

2. Transfer 20 reconstructed oocytes per 100 μL ECM droplet in the embryo culture dish.

3. Incubate reconstructed oocytes with 5% CO_2, 5% O_2, saturated humidity at 38.5 °C for 7–8 days.

4. Replenish 50% of ECM with EFM on day 3 post-activation.

5. Grade blastocysts according to morphology (Fig. 3): Expanded blastocysts with distinctive trophectoderm and inner cell mass (grade I), blastocysts with mild expansion (grade II), and non-expanded blastocysts (grade III).

4   Notes

1. This step is restricted to SCNT using ZP-enclosed oocytes.

2. Micromanipulation systems are usually formed by syringes attached to paraffin oil-filled plastic hoses. Make sure to avoid air bubbles in the micromanipulation system, which interferes with fine control in fluid movement.

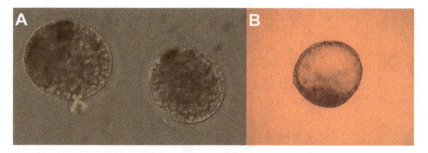

Fig. 3 Cloned blastocysts after horse SCNT. (**a**) Cloned blastocysts obtained from ZP-enclosed oocytes. (**b**) Cloned blastocyst produced with the protocol using ZP-free oocytes. Embryos were imaged with 100× amplification

3. Research with live animals must follow institutional and national guidelines for animal experimentation.
4. Perform this and the following steps with somatic cells in the tissue culture hood.
5. Fibroblasts adapt faster to culture conditions and overgrow other cell types found in the biopsy (e.g., keratinocytes, epithelial cells). The cell line should be restricted to fibroblasts in a few passages.
6. Both immature and mature oocytes can be recovered from live mares as described elsewhere [9].
7. ZP-free oocytes are more sensitive to manipulation than ZP-included oocytes.
8. The oocyte spindle is usually positioned next to the first PB. The removal of the PB and its adjacent cytoplasm allows the enucleation of >90% oocytes.
9. Aspirate more oocyte cytoplasm if the spindle remained in the oocyte. Make sure to not remove large cytoplasm fractions since it diminishes the oocyte developmental competence.
10. An alternative SCNT protocol with zona intact oocytes uses the Piezo drill device [3–6]. The Piezo drill applies subtle vibrations for ZP drilling and oolemma rupturing. Further, the donor cell is also ruptured by additional Piezo pulses, thus releasing its nucleus. Further, the donor nucleus is injected into the enucleated oocyte, which circumvents cell fusion for oocyte reconstruction.
11. Small donor cells are at the G0/G1 stage of the cell cycle.
12. Another approach is to begin oocyte reconstruction after finishing the enucleation of all oocytes.

References

1. Woods GL, White KL, Vanderwall DK, Li GP, Aston KI, Bunch TD et al (2003) A mule cloned from fetal cells by nuclear transfer. Science 301:1063

2. Galli C, Lagutina I, Crotti G, Colleoni S, Turini P, Ponderato N et al (2003) Pregnancy: a cloned horse born to its dam twin. Nature 424:635

3. Hinrichs K, Choi YH, Love CC, Chung YG, Varner DD (2006) Production of horse foals via direct injection of roscovitine-treated donor cells and activation by injection of sperm extract. Reproduction 131:1063–1072

4. Choi YH, Norris JD, Velez IC, Jacobson CC, Hartman DL, Hinrichs K (2013) A viable foal obtained by equine somatic cell nuclear transfer using oocytes recovered from immature follicles of live mares. Theriogenology 79: 791–796

5. Choi YH, Ritthaler J, Hinrichs K (2014) Production of a mitochondrial-DNA identical cloned foal using oocytes recovered from immature follicles of selected mares. Theriogenology 82:411–417

6. Choi YH, Velez IC, Macías-García B, Hinrichs K (2015) Timing factors affecting blastocyst development in equine somatic cell nuclear transfer. Cell Reprogram 17:124–130

7. Gambini A, Jarazo J, Olivera R, Salamone DF (2012) Equine cloning: in vitro and in vivo development of aggregated embryos. Biol Reprod 87:15,1–15,9

8. Gambini A, De Stefano A, Bevacqua RJ, Karlanian F, Salamone DF (2014) The aggregation of four reconstructed zygotes is the limit to improve the developmental competence of cloned equine embryos. PLoS One 9: e110998

9. Vajta G, Korosi T, Du Y, Nakata K, Ieda S, Kuwayama M et al (2008) The well-of-the-well system: an efficient approach to improve embryo development. Reprod Biomed Online 17:73–81

10. Bordignon V, Smith LC (2006) Telophase-stage host ooplasts support complete reprogramming of roscovitine-treated somatic cell nuclei in cattle. Cloning Stem Cells 8:305–317

Chapter 16

A Modified Handmade Cloning Method for Dromedary Camels

Fariba Moulavi and Sayyed Morteza Hosseini

Abstract

Camels play very important economic and sociocultural roles for communities residing in arid and semi-arid countries. The positive impacts of cloning on genetic gain in camel species are indisputable, considering the unique ability of cloning to produce a large number of offspring of a predefined sex and genotype using somatic cells obtained from elite animals, live or dead, and within any age category. However, the current low efficiency of camel cloning seriously limits its commercial applicability. We have systematically optimized technical and biological factors for dromedary camel cloning. In this chapter, we present the details of our current standard operating procedure for dromedary camel cloning, namely, "modified handmade cloning (mHMC)."

Key words Camelids, *Camelus dromedarius*, Modified handmade cloning, Oocyte

1 Introduction

Camels belong to the Camelidae family, the surviving family in the suborder Tylopoda [1, 2]. The even-toed ungulates in this family owe their existence to the Old (OW) and New World (NW) camelids. The OW camelids include two species: dromedary or one-humped (*Camelus dromedarius*) and Bactrian or two-humped (*Camelus bactrianus*). The NW camelids include four species: llama (*Lama glama*), alpaca (*Lama pacos*), guanaco (*Lama guanicoe*), and vicuña (*Vicugna vicugna*). The dromedary camel thrives in the hot dry climates of North Africa, Arabia, and Southern Asia, whereas the Bactrian camels are found in colder and more mountainous regions, such as Southern Russia, China, and Mongolia. The NW camelids are native to the Andes Mountains on the Western side of South America [2].

Supplementary Information The online version contains supplementary material available at https://doi.org/10.1007/978-1-0716-3064-8_16.

Marcelo Tigre Moura (ed.), *Somatic Cell Nuclear Transfer Technology*, Methods in Molecular Biology, vol. 2647, https://doi.org/10.1007/978-1-0716-3064-8_16,
© The Author(s), under exclusive license to Springer Science+Business Media, LLC, part of Springer Nature 2023

Genetic progress in camel breeding programs has been slow due to both limited selection accuracy and intensity of selection, and long generation intervals [3]. For instance, it usually takes 6–7 years for a female dromedary to reach reproductive age and more than 8–12 years to determine its genetic merit for milking ability. The application of assisted reproductive technologies could diminish the impact of such limitations. Since 1990, multiple ovulation and embryo transfer (MOET) has been the only assisted reproductive technology available for multiplying elite animals despite the increasing demand for improved dromedary camel genetics by the camel industry, particularly the United Arab Emirates (UAE) [3, 4]. However, MOET has its own disadvantages, including high costs of hormonal treatments of embryo donors and recipients, variable superovulation responses, risk of disease transmission during mating, and genetic unpredictability of the resulting offspring. In turn, the impact of animal cloning by somatic cell nuclear transfer (SCNT) to genetic gain in camel breeding is indisputable due to its unique ability to produce a large number of cloned offspring from desired genomes using somatic cells obtained from elite animals (both live or dead, and of any age) (Fig. 1). Therefore, the development of an optimized method for efficient cloning would greatly enhance genetic progress in camels [5, 6].

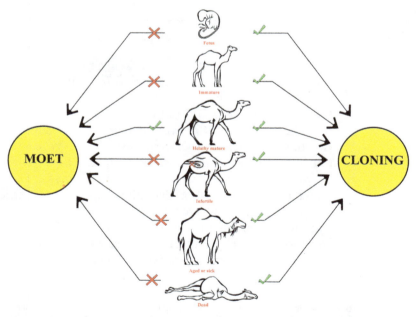

Fig. 1 The impact of animal cloning by SCNT to genetic gain in camel breeding compared to multiple ovulation and embryo transfer (MOET) technology that is widely used in camel. SCNT technology has a unique ability to produce cloned offspring from desired genomes using somatic cells obtained from elite animals without limits for age, health status, and fertility status of the cell donor, whereas MOET requires mature healthy male and female camels

Dromedary camel cloning has been reported by different teams in the UAE [5–8]. The accumulated information from these experiments suggested that improvements toward both technical and biological factors should improve the cloning of dromedary camels. Accordingly, stepwise adjustments allowed to adapt the handmade cloning (HMC) method, originally developed in cattle and sheep [9–11], to the cloning of dromedary camels. Using this novel method named modified HMC (mHMC) for dromedary camels, cloned blastocyst production was significantly increased compared with the standard method of cloning [5, 6]. We then systematically optimized the major biological contributors to dromedary camel cloning, including the conditions for oocyte in vitro maturation (IVM), cell cycle synchronization, cell fusion, oocyte activation, assisted epigenetic reprogramming, and embryo culture system. Importantly, successful vitrification of zona-free cloned blastocysts without a significant loss in viability enabled us to establish a cryobank of cloned embryos for large-scale embryo transfer programs [12]. These modifications enabled us to develop an efficient and consistent mHMC method for large-scale cloning of dromedary camels. The aim of this chapter is to describe our mHMC method in detail for efficient cloning in dromedary camels, which may prove adaptable to other camelid species.

2 Materials

Some methods described in this chapter are similar to the SCNT methods described for sheep cloning by our group [11]. The full detail of chemicals and reagents and their stock preparation are essentially similar to and can be found in the aforementioned study.

2.1 Equipment

1. Two CO_2 incubators: One with high (20%) oxygen tension and the other with low (5%) oxygen tension.

2. Laminar flow cabinet.

3. Stereomicroscope (Olympus SZX series, Japan).

4. Inverted microscope.

5. Electrofusion machine (BTX ECM2001, Harvard Apparatus, USA).

6. Table-top centrifuge.

7. Water bath.

8. Aspiration pump.

9. Liquid nitrogen tank.

10. Warming stages.

286   Fariba Moulavi and Sayyed Morteza Hosseini

2.2 Tools and Consumables

1. Fusion chamber (BTX Microslides, 3.2 mm gap, Harvard Apparatus, USA).

2. Aggregation needle (DN10/N) with handling bulb and cap (BLS, Hungary).

3. Pipette Borosilicate Pasteur (7095D-9, Corning).

4. 100 μm Cell Strainer, Sterile (431752, Corning).

5. Biopsy punch (OD: 8 mm, KruuseTM, Denmark).

2.3 Media and Solutions

Prepare solution using high-grade reagents (*see* **Note 1**), sterile filter (0.22 μm), and store solutions at 4 °C, unless stated otherwise.

1. Biopsy medium (BPM): Phosphate-buffered saline (PBS) with 300 IU/mL penicillin, 300 μg/mL streptomycin, and 2.0 μg/mL amphotericin B.

2. Biopsy washing medium (BWM): Phosphate-buffered saline (PBS) with 100 IU/mL penicillin, 100 μg/mL streptomycin, and 2.0 μg/mL amphotericin B.

3. Cell culture medium (CCM): Dulbecco's modified Eagle's medium/F12 with L-glutamine and phenol red (DMEM/F12; Gibco, Cat# 11320-033) with 10% (v/v) fetal bovine serum (FBS).

4. Cell freezing medium (CFM): DMEM/F12 with 50% (v/v) FBS and 10% (v/v) dimethyl sulfoxide (DMSO).

5. Cell starvation medium (CSM): DMEM-F12 with 0.5% (v/v) FBS and 1 μM rapamycin.

6. Ovum pick-up medium (OPUM): HEPES-buffered medium 199 containing 15 mM HEPES and 5.0 mM $NaHCO_3$ (H-TCM199; Gibco, Cat# 22340) with 1 mg/mL BSA, 3% (v/v) FBS, and 5.0 IU/mL heparin.

7. Oocyte washing medium (OWM): H-TCM199 supplemented with 10% FBS.

8. Ovary transportation solution (OTS): 0.9% (w/v) NaCl (saline solution) supplemented with 100 IU/mL penicillin and 100 μg/mL streptomycin.

9. Ovary storage solution (OSS): Buffer solution containing high-potassium and low-sodium electrolytes and magnesium sulfate, raffinose, lactobionate, antioxidant, and antibiotics (patenting data).

10. Aspiration medium (APM): H-TCM199 supplemented with 10% FBS and 2 IU/mL heparin.

11. In vitro maturation medium (IVMM): Medium 199 containing Earle's salts and L-glutamine without $NaHCO_3$ (Gibco, Cat# 11150-059) supplemented with 25 mM $NaHCO_3$, 10 μg/mL

follicle-stimulating hormone (rhFSH), 10 µg/mL luteinizing hormone (roLH), 1 µg/mL estradiol 17β, 10 µg/mL epidermal growth factor (EGF), 10 µg/mL brain-derived neurotrophic factor (BDNF), 10 µg/mL vascular endothelial growth factor (VEGF), 10 µg/mL insulin-like growth factor-I (IGF-I), 1 µM rapamycin, 1 µg/mL cystine, 5 µg/mL gentamycin, and 5% platelet-rich plasma (PRP) extracted from blood of estrous camels. The production method of PRP can be followed in [13].

12. Denudation solution: H-TCM199 with 10% (v/v) FBS.

13. Hyaluronidase solution: Dissolve 0.075 g hyaluronidase in 7.5 mL H-TCM199. Do not filter and make 25 µL aliquots. Store at – 20 °C for 1 year.

14. Pronase solution: Dissolve 0.05 g protease in 20 mL H-TCM199 and then centrifuge at 700 g, for 5 min. Recover the supernatant and make 500 µL aliquots. Store at – 20 °C for 1 year.

15. Demecolcine solution: Dissolve 5.0 mg demecolcine in 20 mL PBS by pipetting and vortexing. Make 50 µL aliquots. Store at – 20 °C for 2 years.

16. Enucleation medium (ENM): H-TCM199 with 10% (v/v) FBS and 0.4 µg/mL demecolcine.

17. Phytohemagglutinin (PHA) solution: Dissolve 2.0 mg PHA in 4.0 mL H-TCM199. Make 50 µL aliquots. Store at – 20 °C for 1 year.

18. Hypo-osmolar cell fusion buffer (CFB): Add the following sequentially and dissolve well: 3.005 g D-mannitol, 0.0024 g $MgCl_2 \cdot 6H_2O$, 0.012 g HEPES, and 0.05 g fatty acid-free bovine serum albumin (FAF-BSA) to 100 mL ultrapure water. Check and adjust pH and osmolarity to between 7.2 and 7.4 and 200 and 210 mOsm, respectively. Make 1.0 mL aliquots. Store at – 20 °C for 1 year.

19. Post fusion solution (PFS): HTCM199 with 3 mg/mL FAF-BSA, epigenetic cocktail [10 µg/mL vitamin-C (VC) + 10 nM trichostatin A (TSA)], and 10 µg/mL PHA.

20. Ionomycin solution: Pipette 10 µL of ionomycin stock (5 mg ionomycin dissolved in 1340 µL of DMSO and 12.06 mL of absolute ethanol, stored in −20 °C) into 1 mL of activation buffer (HTCM199) with 1 mg/mL FAF-BSA. Keep the ready activation solution in a 1.5 microcentrifuge tube and use within 5 min after preparation.

21. Inactivation solutions: HTCM199 with 30 mg/mL BSA and HTCM199 with 3 mg/mL BSA.

288   Fariba Moulavi and Sayyed Morteza Hosseini

22. 6-Dimethylaminopurine (6-DMAP): Pipette 1 mL mSOFaa into 20 μL of 6-DMAP stock [(0.0652 g 6-DMAP dissolved in 4.0 mL mSOFaa (plus epigenetic cocktail)] and shake the solution thoroughly at 40 °C bath until completely dissolved. Store at −20 °C for 1 year.

23. Synthetic embryo culture medium: A modified formulation of synthetic oviductal fluid with amino acids (mSOFaa) [14] comprised of 107.70 mM NaCl, 7.16 mM KCl, 1.19 mM KH_2PO_4, 0.74 mM $MgSO_4.7H_2O$, 3.30 mM sodium lactate, 1.78 mM $CaCl_2.2H_2O$, 0.33 mM Na-pyruvate, 25 mM $NaHCO_3$, 0.5 mM glucose, 1.0% (v/v) MEM nonessential amino acids (Sigma, M7145, 100X), 2.0% (v/v) BME essential amino acids (Sigma, B6766, 50X), 2.0 mM L-glutamine, 10 μg/mL epidermal growth factor (EGF), 2.77 mM myo-inositol, 0.34 mM tri-sodium citrate, 10 μg/mL brain-derived neurotrophic factor (BDNF), 5 μg/mL lectin, 10 μg/mL vascular endothelial growth factor (VEGF), 10 μg/mL insulin-like growth factor-I (IGF-I), 1 μg/mL cystine, 1 μM rapamycin, 50 μg/mL gentamycin, 5% PRP, 0.01 mM ethylenediamine tetraacetic acid (EDTA), and 4.0 mg/mL FAF-BSA. When amino acids were added to SOF, osmolarity of the medium was maintained at 265–275 mOsmol by adjusting the concentration of sodium chloride and pH was sustained at 7.2–7.4 by adding NaOH or HCl. All slats and additives were embryo culture tested. The embryo culture dishes were preincubated at 38.6 °C with 6% CO_2, 5% O_2, and balance N_2 before using for embryo culture.

24. Epigenetic cocktail: Comprised of trichostatin A (TSA, Sigma, T1952, readymade solution of 5 mM in DMSO) and vitamin C (VC, Sigma, A5960; was dissolved in distilled water to prepare 1 mg/mL stock solution, filtered, and stored at −20 °C). The final concentrations of the epigenetic cocktail were adjusted to 10 nM TSA and 1 μg/mL, respectively, by diluting the stock solutions in post-fusion, post-activation, and initial embryo culture media.

25. Mineral oil (Vitrolife™).

3   The mHMC Setup

3.1 Preparation of Enucleation Micropipette

A unique feature of the mHMC technique is the use of a simple handheld enucleation device. Pasteur pipettes pulled on a flame produce adequate tips with inner diameters for oocyte enucleation (10–20 μm). Following the steps described below, you will be able (after some trial and error) to produce your enucleation device within 1 min (Supplemental Video 1).

1. Prepare commercially available laboratory glass Pasteur pipettes. We prefer Pasteur pipettes made of borosilicate glass (size: 5.75″). Borosilicate glass has a particularly high heating tolerance compared to soda lime, and thus, it is less likely to break or melt during heating and pulling.

2. Hold the Pasteur pipette at both ends and place its tip (2 cm from the opening) into the flame of a Bunsen burner until it softens. Apply a gentle pressure to bend a 45° angle.

3. Draw the pipette out of the flame and immediately pull both ends horizontally to prepare an approximately 10-cm-long straight-drawn pipette.

4. Cut off the extra-long part by bending the pipette. Break the pipette at approximately 4–5 cm above the shoulder to create an internal diameter of about 200–300 μm. Repeat this pulling step if the diameter of the pulled part is still bigger than 200–300 μm.

5. Heat the thin tip of the pulled pipette (about halfway between the bent angle and the narrow tip bend). When it gets soft, remove from the flame and quickly pull it as described previously (*see* **Note 2**). This will end up with one functional micropipette that has a 45° angle.

6. Check the tip of the prepared enucleation devices using a zona-free oocyte under a stereomicroscope. Select pipettes with an internal diameter approximately slightly larger than the cytoplasmic protrusion (~10–20 μm) and with a smooth-tipped opening (*see* **Note 3**).

3.2 Preparation of Donor Cells

1. Restrain the donor animal and perform sedation, if needed, with an IV injection of xylazine hydrochloride (under the supervision of a veterinarian) (*see* **Note 4**).

2. Hold and scrub the back of the ear (the mastoid area) with a soap solution and an impregnated brush to remove any remaining dirt. Shave the area completely with a razor and scrub with 1.0% povidone-iodine. Wash the area thoroughly with BWM, dry with a sterile gauze pad, and spray with 70% (v/v) ethanol. Do not touch the tissue after shaving.

3. Apply the biopsy punch to the center of the shaved area. Rotate the punch with a turning motion in one direction (this will minimize any shearing artifact) with slight downward pressure (the turning force should be higher than this pressure). Use the cutting action of the tool rather than pressure. Once through the skin, remove the punch slowly and straight up.

4. Grasp the edge of the sample and pull it upward and outward using forceps. Cut as low on the stalk (underlying fat) using scissors.

5. Place the biopsy in a sterile 50 mL tube containing 15 mL BPM and transport to the laboratory at room temperature for short distances or beside ice for long (overnight) transportation.

6. Transfer the biopsy to a 100 mm dish containing 10 mL BPM in a laminar flow hood. Shave off any remaining hair with a sterile scalpel blade. Wash the biopsy by serial submerging in three 100 mm dishes containing 20 mL BWM.

7. Cut the biopsy medially to remove cartilage and make small tissue fragments of approximately 4 mm^2 with a sharp razor blade. Transfer tissue fragments to a 25 cm^2 tissue culture flask containing 10 mL CCM without antibiotics. Incubate the culture flask with 6% CO_2, saturated humidity at 37 °C.

8. Check the cell culture flask under low and high magnifications under an inverted microscope at 24 h and 48 h after the onset of culture. Replenish the CCM every other day until the culture reaches 70–80% confluence.

9. Proceed with cell passaging (subculture). Remove the CCM and wash cells three times with PBS at 37 °C to completely remove serum traces.

10. Add 1.0 mL of 2.5% Trypsin-EDTA for 1 min and then discard. Return the flask to the incubator for 5 min (*see* **Note 5**).

11. Add 5.0 mL CCM and pipette it over the surface of the culture flask to provide a single-cell population. Transfer the cell suspension to a 15 mL conical tube and centrifuge at 700 g for 10 min.

12. Discard the supernatant and resuspend cells with 5.0 mL CCM and repeat centrifugation at 700 g for 10 min. Discard the supernatant.

13. Passage cells by adding 1.0 mL CCM and resuspend cells by pipetting. Count cells with a hemocytometer and culture 2.5×10^4 cells/cm^2 with 6% CO_2, saturated humidity at 37 °C.

14. Prepare for cryopreservation by diluting cell pellets (**steps 8–12**) with 10 mL CFM at room temperature. Split cells into 1.5 cryovials and attach them to cryocanes.

15. Place cryocanes at − 70 °C freezer at a 45° angle for 2 h and transfer to a liquid nitrogen tank.

3.3 Cell Cycle Synchronization of Donor Cells

1. Passage cells as described above and culture 2.5×10^4 cells/cm^2 in one well of a four-well dish, add 0.5 mL CCM and culture with 6% CO_2, saturated humidity at 37 °C for 24 h.

2. Remove the CCM and wash cells three times with PBS.

3. Add 1.0 mL CSM and culture with 6% CO_2, saturated humidity at 37 °C for 2 days.

Handmade Cloning in Camels   291

4. Remove the CSM and wash cells three times with PBS. Prepare cells as described above and resuspend the cell pellet with 1.0 mL CSM.

5. Place cells at 4 °C until their use for nuclear transfer.

3.4 Media and Dish Preparation for Handling or Culture of Oocytes

Figure 2a–j illustrates the composition of media and the method of dish preparation for handling oocytes during oocyte aspiration (Fig. 2a), IVM (Fig. 2b), denudation (Fig. 2c), zona-digestion (Fig. 2d), enucleation (Fig. 2e), nuclear transfer (Fig. 2f), artificial activation (Fig. 2g, h), and embryo culture (Figs. 2i, j).

3.5 Preparation of Recipient Oocytes

Ovum pick-up (OPU) is the major source of oocytes for SCNT in dromedary camels. Alternatively, ovaries of slougthered camels can be used for in vitro maturation. Due to limited number of camels slaughtered, very few oocytes are available for in vitro maturation. Therefore, we preserve ovaries in a specially formulated OSS at 10 °C for up to 48 h with negligible effects on oocyte developmental competence. This strategy enabled us to have more oocytes in each SCNT run. However, due to limited number of camels slougthered, and since SCNT runs [6].

3.5.1 Preparation of In Vivo Matured Oocytes

The procedure of donor superovulation and OPU is a standard procedure and can be followed in corresponding studies [6, 7], and is briefly mentioned here:

1. Aspirate >10 mm follicles into 50 mL conical tubes containing 10–15 mL OPUM (Fig. 2a).

2. Transfer ready tubes immediately to the lab, gently agitate the content in the aspiration tube, and pour the aspirated material onto a 100 μM cell strainer. This process will retain cumulus-oocytes complexes (COCs).

3. Backwash the strainer with 10 mL COC washing medium in a 100 mm dish on a warm stage at 38.5 °C. Distribute the aspirated content evenly in the dish by gentle swirling (Fig. 2a).

4. Collect COCs and denuded oocytes with a mouth pipette under a stereomicroscope. The aspirated content is usually bloody and may have coagulated blood particles. Separate COCs from debris, coagulated blood particles, and follicular fluid by washing at least three times in 200 μL COC washing droplets.

5. Transfer the COCs dish into incubator until denudation.

6. Prepare the denudation solution by pipetting 25 μL hyaluronidase stock into 1 mL HTCM199.

7. Prepare denudation dish by making 5 × 100 μL drops of denudation solution in a 6 cm Petri dish; overlay with mineral oil (Fig. 2c).

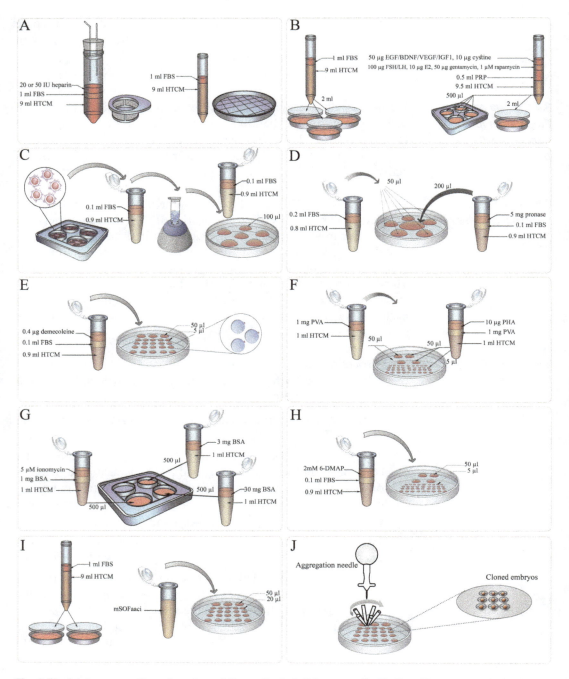

Fig. 2 The basic composition of media and the method of dish preparation for handling oocytes during oocyte aspiration (**a**), IVM (**b**), denudation (**c**), zona digestion (**d**), enucleation (E), nuclear transfer (**f**), artificial activation (**g**, **h**), and embryo culture (**i**, **j**). The media used for post fusion (not shown here), activation for 4 h with 6-DMAP (**h**), and initial embryo culture (for 4 h), were supplemented with epigenetic cocktail (10 μg/mL VC + 10 nM TSA) optimized for assisted epigenetic modification of the reconstructed oocytes before final culture in embryo culture medium

Handmade Cloning in Camels   293

8. Transfer COCs into the first denudation droplet and incubate it for 1 min in incubator.

9. Start denudation of oocytes with large (250 μm), medium (200 μm), and small (150 μm) pulled glass Pasteur pipette in sequence by performing movements of entry and exit from the pipette.

10. By the end of denudation, thoroughly wash denuded oocytes in 200 μL droplets of OWM and incubate there for 30 min before starting *zona pellucida* (ZP) removal.

3.5.2 Preparation of In Vitro Matured Oocytes

Collection and Storage of Abattoir-Derived Ovaries

1. Collect ovaries immediately after slaughter and extrusion of the internal organs from the carcass. Cut ovaries from the reproductive tract with minimum surrounding tissues and place them in a thermos flask containing at least 1 L OTS at room temperature. Transport the thermos flask with ovaries to the laboratory as soon as possible.

2. Trim ovaries from any excessive tissue if destined for long-distance transportation (e.g., overseas). Wash ovaries several times with OTS, gently rub to remove excess blood, and perform a final wash with 1000 mL OTS at room temperature.

3. Transfer ovaries into a sealing plastic bag containing 1 L OSS at room temperature and place the sealed bag in a portable refrigerator adjusted at 10 °C for gradual cooling during transportation.

4. Wash ovaries with OTM after arrival at the laboratory and transfer them to a beaker containing 500 mL OTM.

Follicular Aspiration and IVM

1. Hold one ovary at a time with a sterile absorbing gauze and aspirate all 2–8 mm follicles using a 20-gauge needle of a scalp vein set connected to a 50 mL conical tube containing 10 mL APM and attached to a vacuum pump (Fig. 2a; *see* **Note 6**).

2. Gently agitate the content and pour the entire aspirant onto a 100 μM cell strainer to separate COCs from blood, other cells, and debris (Fig. 2a).

3. Backwash the strainer with 10 mL COC washing medium in a 100 mm dish on a warm stage at 38.5 °C. Distribute the retrieved COCs evenly in the dish by gentle swirling.

4. Retrieve COCs of grades I–III with a mouth pipette. Oocytes with five or more layers of surrounding cumulus cells are graded as class I or II if the cumulus cells are fully compact, and as class III if a slight expansion is observed in the outer layers.

5. Transfer COCs to a 35 mm dish containing 2 mL of H199/10. Swirl the dish and apply gentle pipetting to disperse COCs from each other and debris (Fig. 2b).

294   Fariba Moulavi and Sayyed Morteza Hosseini

6. Wash COCs twice with 2 mL OWM and once with 2 mL IVMM (Fig. 2b).

7. Transfer 30–35 COCs with the minimum amount of medium into a preincubated 500 μL maturation medium prepared in the wells of an IVM 4-well dish (Fig. 2b). Incubate COCs with 6% CO_2, 20% O_2, and saturated humidity at 38.5 °C for 28 h.

3.5.3 Oocyte Denuding

1. Transfer COCs from the maturation dish 28 h after IVM to a microtube containing 1.0 mL denudation solution and vortex it for 1 min (Fig. 2c). Dromedary camel oocytes following IVM are easily denuded during pickup from IVM dish or following a short vortexing without hyaluronidase.

2. Spin down the microtube briefly.

3. Transfer oocytes from the microtube to a 35 mm dish. Swirl the dish to collect the oocytes in the center, pick up denuded oocytes, and transfer them to the first 100 μL OWM droplet (Fig. 2c). Wash oocytes in at least three 200 μL OWM droplets to remove remaining cumulus cells.

4. Select oocytes with a polar body (PB) or undergoing extrusion.

5. Place denuded oocytes in 200 μL droplets and incubate for 30 min before starting ZP removal (Fig. 2c).

3.5.4 Removal of the Zona Pellucida

1. Transfer 50–100 denuded oocytes to a 200 μL pronase droplet in the ZP digestion dish and gently rotate the dish by hand on the warm stage for 1–2 min (Fig. 2d). Decrease light intensity and adjust the contrast of the stereomicroscope to focus on the ZP. The oocytes pose no concerns since they lose their shapes during zona removal (*see* **Note 7**). However, they will retain a spherical shape soon after pronase treatment.

2. Do not wait for complete ZP removal. Once the ZP starts dissolving, pick up the oocyte with the minimum volume possible of the solution. Wash at least three times in 200 μL washing solution droplets.

3. Incubate ZP-free oocytes in 200 μL washing solution droplets for 15 min before oocyte enucleation (Fig. 2d).

3.6 Somatic Cell Nuclear Transfer

3.6.1 Oocyte Enucleation Assisted by Demecolcine Treatment

1. Wash ZP-free oocytes in 100 μL ENM droplets, transfer 2–3 oocytes per 10 μL ENM droplet, and incubate for 1 h (Fig. 2e; *see* **Note 8**).

2. Pick oocytes with a clear cytoplasmic protrusion under the stereomicroscope. Discard oocytes without the protrusion. Usually, more than 95% of treated oocytes have a clear cytoplasmic protrusion of MII spindle following demecolcine treatment. If not, check the IVM system.

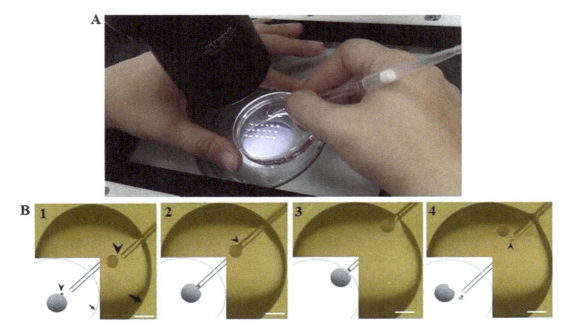

Fig. 3 Oocyte enucleation setup and hand position (**A**). Steps for enucleation of oocytes in mHMC shown in real and schematic (inserts) images (**B** 1–4). For further description, refer to the oocyte enucleation section

3. Place the enucleation dish on warm plate of the stereomicroscope and hold the enucleation dish in your left hand with your thumb and index finger (for right-handed people) and hold the enucleation device in right hand in the same focus as for the oocytes (Fig. 3a).

4. Roll the oocyte and place the cytoplasmic protrusion at 3 o'clock position in a clock's face (Fig. 3b; Supplemental Video 2).

5. Place the pipette tip adjacent to the cytoplasmic protrusion. Ensure that the micropipette opening shares the same focus as the cytoplasmic protrusion (Fig. 3b1).

6. Allow the cytoplasmic protrusion to enter the micropipette via capillary and apply gentle suction by mouth if needed (Fig. 3b2).

7. Remove the pipette (while holding the oocyte) from the droplet and place it in the mineral oil. This movement will split the oocyte into two parts. The cytoplasmic portion held by the pipette will contain the oocyte chromosome mass and the enucleated oocyte (i.e., cytoplast) will remain in the enucleation droplet (Fig. 3b3).

8. Discard the removed cytoplasmic protrusion in an enucleation droplet and check whether enucleation occurred (the enucleated oocyte lacks a cytoplasmic protrusion). Repeat the process with all the remaining oocytes (Fig. 3b4).

296 Fariba Moulavi and Sayyed Morteza Hosseini

3.6.2 Oocyte Reconstruction

1. Pick up 10–20 cytoplasts and wash twice in 50 μL HTCM199 with 1 mg/mL PVA droplets in an oocyte reconstruction dish (*see* **Note 9**; Fig. 2f).

2. Transfer cytoplasts to the PHA droplet. Pipette gently each cytoplast in PHA to make them sticky. Distribute cytoplasts apart in the PHA droplet (Fig. 2f; *see* **Note 10**).

3. Resuspend donor cells and place 1–3 μL of the cell suspension into an HTCM199/PVA droplet under a stereomicroscope to obtain an approximate population of 100–200 cells per droplet.

4. Pick up 20–30 donor cells with a mouth pipette and place them in PHA droplets with cytoplasts. Deposit cells apart from cytoplasts.

5. Allow cells to settle for a few minutes and adjust the contrast filter of the stereomicroscope to visualize the cells.

6. Pick up five cytoplasts and drop them on single donor cells. Give preference to small well-rounded donor cells. Push and press gently the cytoplast onto the donor cell to sustain attachment, if necessary. Repeat this process with the remaining cytoplasts.

7. Transfer cell couplets to a second PHA droplet, while checking for the presence of a single donor cell in each cytoplast. Incubate cell couplets in the PHA droplet for 2–3 min.

8. Wash cell couplets in 50 μL HTCM199/PVA droplets and transfer them individually into 5 μL HTCM199/PVA droplets to prevent them from sticking together (Fig. 2f).

3.7 Cell Fusion

Cell couplets are fused 35–37 h after the onset of IVM. We also suggest using hypo-osmolar (200–210 mOsm) fusion buffer to increase fusion efficiency.

1. Turn on the fusion machine (*see* **Note 11**) and set parameters as follows: pre-fusion alignment (AC) of 15 V (1,000 Hz), fusion pulse (DC) with two pulses of 600 V (20 μs/each and 1 s delay), and post-fusion alignment of 15 V for 9 s).

2. Place the fusion chamber on the stereomicroscope. Attach live (red) wire to northern and ground (black) to southern wire in the fusion chamber. Cover fusion chamber with 500 μL CFM between electrodes at the center of the microscope field of view.

3. Add 100 μL CFM on the lid of a 60 mm Grainer dish and equilibrate 10–20 cell couplets for 1–2 min.

4. Transfer cell couplets over regular distances on the electrodes. Roll each cell couplet with a mouth pipette until the contact plane between the cytoplast and the donor cell is parallel to electrodes (*see* **Note 12**).

Handmade Cloning in Camels   297

5. Apply the AC current for 5–10 s to further align cell couplets.

6. Apply the DC fusion pulses and let fused couplets drop from the electrodes.

7. Remove fused couplets gently with a mouth pipette. Wash them in 100 μL droplets of PFS medium droplets and incubate them individually in PFS microdroplets until activation.

8. No need not check for fusion events and proceed to the next step as soon as possible.

3.8 Oocyte Activation

1. Prepare reconstructed oocytes 2 min before oocyte activation. Wash fused couplets in 50 μL PFS medium and keep them in one droplet.

2. Place 500 μL ionomycin solution into one well of a four-well dish. Place inactivation solutions in the two remaining wells (Fig. 2g).

3. Incubate reconstructed oocytes in ionomycin solution for 1 min. Take care to avoid reconstructed oocytes from aggregating.

4. Wash reconstructed oocytes and incubate in inactivation solution (HTCM199 with 30 mg/mL BSA) for 5 min (Fig. 2g).

5. Transfer activated oocytes into post-activation medium (HTCM199 with 3 mg/mL BSA) and incubate for 5 min (Fig. 2g).

6. Wash reconstructed oocytes three times in 50 μL 6-DMAP droplets in a dish and then culture them individually in 5.0 μL 6-DMAP droplets with 6% CO_2, saturated humidity at 38.5 °C for 4 h (Fig. 2h).

3.9 Embryo Culture

ZP-free embryos need to be cultured in wells to avoid disaggregation of blastomeres at pre-compaction stages and also to avoid aggregation of embryos together.

1. Using nontreated culture dishes for embryo culture is necessary to avoid attachment of zona-free embryos to the culture dish. Place three washing (50 μL) and nine culture (20 μL) droplets of mSOFaaci in a 60 mm culture dish and cover with prewashed embryo-tested paraffin oil (Fig. 2i).

2. Sterilize the aggregation needle tip with drips of 70% ethanol. Allow it to dry before washing with drips of embryo culture-grade water and then let it dry completely.

3. Support the bottom of the embryo culture dish with a thick glass microscope filter.

4. Make 10 wells in each 20 μL mSOFaaci droplet. Press the aggregation needle tip into the plastic through the paraffin oil and culture medium, while making a circular movement with

the free end of the needle held by your hand. This movement makes tiny depressions (i.e., wells) of approximately 300 μm in internal diameter and with smooth border (Fig. 2j).

5. Incubate the prepared dish in 6% CO_2 and 5% O_2, at 38.5 °C for at least 2 h before embryo culture.

6. Embryo culture takes place in two steps: **Step 1** in the presence of the epigenetic cocktail for 4 h and **step 2** in the absence of epigenetic cocktail for the rest of the culture period.

7. Wash the reconstructs three times in 50 μL mSOFaaci plus epigenetic cocktail droplets.

8. Place each activated oocytes in a well, transfer the embryo culture dish plus epigenetic cocktail gently to the incubator and culture in 6% CO_2, 5% O_2, and saturated humidity at 38.5 °C for 4 h.

9. Wash the reconstructs three times in 50 μL mSOFaaci droplets.

10. Place each activated oocytes in a well, transfer the embryo culture dish gently to the incubator and culture in 6% CO_2, 5% O_2, and saturated humidity at 38.5 °C.

11. Do not check embryonic development until days 5.5–6 post-activation when embryos are checked for vitrification or transfer to the recipients (*see* **Note 13**; Fig. 4).

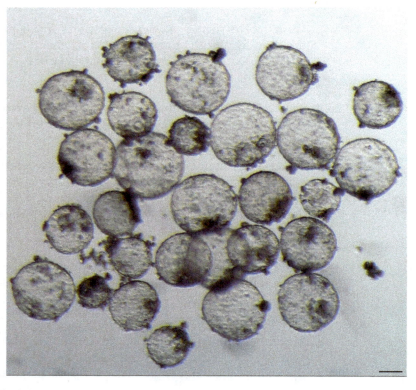

Fig. 4 A batch of day 6 embryos produced by mHMC using in vitro matured oocytes. Bar represents 100 μm

Handmade Cloning in Camels   299

3.10 Vitrification of Embryos

Due to limits in the source of abattoir-derived oocytes and also in performing OPU on a large scale, the number of recipient oocytes in each SCNT run is also limited. Fresh transfer of the resultant few clone blastocysts requires keeping recipients throughout the breeding season for ET sessions. To avoid this, our routine plan is to stock embryos in liquid nitrogen tank for bulk transfer during the best season time. The method of blastocyst vitrification was described previously [14] and supports high survival of vitrified-warmed blastocysts with comparable, even higher, pregnancy outcomes.

3.11 Embryo Transfer

Blastocysts of morphological grades 1 and 2 are selected for transfer into recipients according to the established criteria for morphological grading of embryos developed by the International Embryo Transfer Society (IETS). For ZP-free embryos, however, a systematic study needs to be conducted to understand how meaningful the morphological grading criteria of zona-intact fertilized embryos are for ZP-free cloned embryos, especially with regard to their survival after embryo transfer [15]. In our routine clone-ET system that supports high rates of initial pregnancy and delivery, we transfer grade 1 and 2 blastocysts on day 5.5–6 of the culture into the synchronized recipient she camels on day 5 of ovulation. The standard method for preparation and synchronization of the recipients can be found in [14].

1. Load one grade 1 or two grade 2 blastocysts in the ET medium (mSOFaa) per 0.25 mL straw (IMV, France) and transfer nonsurgically into the uterine lumen.

2. Perform initial pregnancy diagnosis using transrectal ultrasonography between days 25 and 25 of ET. The diagnosis is based on the presence of three endpoints: embryonic vesicle, embryo proper, and embryonic heartbeat.

3. Pregnancies can be examined by a second round of ultrasonography at days ~70–90 after ET.

4. The gestation length in cloned pregnancies varies widely, from 345 to 400 days post-ET.

5. Monitor pregnant camels approaching the expected delivery date with the assistance of a veterinarian and trained herd shepherd. Allow the camel to calve naturally and provide manual assistance if necessary.

Current Success Rate and Future Directions

In our experience, the mHMC has proved to be a viable alternative to the traditional method. Its simplicity, ease of implementation, low cost, and high efficiency are important features that will encourage its broader application in both research and commercial applications. Use of the mHMC decreases costs radically as the enucleation and nuclear transfer devices are each reproducible within as short a time period as 1 min using simple equipment such as a glass Pasteur pipette and a laboratory flame burner [5]. This technique is facile enough for a person with no previous experience in micro-manipulation to learn the basic routines of the procedure in 3–4 weeks, which is in stark contrast to the several months typically required to master the traditional method.

The overall efficiency of the in vitro blastocyst development reached to 90–100% when using OPU-derived oocytes, clearly pointing to significant improvements (unpublished data). It is noteworthy that this high rate of blastocyst development was achieved when we used a static embryo culture medium containing all nutrients and growth factors from the beginning of culture and did not disturb the embryos throughout culture period compared to the biphasic embryo culture system used during our first experiments [5, 6]. One first conclusion is that 100% clone blastocyst development may be achievable in dromedary camels when using in vivo matured oocytes as recipient oocytes and an optimized cloning protocol. The results also indicate that current IVM practice in dromedary camels continues to remain inefficient, and future studies are required to optimize maturation and produce more competent recipient oocytes in vitro.

Following transfer of vitrified-warmed embryos derived from in vivo matured oocytes with our established method the overall efficiencies of the initial pregnancy/total recipients, development to term/total pregnancies, and weaving/total pregnancies were 42.1%, 39.7%, and 37.7%, respectively [5, 6, 14, and unpublished data]. In our experience, approximately 6.6% of the transferred embryos developed into viable offspring. This overall efficiency of 6.6% compares very unfavorably with that of MOET, where approximately 40–60% of freshly transferred embryos develop into healthy calves at weaning [3]. This great difference between the SCNT and MOET with regard to dromedary camels is not surprising in view of the similar differences observed between in vivo developmental fate of SCNT and IVF embryos in other species as well [15].

In dromedary camels, most of the early embryonic deaths in natural mating and MOET programs usually occur during the first 2 months of pregnancy, with rates ranging from 30% to 40% [16]. For clone pregnancies, in our experience, the frequency of pregnancy losses until day 90 was higher than those in the remainder of the gestation period, which is consistent with other studies [7]. Importantly, in vivo development of dromedary camel clones does not seem to exhibit the same complications that have been reported for cattle, namely, large offspring syndrome, hydrops, and advanced pregnancy failure [15]. This important advantage of a

Handmade Cloning in Camels   301

good pregnancy prognosis, especially for camels with a very long gestation period (≥ 1 year), increases the economic value of camel cloning in large-scale applications. We successfully used the mHMC system presented here for the production of dromedary camel clones at thus far the largest scale from a single laboratory. Since then, this SCNT system is actively used for the production of clones from several other cell lines in a commercial plan. This innovative technique can accelerate cloning technology in camels, and we expect it to be readily applicable to other camelids as well.

4 Notes

1. The efficiency of SCNT depends on the reagents, glassware, and disposable supplies. All items should be of tissue grade. The quality of several reagents varies widely (e.g., BSA, FBS). It is paramount to test the quality of each new lot or reagent before routine use.

2. An alternative is to bend the shoulder during softening, which provides a 45° angle between the stem and the pipette tip. This fact makes enucleation pipettes more convenient for handling during oocyte enucleation.

3. Handheld pipettes need a blunt tip end with a smooth surface to prevent cell damage. The stereomicroscope assists visualizing the pipette tip. Laboratories that apply pipette handles may rely on commercially available tips or pipettes from hematocrit tubes. Make sure to use safety glasses while preparing pipettes.

4. Perform experiments with live animals under institutional and national guidelines.

5. Use differential trypsinization to remove epithelial cells at the first passage. Fibroblasts will slough off faster than epithelial cells and should be removed carefully with a Pasteur pipette. We typically use primary fibroblast cell lines at passages 2 and 3.

6. The pressure on the vacuum pump is set to 100 mm Hg via Tygon tubing containing a 65 mm 0.2 μm PTFE membrane filter in the vacuum line to protect the pump from accidental fluid aspiration. Different experimental conditions or unintended results (e.g., excessive oocyte denuding) may require adjustments of these conditions.

7. Do not wait until the ZP dissolves completely. Rather, remove oocytes from pronase solution when observing ZP thinning and release them by gentle pipetting in neutralizing droplets.

8. ZP-free oocytes may stick to the inner surface of enucleation pipettes. The use of media with protein supplementation (e.g., FBS, BSA, PVA) prevents this from happening.

9. FBS inhibits or weakens PHA-mediated cell attachment. Therefore, avoid carryover of FBS-containing media into PHA droplets while transferring cytoplasts and donor cells.

10. Cytoplasts and donor cells become very sticky in PHA solution. Avoid cytoplast and cell aggregation by distributing them evenly in each droplet and while inside the mouth pipette.

11. There are fusion machines from various companies (e.g., BTX, Cryologic). Adjust fusion parameters accordingly, preferably using a graphic pulse analyzer. These parameters are based on using a BTX-2001 (BTX ECM2001, Harvard Apparatus, USA) fusion machine and BTX fusion chamber with a 3.2 mm gap between electrodes.

12. The components of H-TCM199 may increase conductivity and cause cell lysis after fusion. Minimize the carryover of H-TCM199-containing media to the fusion chamber.

13. Place the CO_2 incubator destined for embryo culture in a vibration-free area and avoid opening the incubator. Handling the culture dish may remove SCNT embryos from their wells.

Acknowledgments

The authors extremely grateful to the General Director of CARTC, for the moral and unconditional support. The authors are also grateful to all participants of the reproduction group at CARTC, including embryology department and clinical department. This work was supported by grants from Zaabeel Office, Government of Dubai, under regulation by H. H. Sheikh Mohammed Bin Rashid Al Maktoum, vice president and prime minister of the UAE and ruler of Dubai.

References

1. Agnew D (2018) Camelidae. In: Pathology of wildlife and zoo animals. Academic Press, pp 185–205

2. Fowler M (2010) Medicine and surgery of camelids, 3rd edn. Wiley-Blackwell, Ames. General Biology and Evolution, pp 3–16

3. Skidmore JA (2019) The use of some assisted reproductive technologies in old world camelids. Anim Reprod Sci 207:138–145

4. Nagy P, Skidmore JA, Juhasz J (2013) Use of assisted reproduction for the improvement of

milk production in dairy camels (*Camelus dromedarius*). Anim Reprod Sci 136:205–210

5. Moulavi F, Hosseini SM (2019) Development of a modified method of handmade cloning in dromedary camel. PLoS One 14:e0213737

6. Moulavi F, Asadi-Moghadam B, Omidi M, Yarmohammadi M, Ozegovic M, Rastegar A, Hosseini SM (2020) Pregnancy and calving rates of cloned dromedary camels produced by conventional and handmade cloning techniques and in vitro and in vivo matured oocytes. Mol Biotechnol 62:433–442

7. Wani NA, Wernery U, Hassan FAH, Wernery R, Skidmore JA (2010) Production of the first cloned camel by somatic cell nuclear transfer. Biol Reprod 82:373–379

8. Son YB, Jeong YI, Jeong YW, Olsson PO, Hossein MS, Cai L, Kim S, Choi EJ, Sakaguchi K, Tinson A, Singh KK, Rajesh S, Noura AS, Hwang WS (2022) Development and pregnancy rates of *Camelus dromedarius*-cloned embryos derived from in vivo- and in vitro-matured oocytes. Anim Biosci 35: 177–183

9. Vajta G, Lewis IM, Hyttel P, Thouas GA, Trounson AO (2001) Somatic cell cloning without micromanipulators. Cloning 3:89–95

10. Hosseini SM, Hajian M, Moulavi F, Asgari V, Forouzanfar M, Nasr-Esfahani MH (2013) Cloned sheep blastocysts derived from oocytes enucleated manually using a pulled Pasteur pipette. Cell Reprogram 15:15–23

11. Hosseini SM, Moulavi F, Nasr-Esfahani MH (2015) A novel method of somatic cell nuclear transfer with minimum equipment. In: Cell Reprogramming. Humana Press, New York, NY, pp 169–188

12. Moulavi F, Soto-Rodriguez S, Kuwayama M, Asadi-Moghaddam B, Hosseini SM (2019) Survival, re-expansion, and pregnancy outcome following vitrification of dromedary camel cloned blastocysts: a possible role of vitrification in improving clone pregnancy rate

by weeding out poor competent embryos. Cryobiology 90:75–82

13. Moulavi F, Akram RT, Khorshid Sokhangouy S, Hosseini SM (2020) Platelet rich plasma efficiently substitutes the beneficial effects of serum during in vitro oocyte maturation and helps maintain the mitochondrial activity of maturing oocytes. Growth Factors 38:152–166

14. Holm P, Booth PJ, Schmidt MH, Greve T, Callesen H (1999) High bovine blastocyst development in a static in vitro production system using SOFaa medium supplemented with sodium citrate and myo-inositol with or without serum-proteins. Theriogenology 52: 683–700

15. Hosseini SM, Dufort I, Nieminen J, Moulavi F, Ghanaei HR, Hajian M, Jafarpour F, Forouzanfar M, Gourbai H, Shahverdi AH, Nasr-Esfahani MH, Sirard MA (2015) Epigenetic modification with trichostatin A does not correct specific errors of somatic cell nuclear transfer at the transcriptomic level; highlighting the non-random nature of oocyte-mediated reprogramming errors. BMC Genomics 17:16

16. Tibary A, Anouassi A (1997) Reproductive management of Camelidae. In: Theriogenology in Camelidae. Institute Agronomique et Veterinaire Hassan II, Rabar, Maroc, pp 459–479. (Chapter XII)

Chapter 17

Derivation of Bovine Primed Embryonic Stem Cells from Somatic Cell Nuclear Transfer Embryos

Delia A. Soto, Micaela Navarro, and Pablo J. Ross

Abstract

Derivation of bovine embryonic stem cells from somatic cell nuclear transfer embryos enables the derivation of genetically matched pluripotent stem cell lines to valuable and well-characterized animals. In this chapter, we describe a step-by-step procedure for deriving bovine embryonic stem cells from whole blastocysts produced by somatic cell nuclear transfer. This simple method requires minimal manipulation of blastocyst-stage embryos, relies on commercially available reagents, supports trypsin passaging, and allows the generation of stable primed pluripotent stem cell lines in 3–4 weeks.

Key words Blastocyst, Derivation, Embryonic stem cells, Inner cell mass, Pluripotency, Somatic cell nuclear transfer

1 Introduction

Embryonic stem cells (ESCs) harbor indefinitely the developmental potential of the pluripotent epiblast. Under appropriate culture conditions, ESCs exhibit capacities of unlimited self-renewal and differentiation toward tissues representative of the three germ lineages and the germline [1]. Derivation of ESC was successfully achieved from in vivo and in vitro produced embryos in rodents, primates, and humans [1], while in livestock progress has been challenging. ESCs from ungulates species represent an important tool for animal agriculture with potential utility for genetic selection, introduction of multiple and complex genetic modifications, and understanding cell fate decisions and pluripotency [2]. Additionally, by combining somatic cell nuclear transfer (SCNT) with ESC derivation, genetically matched pluripotent cell lines to valuable animals can be derived and maintained indefinitely.

Delia A. Soto and Micaela Navarro contributed equally to this work.

Marcelo Tigre Moura (ed.), *Somatic Cell Nuclear Transfer Technology*, Methods in Molecular Biology, vol. 2647,
https://doi.org/10.1007/978-1-0716-3064-8_17,
© The Author(s), under exclusive license to Springer Science+Business Media, LLC, part of Springer Nature 2023

306   Delia A. Soto et al.

Over the last 30 years, many attempts have been made to capture in vitro the pluripotent potential of the bovine inner cell mass (ICM) [3]. However, most presumed bovine ESC (bESCs) lines presented poor derivation efficiency, survived a limited number of passages in culture, lost expression of pluripotency marker genes over time, and lacked capacity for multilineage commitment in teratomas. Recently, a culture system that relies on custom mTeSR1 medium supplemented with fibroblast growth factor 2 (FGF2) and the Tankyrase/Wnt inhibitor IWR-1 succeeded in deriving bona fide primed bESCs [4]. bESC lines from bovine embryos of different sources and genetic backgrounds were established by plating isolated ICMs or zona-free whole blastocysts on gamma-irradiated mouse embryonic fibroblasts (MEF). Since FGF2 is involved in maintaining self-renewal of ESCs [5] and IWR-1 has been related to trophectoderm development suppression [6, 7], the simultaneous supplementation of this growth factor and small molecule favored the growth of ICM cells over the extraembryonic lineages, allowing efficient propagation of the pluripotent ICM-derived bESCs from whole bovine blastocysts. However, bESC derivation and culture was based on a custom-made base medium (mTeSR1 devoid of growth factors), which is cumbersome to prepare, costly, and largely inaccessible to most research laboratories. Hence, we describe here a modification of the original protocol for bESC derivation and culture, in which custom-made mTeSR1 medium was replaced by N2B27 base medium supplemented with 1% low fatty acid bovine serum albumin (BSA) and 20 ng/mL activin A. Since all components are commercially available, this technology can be easily adopted by different laboratories. Using this novel media and following the step-by-step protocol described below, bESCs can be efficiently derived from embryos from in vitro fertilization (IVF), SCNT, and parthenogenesis in only 3 weeks, are simple to propagate, and show long-term pluripotency and genomic stability.

2   Materials

High-quality stem cell-certified reagents must be used for bESC derivation and maintenance.

2.1   Cell Culture

1. Biosafety cabinet class I or II for cell culture.
2. CO_2 incubator.
3. Centrifuge.
4. Water bath.
5. Inverted microscope.
6. Liquid nitrogen tank.

Primed Embryonic Stem Cells from Cloned Bovine Embryos   307

7. Autoclave.

8. Sterile storage bottle.

9. Filter system 0.22 μm PES.

10. 27-gauge hypodermic needle.

11. Irradiated CF-1 Mouse Embryonic Fibroblasts (Gibco).

12. Gelatin stock: 10 mg/mL porcine skin gelatin in double-distilled water, autoclave, aliquot, and store at 4 °C for up to 3 weeks.

13. MEF medium: DMEM supplemented with 1% GlutaMAX Supplement, 1% penicillin-streptomycin, 1% MEM nonessential amino acid solution, and 10% fetal bovine serum. Filter sterilize and store at 4 °C for up to 1 month.

14. NBFR medium: 1:1 neurobasal and DMEM/F12 medium, 0.5% N-2 supplement, 1% B-27 supplement, 1% MEM nonessential amino acid solution, 1% GlutaMAX Supplement, 1% penicillin-streptomycin, 20 ng/mL activin A, 10 mg/mL bovine serum albumin (BSA) 20 ng/mL FGF, 2.5 μM IWR-1 stock, and 0.1 mM 2-mercaptoethanol. Store at 4 °C for up to 3 weeks (*see* **Note 1**).

15. ROCK Inhibitor Y-27632 stock: 10 mM ROCK inhibitor Y27632 in sterile double-distilled water, aliquot, and store at −80 °C.

16. 100× antimycotic/antibiotic (JR Scientific).

17. 1× TrypLE select enzyme (Gibco).

18. Dimethyl sulfoxide (DMSO) (Sigma).

19. Mr. Frosty freezing container (Nalgene) or equivalent.

2.2 Handling of Blastocysts for ESC Derivation

1. Thermoplates.

2. Heating block.

3. Stereomicroscope.

4. Microdispenser or mouth pipette.

5. Filter system 0.22 μm with PES membrane.

6. SOF-HEPES medium: 107.7 mM NaCl, 7.16 mM KCl, 1.19 mM KH_2PO_4, 0.49 mM $MgCl_2 \cdot 6H_2O$, 5.3 mM Na lactate, 1.71 mM $CaCl_2 \cdot 2H_2O$, 0.5 mM fructose, 21 mM HEPES, 1% MEM nonessential amino acid solution, 0.5% BME amino acids solution, 4 mM $NaHCO_3$, 0.33 mM Na pyruvate, 1% GlutaMAX Supplement, 1 mg/mL BSA, 5 μg/mL gentamicin, and ultrapure water. Adjust pH to 7.3–7.4 and osmolarity to 280 ± 10 mOsm. Filter sterilize (0.22 μM) and store at 4 °C for up to 1 month.

7. Pronase solution: 2 mg/mL pronase in SOF-HEPES, aliquot, and store at −20 °C. Spin down before use.

308   Delia A. Soto et al.

2.3 Immuno-fluorescence Staining

All solutions should be prepared fresh for every use and store short term at 4 °C.

1. Inverted fluorescent microscope.

2. Nutator.

3. Paraformaldehyde 4% in PBS (Santa Cruz Biotechnology).

4. Dulbecco's phosphate-buffered saline (DPBS) with calcium and magnesium.

5. Hoechst 33342 solution: 1 mg/mL Hoechst 33342 in DPBS calcium and magnesium free, aliquot, and store protected from lights at −20 °C. Spin down before use.

6. Washing solution: 3% Triton X-100 in DPBS calcium and magnesium free.

7. Blocking solution: 3% normal donkey serum in washing solution.

8. Antibody solution: 1% normal donkey serum in washing solution.

9. Primary antibody solution: 0.6 μg/mL goat anti-OCT4 antibody (Santa Cruz Biotechnology), 1:300 rabbit anti-SOX2 antibody (Biogenex) in antibody solution.

10. Secondary antibody solution: 20 μg/mL donkey anti-goat IgG Alexa Fluor 568 antibody, 20 μg/mL donkey anti-rabbit IgG Alexa Fluor 488 antibody, and 10 μg/mL Hoechst 33342 solution in antibody solution. Keep protected from the light.

11. Parafilm or equivalent.

3   Methods

In this chapter, we describe a detailed procedure for deriving bESC from SCNT embryos; however, this protocol can be indistinctly used for deriving bESCs from IVF or parthenogenetic embryos. Detailed materials and methods for producing bovine SCNT embryos are described elsewhere [8]. To avoid contamination, all procedures should be performed inside a class I or class II biosafety cabinet. bESC derivation is facilitated by having a stereomicroscope inside a biosafety cabinet.

3.1 Plating of MEF Feeders

1. One day before starting ESC derivation, coat a 48-well dish with 150 μL 0.1% gelatin solution and incubate at room temperature (RT) for 30 min (*see* **Note 2**).

2. After incubation, remove the surplus of gelatin and rinse coated wells with 300 μL DPBS calcium and magnesium free just before plating MEF feeders.

Primed Embryonic Stem Cells from Cloned Bovine Embryos   309

3. Thaw MEF feeders in a 37 °C water bath until a small piece of ice is still visible.

4. Transfer MEF feeders from the cryovial to a 15 mL centrifuge tube and dilute cryopreservation medium slowly by adding 9 mL of MEF medium.

5. Centrifuge for 5 min at 200 × g and discard the supernatant.

6. Resuspend cell pellet in 12 mL of MEF medium and add 250 μL of the cell suspension to each well of the gelatin-coated 48-well plate, plating ~2.0×10^6 cells per plate (*see* **Note 3**).

7. Culture MEF feeders at 37 °C and 5% CO_2 for 24 h.

3.2 Bovine ESC Derivation from SCNT Embryos

1. Prepare NBFR media and depending on how many embryos will be plated, keep a working aliquot at RT. Store the rest at 4 °C (*see* **Note 4**).

2. Prepare wells containing MEF feeders for blastocyst plating by removing MEF media, rinsing two times with 300 μL DPBS calcium and magnesium free, and adding 250 μL of RT NBFR media, 10 μM ROCK Inhibitor Y-27632 and 1% antibiotic/antimycotic solution. Leave the prepared plate at 37 °C and 5% CO_2 until use.

3. Place SOF-HEPES medium and the pronase solution in a heating block at 38.5 °C.

4. Remove day 7 blastocysts from the incubator (*see* **Note 5**) and using a microdispenser wash them in three drops of SOF-HEPES to eliminate embryo culture medium and mineral oil (*see* **Note 6**). Incubate embryos in pronase solution until zona pellucida starts to look thinner and separates from the embryo (~3 min) (*see* **Note 7**).

5. After pronase incubation, wash blastocysts in SOF-HEPES drops moving them vigorously to remove pronase. Zona pellucida will slough from the embryo during washes (*see* **Note 8**).

6. Once zona pellucida has been removed, transfer zona-free blastocysts through 2–3 washes of NBFR medium.

7. Plate a single zona-free blastocyst per well containing MEF and NBFR medium plus supplements.

8. Culture plated embryos at 37 °C and 5% CO_2 for 24 h (*see* **Note 9**). At this time, the culture is considered passage zero (P0).

9. Blastocysts that have failed to adhere to the feeder during the 24 h of initial culture should be physically pressed against the bottom of the culture plate. Carefully press the blastocyst on an area away from the ICM with a 27-gauge needle against the culture plate. Slowly move away the needle from the blastocyst avoiding dragging the embryo along with the needle.

310   Delia A. Soto et al.

10. Check blastocysts after 24 h of forced plating and discard embryos that failed to adhere to MEF feeders.

11. After 48 h of blastocyst plating, carefully refresh half of NBFR media under a stereomicroscope. Thereafter, NBFR media should be changed daily (*see* **Note 10**).

3.3 Passaging of Bovine bESCs

After 7 days in culture, dissociate and passage (sub-culture) outgrowths onto fresh MEF feeders.

1. Thaw MEF feeders and plate as indicated in Subheading 3.1.

2. Passage outgrowths by removing NBFR media and washing with 300 μL DPBS.

3. Incubate cells with 150 μL TrypLE Express for 3 min at 37 °C (*see* **Note 11**).

4. Stop trypsin enzymatic reaction by adding 300 μL NBFR media, pipet up and down a few times, and transfer to a 1.5 mL microcentrifuge tube.

5. Centrifuge for 5 min at 200 × g and discard the supernatant.

6. Resuspend cell pellet in 250 μL NBFR media supplemented with 10 μM ROCK Inhibitor Y-27632.

7. Seed the cell suspension into the previously prepared wells and culture cells at 37 °C and 5% CO_2 (*see* **Note 12**).

8. Repeat passaging until bESC lines are established after 3–5 passages (*see* **Note 13**). Once multiple bESC colonies appear in culture (Fig. 1), start splitting at a ratio of 1:5–1:10 every 3–4 days (*see* **Note 14**). At this point, cryopreservation or evaluation of pluripotency can be performed (*see* **Note 15**).

3.4 Cryopreservation of Bovine ESCs

1. Add 500 μL of NBFR medium containing 20% DMSO to each appropriately labeled cryovial (*see* **Note 16**).

2. Dissociate cells as described in Subheading 3.3.

3. After centrifugation, discard supernatant and resuspend cell pellet in 500 μL NBFR medium.

4. Transfer cell suspension to the prepared cryovial and place in a cell-freezing device.

5. Place cell-freezing device at −80 °C overnight and then transfer cryovials into a liquid nitrogen tank for long-term storage.

6. Thaw bESC by removing cryovial from liquid nitrogen tank and incubating in a water bath at 37 °C until a small piece of ice is left.

7. Transfer cryovial to the biosafety cabinet, place thaw cells in a 15 mL centrifugation tube, and slowly add 9 mL NFBR medium to dilute cryoprotectant.

Primed Embryonic Stem Cells from Cloned Bovine Embryos 311

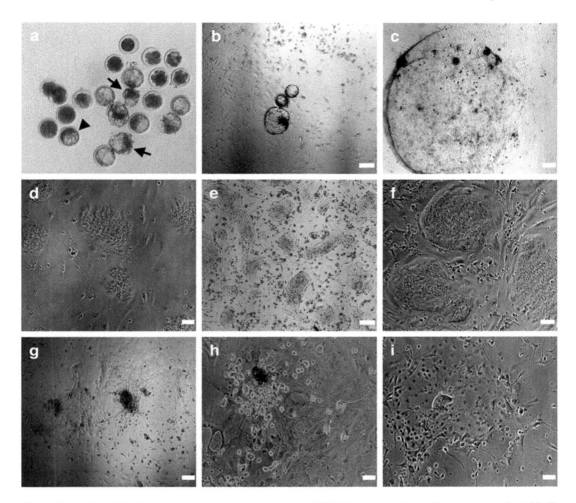

Fig. 1 Derivation of bovine primed embryonic stem cells (bESCs) from somatic cell nuclear transfer (SCNT) embryos. (**a**) SCNT bovine embryos with different grading [9]. Only grade 1 quality blastocysts should be selected for bESC derivation. Arrows indicate good-quality embryos, whereas the arrowhead points to a poor-quality embryo (**b**) Hatching SCNT blastocyst plated in ESC derivation conditions. (**c**) Blastocyst outgrowth after 7 days of embryo attachment to the mouse embryonic fibroblast feeder layer. (**d**) Compact colonies of bESCs established at passage 4, showing high nucleus-to-cytoplasm ratio. (**e**) Confluent bESC culture at passage 8. (**f**) bESC cell line maintains morphology and proliferation rate at passage 15. (**g–i**) Blastocyst outgrowths that failed to establish a cell line. No visible bESC colonies and trophectoderm-like cells were maintained for several passages. Scale bars 100 μm

8. Centrifuge for 5 min at 200 × g and discard the supernatant.
9. Resuspend cell pellet in NBFR medium supplemented with 10 μM ROCK Inhibitor Y-27632 and seed onto MEF feeders.

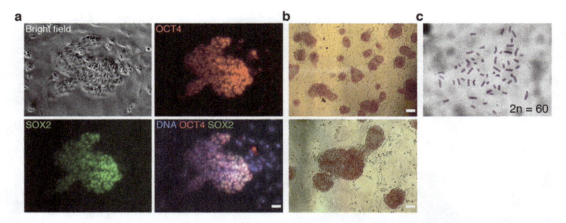

Fig. 2 Characterization of bovine primed embryonic stem cells (bESCs) derived from somatic cell nuclear transfer (SCNT) embryos. (**a**) Immunofluorescence analysis showing detection of OCT4 and SOX2 proteins in bESCs. (**b**) bESC colonies are alkaline phosphatase-positive compared with mouse embryonic fibroblast feeder cells. (**c**) bESCs have normal chromosome number ($2n = 60$). Scale bars 100 μm

3.5 Immunofluorescence Staining

An option for routine pluripotency analysis is immunostaining for core pluripotency transcription factors *OCT4* (also known as *POU5F1*) and *SOX2* (Fig. 2).

1. Grow bESCs until they reach 60–70% confluency (*see* **Note 17**).
2. Fix cells by discarding culture media, rinsing two times with DPBS containing calcium and magnesium and incubating cells in 4% paraformaldehyde solution for 10 min.
3. Remove fixative, dispose appropriately, and rinse cells three times with DPBS containing calcium and magnesium (*see* **Note 18**).
4. Remove DPBS, add blocking solution, and incubate for 30 min.
5. After blocking, discard blocking solution and incubate in primary antibody solution for 1 h.
6. Wash three times for 10 min in washing solution.
7. Remove washing solution and incubate for 1 h in secondary antibody solution, protected from the light.
8. Wash three times for 10 min in washing solution.
9. After the final wash, add DPBS containing calcium and magnesium and seal plate with parafilm (*see* **Note 19**).
10. Image cells using an inverted fluorescence microscope (*see* **Note 20**).

4 Notes

1. BSA low free fatty acids should be used. A 100 mg/mL BSA stock solution can be prepared by stirring BSA at low speed in DMEM/F12 medium. Avoid making bubbles, filter sterilize, and store at 4 °C for up to 1 month.

2. Coat cell culture dishes by diluting gelatin stock at 1 mg/mL in DPBS. Gelatin-coated dishes can be prepared in advance and store at 4 °C for up to 1 week.

3. Spread MEF feeders evenly by moving the culture plate once back and forth and once left and right.

4. Do not warm up NBFR medium to 37 °C. It should be used at room temperature since FGF2 degrades rapidly at 37 °C.

5. It is critical to work with high-quality embryos. We observed that embryo quality is a critical factor defining success or failure of the procedure. Blastocysts used in ESC derivation should be selected carefully; embryos with a large and well-defined ICM are ideal.

6. It is convenient to use stereomicroscopes coupled to thermo-plates set at 38.5 °C during handling of embryos.

7. Avoid prolonged pronase incubation as the enzyme can damage the embryo.

8. Hatched blastocysts should not be subjected to pronase treatment. Because SCNT embryos usually hatch from the scar produced during micromanipulation, it is preferable to let them finish hatching during embryo culture or in NBFR derivation conditions. We have seen that blastocysts continue development when cultured on MEF in NBFR medium.

9. To facilitate blastocyst adherence to the monolayer of MEF feeders, avoid moving the culture plate during the first hours of derivation.

10. To contain potential mycoplasma contamination propagated from embryos, NBFR media can be supplemented with 2.5 µg/mL of plasmocin (In Vivo Gen) during ESC line derivation (P0-P3).

11. Even though NBFR-bESCs can tolerate enzymatic disaggregation, care should be taken during initial passages. ICM cells are fragile, thus we do not recommend mechanical disruption (excessive pipetting) until later passages, when pluripotent cells have been sufficiently expanded. After passage 3 or 4, mechanical disruption of bigger clumps is recommended for a more homogeneous culture density.

314   Delia A. Soto et al.

12. Until colonies are observed, the whole content of a well should be reseeded into a new well with fresh MEF feeders. Do not split the cell suspension.

13. bESC colonies should appear between the third and fifth passages. If no colonies are seen after passage 5, cells should be discarded.

14. bESCs should be subcultured reaching a maximum confluency of 80%. Over-confluency can induce differentiation of bESCs.

15. Time of establishment and proliferation rate of bESC lines can vary. This may be related to the blastocyst stage or ICM developmental capacity upon ESC line derivation.

16. To improve cell survival during cryopreservation, ROCK Inhibitor Y-27632 (10 μM) can be added to the cryopreservation medium, especially throughout the first few passages.

17. Preferably perform immunostaining in four-well plates, thus all washes and incubations are done using only 400–500 μL of solutions. Immunostaining incubations should be performed at RT on a nutator at minimum speed. Extreme care should be taken when adding and removing solutions from wells, and bESC colonies easily detach from the plate.

18. Fixed cells can be stored in DPBS with calcium and magnesium at 4 °C for up to 1 week.

19. Avoid drying out cells during solution changes and washes as this can result in nonspecific antibody binding.

20. Cells can be imaged immediately or within a week if plate is stored at 4 °C sealed and protected from lights.

Acknowledgments

This work was supported by Chancellor's Fellow award and USDA/NIFA/AFRI W-4171 multistate project to PJR. D.A.S. was supported by CONICYT within the funding program Becas Chile. M.N was supported by the National Scientific and Technical Research Council (CONICET) grants from Argentina.

References

1. Nichols J, Smith A (2009) Naive and primed pluripotent states. Cell Stem Cell 4:487–492

2. Goszczynski DE, Cheng H, Demyda-Peyrás S, Medrano JF, Wu J, Ross PJ (2019) In vitro breeding: application of embryonic stem cells to animal production. Biol Reprod 100:885–895

3. Navarro M, Soto DA, Pinzon CA, Wu J, Ross P (2019) Livestock pluripotency is finally captured in vitro. Reprod Fertil Dev 32:11–39

4. Bogliotti YS, Wu J, Vilarino M, Okamura D, Soto DA, Zhong C et al (2018) Efficient derivation of stable primed pluripotent embryonic

stem cells from bovine blastocysts. Proc Natl Acad Sci 115:2090–2095

5. Levenstein ME, Ludwig TE, Xu R-H, Llanas RA, VanDenHuevel-Kramer K, Manning D et al (2006) Basic fibroblast growth factor support of human embryonic stem cell self-renewal. Stem Cells 24:568–574

6. Tribulo P, Leão BCS, Lehloenya KC, Minot GZ, Hansen PJ (2017) Consequences of endogenous and exogenous WNT signaling for development of the preimplantation bovine embryo. Biol Reprod 96:1129–1141

7. Wang C, Han X, Zhou Z, Uyunbilig B, Huang X, Li R et al (2019) Wnt3a activates the WNT-YAP/TAZ pathway to sustain CDX2 expression in Bovine Trophoblast stem cells. DNA Cell Biol 38:410–422

8. Ross PJ, Cibelli JB (2010) Bovine somatic cell nuclear transfer. Methods Mol Biol 636:155–177

9. Bó GA, Mapletoft RJ (2013) Evaluation and classification of bovine embryos. Anim Reprod 10:344–348

Correction to: Cloning by SCNT: Integrating Technical and Biology-Driven Advances

Marcelo Tigre Moura

Correction to:
Marcelo Tigre Moura (ed.), *Somatic Cell Nuclear Transfer Technology*,
Methods in Molecular Biology, vol. 2647,
https://doi.org/10.1007/978-1-0716-3064-8_1

The book was inadvertently published with an incorrect tagging of the Editor's name Prof. Marcelo Tigre Moura.
It has now been corrected as below.
Family name: Moura
First name: Marcelo Tigre

The updated original version of this chapter can be found at
https://doi.org/10.1007/978-1-0716-3064-8_1

Marcelo Tigre Moura (ed.), *Somatic Cell Nuclear Transfer Technology*, Methods in Molecular Biology, vol. 2647,
https://doi.org/10.1007/978-1-0716-3064-8_18,
© The Author(s), under exclusive license to Springer Science+Business Media, LLC, part of Springer Nature 2023

INDEX

A

Actinomycin D 17
Amphibians
    cloning 10
    embryology.. 5

C

Cell cycle synchronization
    cell contact inhibition 232, 241
    quiescence.. 232
    serum starvation 11, 232, 241, 275
Cell fusion
    electrofusion 6, 199, 204, 207, 228,
        247, 253, 264, 265, 278
    Sendai virus (SV)... 6–8
Cell identity
    cumulus cells 21, 113,
        114, 158, 160, 161, 165, 174, 175, 187, 193,
        202, 203, 220, 223, 234, 235, 241, 251, 276,
        293, 294
    fibroblasts 13, 21, 90, 92, 108–114,
        137, 139, 160, 161, 165, 167, 174, 230, 232,
        233, 246, 250, 301
    lymphocytes 7, 15
    neurons 15, 114
    sertoli cells 20, 21, 114
Chromatin
    DNA hypersensitive sites 23, 24
    euchromatin 4, 38, 51
    heterochromatin.. 4, 38, 65
    high-order configuration 24, 25
    nucleosome... 24, 39
    topologically associating domains (TADs) 24, 25
Cis-regulatory elements (CREs)
    enhancers 3, 24
    promoters 2, 24
Cloning
    efficiency 6–8, 10, 13,
        19–23, 25, 26, 49–51, 60, 91, 105–107, 110,
        112–114, 151, 152, 170, 198, 211, 226, 227,
        242, 246

embryo..8–12, 60, 69,
    90, 91, 169, 170, 186
    handmade (HMC) 18, 183–195,
        206, 245–256, 278, 283–302
    neonatal care 240
    reproductive 15, 16, 67, 69
    somatic cell 6, 19, 105, 108, 151,
        184, 225–242
    therapeutic 15, 16, 67, 69
Cryopreservation
    cryoprotectant 310
    thawing 186, 248
    vitrification.................................... 184, 285, 298, 299
    warming 213, 220, 285
Cycloheximide 172, 190, 201, 208, 274
Cytochalasin B (CB) 7, 9, 152, 155,
    171, 186, 201, 208, 214, 229, 230, 234, 241,
    252, 261, 263, 273
Cytoplast
    demi 250, 252, 253

D

Demecolcine 17, 200, 203, 204, 207,
    273, 287, 294, 295
Development
    developmental potential 3, 10, 17,
        18, 21, 26, 46, 105, 107, 125, 170, 235, 238,
        259, 305
    full-term 6, 10, 14, 15, 21, 23, 26,
        48, 50, 51, 67, 71, 72, 108, 111, 113–115, 125,
        164, 191, 212, 226, 232, 246, 269
    gastrulation 2, 5, 19
    implantation 62, 115
    post-implantation................... 46, 106, 112, 113, 115
    pregnancy 10, 300
6-Dimethylaminopurine (6-DMAP) 172, 214,
    222, 238, 248, 253, 256, 261, 265, 266, 274,
    288, 292, 297
D-mannitol 186, 201,
    248, 261, 287
DNA damage
    pathway... 1

Marcelo Tigre Moura (ed.), *Somatic Cell Nuclear Transfer Technology*, Methods in Molecular Biology, vol. 2647,
https://doi.org/10.1007/978-1-0716-3064-8,
© The Editor(s) (if applicable) and The Author(s), under exclusive license to Springer Science+Business Media, LLC, part of Springer
Nature 2023

SOMATIC CELL NUCLEAR TRANSFER TECHNOLOGY
Index

Donor cells 5–15, 18–22, 24,
    25, 46, 50, 71, 90, 95, 105, 106, 110–112, 114,
    137, 152, 156, 158, 160, 161, 163, 165, 167,
    169, 170, 174–177, 179, 180, 186–187, 189,
    190, 193, 198, 202–204, 215, 222, 226, 227,
    232, 236, 237, 241, 242, 246, 248, 250, 252,
    260, 269, 270, 275, 277–280, 289–290, 295,
    300, 302

E

Embryo
    blastocyst 6, 9–11, 13, 21, 45, 46,
        49, 50, 61, 62, 65–67, 69–73, 109, 111, 112, 122,
        177, 191, 199, 205, 207, 227, 239, 254, 285,
        299, 306, 309–312
    blastula 7, 8
    cleavage 10, 25, 177, 222
    culture 13, 156, 157, 163, 173,
        177, 186–187, 190, 191, 198, 204, 205, 214,
        215, 222, 238, 246, 248, 254–256, 270, 273,
        274, 277, 279, 285, 288, 291, 292, 297, 298,
        300, 302, 309, 313
    epiblast 61–64
    gastrula 8
    grading 191, 279, 299, 311
    inner cell mass (ICM) 11, 42,
        59–63, 65–69, 71, 72, 122
    morula 9, 10, 13, 15, 16, 23, 62
    primitive endoderm (PE) 61–64
    pronucleus 5, 65
    transfer 13, 21, 106, 108,
        111–113, 164, 177, 178, 191–194, 198,
        205–206, 208, 239, 240, 242, 254, 284, 285,
        299–301
    trophectoderm (TE) 49, 59–63, 65–72
    zygote 2, 16, 21, 95, 135
Embryonic genome activation (EGA) 2, 16,
    20–25, 27, 114, 198, 227, 260
Epigenetics
    definition 37
    DNA methylation 3, 18, 37–45,
        47, 49, 50, 198, 226
    histone post-translational modifications 4, 37, 47,
        64, 65, 226
    imprinting 40, 44
    non-coding RNAs (ncRNAs) 4, 37, 40,
        47, 226
    X-chromosome inactivation (XCI) 44–48

F

Fusion couplets 12, 177, 190,
    229, 237, 238, 279, 295, 297

G

Genomics
    DNA accessibility 4, 39
    DNA methylation maps 16, 21, 23, 25
    multi-omics 25, 26
    nucleosome occupation 4
    three-dimensional chromatin conformation 24
    transcriptome 16, 20, 22,
        23, 25, 26, 69, 106

H

Hoechst 33342 .... 207, 214, 220, 229,
    230, 252, 254, 261, 263, 273, 308

I

Ionomycin 201, 204, 208, 214, 222, 238,
    261, 265, 266, 273, 279, 287, 297

L

Livestock 6, 10, 13, 14, 17, 18,
    25, 48, 90, 105–107, 112, 113, 121–127, 130,
    133–142, 225, 305

M

Micromanipulation
    holding pipette 158, 172, 173,
        199, 216, 229–231, 260, 263, 264
    injection pipette 173, 199, 216,
        223, 229–231, 240, 264, 273
    inverted microscope 173, 199,
        223, 263, 270
    micromanipulators 18, 154, 157, 158,
        166, 170, 173, 176, 184, 198, 199, 206, 215,
        229–231, 245, 260, 263, 272, 278
    nuclear transfer 231
    ooplasm transfer 259–267
    piezo unit 6, 18, 157, 158,
        162, 163, 166, 216
    pol-scope microscope 6, 17
    pronuclear microinjection 124, 137
Mitochondria
    bottleneck 83, 88, 89, 91
    heteroplasmy 83, 91, 93
    mtDNA inheritance 83–95, 241
    replacement therapy (MRT) 93–95
Mitomycin C 17

N

Nuclear reprogramming
    abnormalities 14, 49, 105, 106, 113

definition 7
epigenetic memory 19–22, 24, 226
induced pluripotent cells
    (iPSCs) 16, 17, 51, 68, 71, 86, 89
nuclear swelling 11
premature chromosome condensation
    (PCC) 11, 12
reprogramming resistant regions
    (RRRs) 21–24, 226, 227

O

Oocyte
    activation 8–10, 12, 13, 152,
        156, 158, 161, 163, 172, 177, 190, 193, 198,
        204, 222, 253, 265, 273, 274, 279, 285, 297
    bisection 10, 18, 188
    denuding 175, 188, 203,
        207, 220, 229, 234, 241, 250, 251, 262, 263,
        276, 291–294, 301
    enucleation 5, 6, 9–11, 17, 18,
        21, 44, 46, 50, 90, 92, 94, 105, 111, 152,
        161–163, 175–177, 180, 188, 198, 204, 207,
        212, 216, 219–222, 229, 230, 233, 234,
        236–238, 245, 250–252, 255, 263, 266, 269,
        277, 278, 280, 289, 294, 295
    maturation 228, 233
    oolemma 18, 161, 162, 222, 223
    polar body 176, 188,
        203, 204, 207, 220, 234, 236, 262, 276, 294
    protrusion 6, 18, 252, 278, 294, 295
    reconstruction 8, 17, 18, 90,
        109, 114, 175, 176, 204, 220–222, 232, 236,
        237, 241, 252, 263, 264, 280, 295
    spindle 17, 94, 153,
        155, 161, 162, 166, 176, 203, 219, 221, 236,
        252, 255, 266, 278, 280, 294
    staining 207, 230, 234, 250, 262, 263

P

Phytohemagglutinin (PHA) 186, 188,
    189, 193, 248, 252, 270, 287, 295, 302
Pluripotency
    assays 108
    definition 106
    tetraploid complementation 15

R

Reproductive technologies
    intracytoplasmic sperm injection 45, 212
    in vitro fertilization (IVF) 20, 21, 45,
        46, 67–73, 109, 111, 246, 269, 300, 306, 308

parthenogenesis 306

S

Somatic cell nuclear transfer
    interspecies SCNT 91, 113, 259–267
    recloning 14, 15
Species
    Bison bison 260, 262
    Bos taurus 69, 70, 108–109, 262
    Bubalus bubalis 69, 70, 109–110, 246
    Camelus dromedarius 283
    Canis lupus familiaris 113
    Capra hircus 112–113
    Cervus elaphus 113
    Drosophila melanogaster 8
    Equus caballus 111–112
    Macaca mulatta 16, 69
    Misgurnus d L. 8
    Mus musculus 69, 113–114
    Oryctolagus cuniculus
    Ovis aries 112–113
    Rana pipiens 5–7
    Sus scrofa 69, 70, 110–111
    Xenopus laevis 6–8, 17, 22, 24
Stem cells
    adult 106
    embryonic 6, 11, 15, 16, 26, 47,
        60–62, 67, 69–71, 90, 93, 94, 160, 161, 305–314
    fetal 11, 111–113
    mesenchymal 107–114
    somatic 7, 106, 107
Strontium chloride (SrCl$_2$) 152, 155–157,
    201, 208
Sucrose 17, 203, 204, 207

T

Totipotency
    definition 10
Transcription factors
    binding sites 3
    motifs 22, 24
Transgenesis
    CRISPR-Cas9 .... 132, 136–138, 140–142
    disease models 14
    gene editing 14, 132, 138, 184
    gene targeting 124, 125, 137
    genetically edited organisms 133–135
    genetically modified organisms 127–130,
        133, 134
    homologous recombination 124, 134, 135
    short-interfering RNAs 21

SOMATIC CELL NUCLEAR TRANSFER TECHNOLOGY
Index

Transgenesis (*cont.*)

    TALENs.. 131, 132, 136,
        138, 140, 141

    transgenic cells.. 14

    zinc-finger nucleases 131, 132, 135,
        136, 138, 140

Transposable elements (TE)................................. 40, 121

U

Ultraviolet irradiation 207, 230

X

X-ray.. 8

Z

Zona pellucida

    zona enclosed 255, 275

    zona-free 250

www.ingramcontent.com/pod-product-compliance
Lightning Source LLC
Chambersburg PA
CBHW080107030225
21296CB00006B/253